WITHDRAWN
LIBRARY
WARWICKSHIRE COLLEGE
D0301640

TREE DISEASES AND DISORDERS

TREE DISEASES AND DISORDERS

Causes, Biology, and Control in Forest and Amenity Trees

HEINZ BUTIN

Biologische Bundesanstalt für Land- und Forstwirtschaft
Institut für Pflanzenschutz im Forst,
Braunschweig, Germany

Edited by David Lonsdale from
a translation by Robert Strouts

With 460 individual
illustrations in 123 figures
and two fungal spore charts

OXFORD NEW YORK TOKYO
OXFORD UNIVERSITY PRESS
1995

Oxford University Press, Walton Street, Oxford OX2 6DP

Oxford New York
Athens Auckland Bangkok Bombay
Calcutta Cape Town Dar es Salaam Delhi
Florence Hong Kong Istanbul Karachi
Kuala Lumpur Madras Madrid Melbourne
Mexico City Nairobi Paris Singapore
Taipei Tokyo Toronto

and associated companies in
Berlin Ibadan

Oxford is a trade mark of Oxford University Press

Published in the United States
by Oxford University Press Inc., New York

Originally published in German as Krankheiten der Wald- und Parkbäume *by Georg Thieme Verlag, Stuttgart, 1983*

First published in English by Oxford University Press 1995 as Tree Diseases and Disorders.

A catalogue record for this book is available from the British Library

Library of Congress Cataloging in Publication Data
Butin, Heinz.
[Krankheiten der Wald- und Parkbäume. English]
Tree diseases and disorders : causes, biology, and control in forest and amenity trees / Heinz Butin ; edited by David Lonsdale from a translation by Robert Strouts.
Revised translation of the 2nd German edition.
Includes bibliographical references and index.
1. Trees–Diseases and pests–Europe. 2 Trees–Wounds and injuries–Europe. 3. Trees–Diseases and pests. 4. Trees–Wounds and injuries. 5. Ornamental trees–Diseases and pests–Europe. 6. Ornamental trees–Wounds and injuries–Europe. 7 Ornamental trees–Diseases and pests. 8. Ornamental trees–Wounds and injuries.
I. Lonsdale, D. II. Title.
SB764.E85B8713 1995 634.9'6–dc20 94–45537
ISBN 0 19 854932 6 (Hbk)

Typeset by
Light Technology Ltd., Fife, Scotland
Printed in Great Britain by
Antony Rowe Ltd, Chippenham, Wiltshire

Preface

The great interest stimulated by the 2nd German edition of *Krankheiten der Wald- und Parkbäume* has encouraged the author and publishers to issue this English translation. The essential features of the German text have been retained both in concept and contents. However, the diseases now covered include a number that, although of minor importance in Germany, have more significance in other countries. Thus, the book is of practical value not only in Europe, but around the world. It contains information on important diseases that do not occur in Europe, especially those that are scheduled for quarantine control. As far as the range of host plants is concerned, most of those dealt with are forest trees, in keeping with the title of the German edition. Nevertheless, the reader will also find details of diseases of ornamental tree species that are found in streets or parks.

The main layout of the book is based on the various types of tree diseases and disorders, as defined by the parts of the tree affected. So, for example, there are chapters dealing with leaf diseases, bark damage and damage to the woody tissues of the main stem. Within this framework, the information is presented according to the causal processes that are recognized in classic forest pathology, including both external agents and botanical factors within the tree. Thus, there are sub-sections dealing with conditions caused by viruses, bacteria, fungi, and parasitic flowering plants, together with abiotic injuries and growth abnormalities. The chapters are subdivided according to the particular pathological conditions that occur among the different tree species covered.

The aim of the book is to provide the practitioner with an introduction to the diagnosis of tree diseases, and so the emphasis is on the description of symptoms and of the causes of damage. However, information is additionally provided on the biology and economic significance of the particular causal agents, together with notes on measures for the prevention and control of the principal diseases.

As for the illustrations, to which six new ones have been added, these are not intended to serve merely as decorative additions; much more, they are an important diagnostic tool in that they show many features which are not noted in the text. The illustrations have not been provided with scales as dimensions are included in the text. This applies particularly to spore sizes.

For the benefit of readers who may seek further information, literature references are given in brackets at appropriate places. Since the forest pathological literature is very extensive, these are limited to works which either give a comprehensive account of the problem under consideration, or can be recommended for use in further diagnostic investigations. Particular attention has been given to the more recent literature.

My thanks to those who have helped me in the preparation of this book go first to the graphic artist, Angelika Krischbin, for the artwork; she sensitively and carefully prepared the majority of the drawings. The foundations for the illustrations were laid at that time by Richard Kliefoth. I owe particular thanks to Dr. David Lonsdale and Robert Strouts not only for dealing with the translation of the German text, but also for their many technical comments and suggestions for improvements. Finally, I am indebted to the publishers, Oxford University Press, for being so accommodating in the production and completion of the book.
Now that the second German edition of *Krankheiten der Wald- und Parkbäume* is available in this revised and translated form, it is to be hoped that it will meet the needs of a wider readership, whether as a textbook, reference work, or as an introduction into the diagnosis of diseases and disorders of forest and ornamental trees.

Braunschweig, Autumn 1993 Heinz Butin

Contents

1 *Damage to Flowers and Seeds*

Flowers and seeds, in common with all plant tissues, can be damaged either by biotic or abiotic agents which, in classic plant pathology, are distinguished as the two principal causes of disease; either 'parasitic' or 'non-parasitic' disease, respectively.

Among the most important abiotic factors which can damage **flowers** or inflorescences are extremes of weather, in particular late frosts which principally affect the early flowering trees like walnut, Sweet chestnut, beech, and oak. In these species, the damage is usually noticed only because of the failure of the seed crop but there are others, especially ornamentals, which can show striking changes directly after the first frosty night: in magnolias, for example, the normally white petals become an unsightly brown and hang down limply. Other weather events which can damage the flowers are hailstorms and sudden, dry winds (cf. p. 47).

Biotic agents tend to cause less conspicuous damage to flowers or inflorescences. They are nearly always fungi which are strongly host specific and which induce characteristic changes in flowers or inflorescences. Among the more frequently occurring species are:

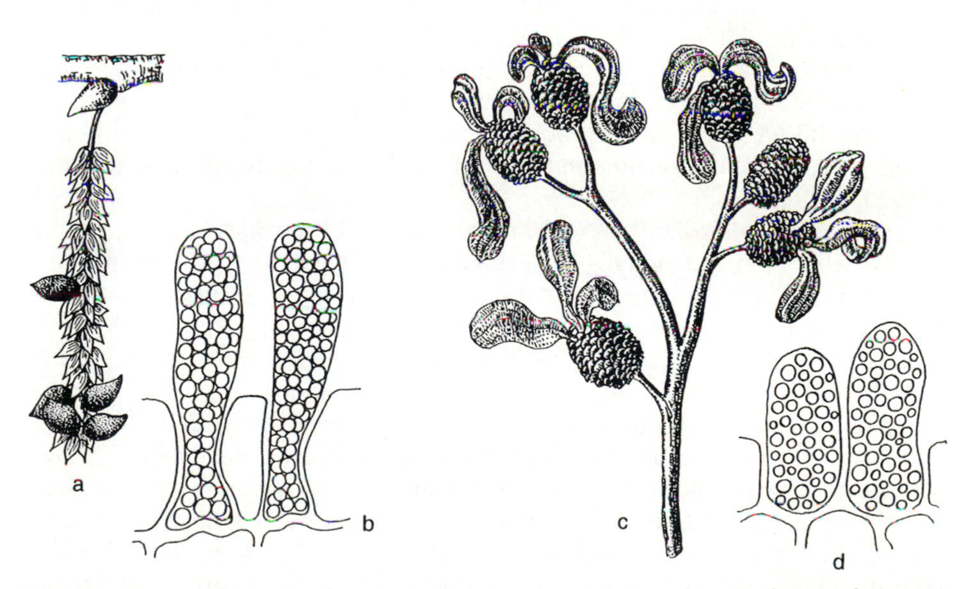

Fig. 1 Damage to flowers and seeds. **a** *Populus tremula* infected by *Taphrina johanssonii*, **b** asci of *T. johanssonii*; **c** a female catkin of *Alnus incana* infected by *Taphrina amentorum*, **d** asci (a after Dennis 1978; b,d after Mix 1969; c after Hartig 1900)

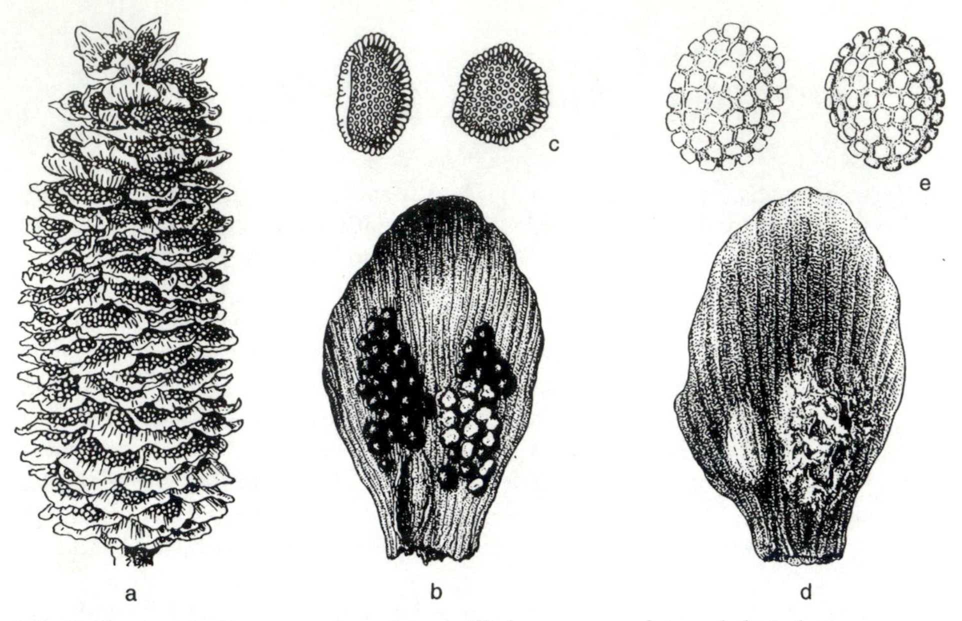

Fig. 2 Damage to flowers and seeds. **a-c** *Thekopsora areolata*: **a** infected spruce cone, **b** cone scale with aecidia, **c** aecidiospores; **d,e** *Chrysomyxa pyrolatum*: **d** cone scale with aecidia, **e** aecidiospores (a after Ferdinandsen and Jørgensen 1938/39)

- *Taphrina amentorum* (Sadeb.) Rostrup, which affects alder. This ascomycetous fungus causes the formation of reddish, tongue-like outgrowths on the female catkins of *Alnus incana* and other alder species (Fig. **1a,b**).
- *Taphrina johanssonii* Sadeb. attacks the female flowers of *Populus tremula* and related species, causing individual fruits to develop in the form of inflated, blister-like, golden yellow, sterile capsules; the asci are 60–140 μm long, and, as seen in section under the microscope, hardly attenuated at the base (Fig. **1c,d**).
- *Taphrina rhizophora* Johansson similarly causes blister-like, golden yellow deformations on individual pistils on the female catkins, though only on *Populus alba*; the asci are 80–200 μm long and have a root-like attenuation at the base [137].
- *Thekopsora areolata* (Fr.) Magnus (Syn. *Pucciniastrum areolatum* [Fr.] Otth) is a heteroecious rust fungus on cones and young shoots of Norway spruce (haplontic-state host) and leaves of *Prunus* species (dikaryotic-state host). On spruce, the mycelium of the rust first permeates the whole female inflorescence; then aecidia develop in summer on the upper and lower sides of enlarged, protuberant cone scales. These are globular with finely warty, polyhedral aecidiospores 21–28 × 17–20 μm in size (Fig. **2a,c**). Usually only spermogonia are produced on infected shoots (these shoots are recognizable by being slightly curved and with necrotic bark on one side) [168]. The next stage of development occurs on the alternate host; *Prunus padus* or

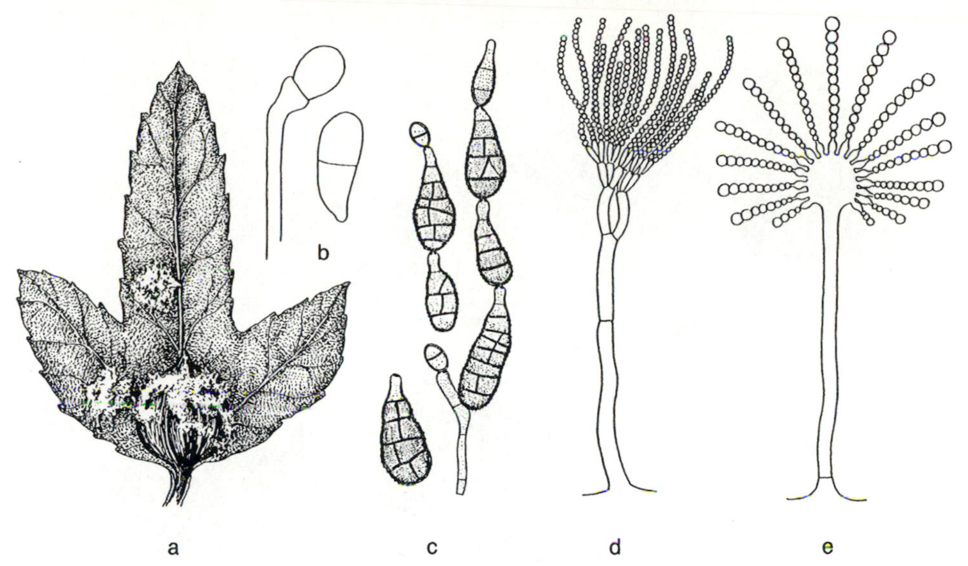

Fig. 3 Moulds on seeds and fruit. **a,b** *Trichothecium roseum* on hornbeam: **c** *Alternaria alternata*; **d** *Penicillium* sp. **e** *Aspergillus* sp. (c after Ellis 1971)

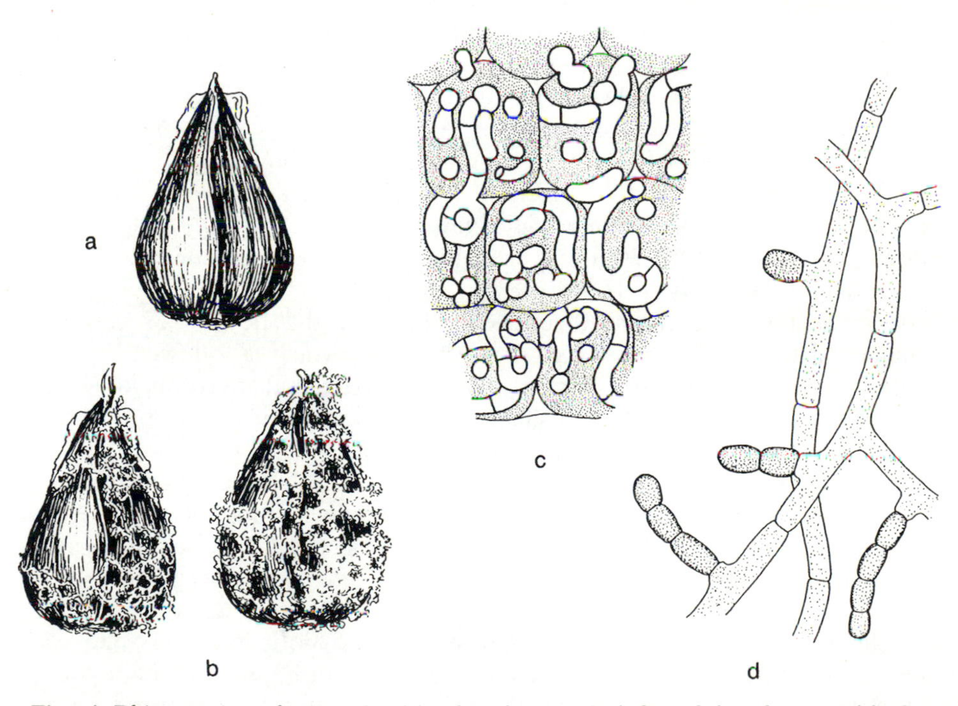

Fig. 4 *Rhizoctonia solani*. **a** healthy beech nut, **b** infected beech nut with fungal mycelium, **c** hyphae in the seed-leaf tissue, **d** mycelium with chlamydospores, in artificial culture

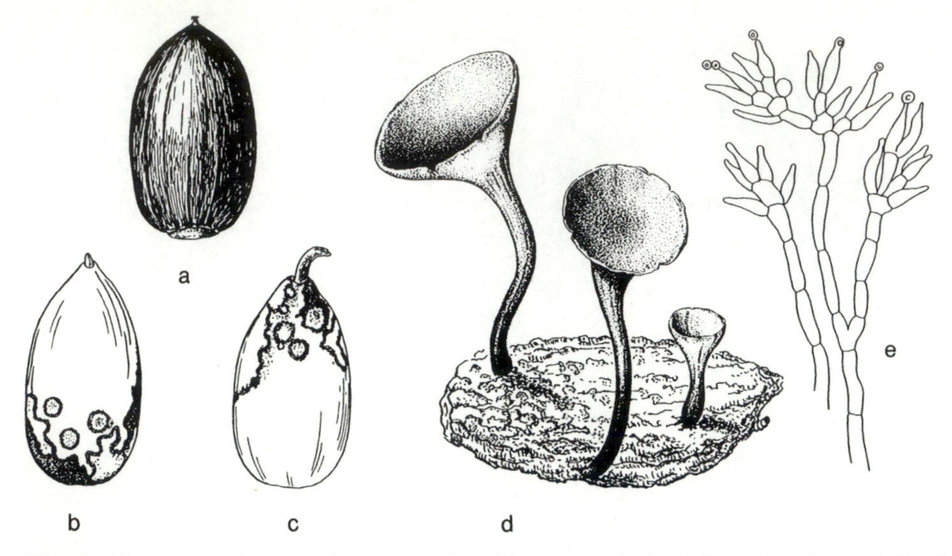

Fig. 5 *Ciboria batschiana*. **a** healthy acorn with seed coat, **b,c** infected seeds, **d** apothecia on mummified acorn, **e** conidial state (b,c after Delatour 1978)

P. serotina, whose leaves in summer show dark red spots (uredosori) on the underside. The disease is of economic significance only in seed orchards where it can result in the failure of the seed crop.

- *Chrysomyxa pyrolatum* (Schwein.) Winter is also a heteroecious rust fungus and develops on the cones of Norway spruce (haplontic-state host) and the leaves of *Pyrola* species (dikaryotic-state host). The yellow to golden yellow aecidia of this rust appear in the form of 1–2 mm blister-like swellings on the outside of the cone scales. The aecidiospores are coarsely warty, roundish to elliptic, and 25–36 × 20–30 μm in size (Fig. **2d,e**). Diseased cones produce no, or only a few seeds [17].
- *Monilia laxa* (Ehrenb.) Sacc. causes a blight of the flowers of various ornamental trees and shrubs, e.g. *Prunus triloba*, which is followed by the death of whole shoots; grey fungal mats form here and there on infected parts of the plant, carrying lemon-shaped conidia borne in simple or branched chains; infected parts should be removed immediately.
- *Botrytis cinerea* Pers. can, in wet years, lead to the wilting and complete destruction of flowers of various woody ornamentals (*Magnolia, Syringa*, ornamental forms of *Prunus*); otherwise it is a non-host-specific weak parasite on shoots of conifers (cf. Fig. **53c**).

Seeds, like flowers, can be damaged both by abiotic and biotic factors. Examples of abiotic damaging agents include high temperatures during seed extraction or the improper use of seed disinfectants during seed preparation. Another factor, inherent in the seed itself, is ageing which can result in the failure of germination, despite optimal seed treatment. The longevity of seed depends largely on species, ranging from a few weeks (e.g. poplar, willow) to several years (e.g. spruce, pine, robinia). However, it is possible to prolong the germinability by using certain methods of seed preparation (e.g. drying the seed) or storage (maintaining low temperatures). Non-germination of seed due to ageing should not be confused with dormancy which can be physiologically induced and which, with some species, can be artificially overcome by chilling [167].

Biotic damage to seed is almost exclusively caused by fungi, among which two biological groups can be recognized. One of these includes numerous non-specific 'moulds' which, when the air humidity is high, infest only the outer seed coat, or can penetrate to the interior of the seed or fruit only after the seed coat has been damaged, e.g. by insects [136]. Among the more commonly occurring are imperfect species of the genera *Alternaria, Fusarium, Penicillium,* and *Trichothecium* (Fig. **3**).

The other group of seed fungi consists of specialists which can attack intact seed and cause internal rotting. This group includes some species of the genera *Rhizoctonia* and *Ciboria* which can infect the seed or fruits of various tree species, though causing economically significant damage only on beech nuts and acorns. The following are descriptions of the two most important seed pathogens:

- *Rhizoctonia solani* Kühn, anamorph of the basidiomycete *Thanatephorus cucumeris* (Frank) Donk, infects the seed predominantly from the soil. An attack can be recognized externally from a loosely attached woolly white mycelium on the seed surface or on the outer shells of 'seeds' such as those of beech, which are fruits in strict botanical terms. If the interior of the seed is examined, numerous 'fat' fungal hyphae will be found in the pale brown starch-containing cells of the cotyledons. A positive identification can be obtained by isolating the fungus in pure culture. On suitable nutrient media, a white, uniform mycelial growth develops with chlamydospores arranged in chains. In older cultures, dark sclerotia, 3–5 mm in diameter are formed—mostly on the edge of the colony (Fig. **4**). The fungus occurs in an epidemic form mainly after cold, wet autumn and winter months. To prevent heavy losses, it can be helpful in areas of natural regeneration to cultivate the ground. Serious losses in the harvested seed can be avoided by collecting early, by laying out catching nets, or by dry storage. Finally, it is possible to use fungicides, although this only deals with the spores adhering to the exterior of the seed coat. Heat treatment appears to be an effective method of disinfecting seed which is already colonized (24 hours at 36–38°C and 100% r.h.) [152].
- *Ciboria batschiana* (Zopf) Buchwald (Syn. *Stromatinia pseudotuberosa* [Rehm] Boud.) is the cause of a black rot of acorns and which on several occasions has been blamed for the almost complete loss of sowings of oak. This fungus, too, develops best in damp, cool weather. It infects the acorns on the ground in the form of ascospores which are released in the autumn. They produce hyphae which enter the seed either through the apex or the base of the acorn. The first symptom of attack is the formation of dark spots on the outer skin of the seed while, beneath the skin, small, orange-yellow spots

with dark margins develop. At a later stage, which is accompanied by an unpleasant smell of decay, the cotyledons become brown and porous. In the following autumn, the fruit bodies of the fungus break out from the acorns which have in the meantime become shrunken, mummified, and completely black. They are apothecia, mainly occurring in clusters, each with a cinnamon-brown, bowl- to funnel-shaped disc 0.5–2 cm in diameter, and a blackish stalk tapering downwards. They produce ascospores which are formed in cylindrical asci. Prior to ascospore formation, microconidia (*Rhacodiella castaneae* Peyr.) develop. These are not viable, and so have no epidemiological role; they can, however, be used for diagnostic purposes (Fig. **5**).

As infection spreads from old acorns lying on the ground, the collection of acorns should commence as early as possible in the autumn. During seed storage, attack can be prevented by suitable storage conditions [167] or by the use of fungicides [20]. For seed that is already infested, hot-water disinfection (24 hours at 38°C) appears to be an effective treatment [55].

2 *Damage to Seedlings and Young Plants*

Nonparasitic Seedling Disorders

Frost is one of the more frequent abiotic causes of damage to seedlings and young plants. Frost damage may occur in almost any month, but is most prevalent in winter. The symptoms are the browning of needles or leaves or the death of whole plants. Susceptibility is largely determined by species and provenance. The importance of provenance is shown particularly clearly by Douglas fir which embraces a broad spectrum from highly resistant (*glauca* and *caesia* forms) to very frost-susceptible (some *viridis* provenances). Frost hardiness can also be influenced to some extent by fertilization.

A kind of frost damage peculiar to seedlings is 'frost lift' (also known as 'frost heave'). This phenomenon results from ice formation in the soil; the resulting expansion and breaking up of the soil pushes small plants right out of the ground. If the soil then thaws and settles back again, the seedlings fall over and dry up. This kind of winter damage happens chiefly in the absence of snow cover. Ways of avoiding frost damage in the forest nursery are late spring sowing, protecting the beds with brushwood or plastic netting, and ensuring well-balanced nutrition, taking care to supply sufficient potash, and to avoid too much nitrogen.

Heat damage occurs when the uppermost layer of soil is heated to a lethal temperature by insolation. For conifer seedlings, the critical point lies between 45 and 55°C [141]. A characteristic feature of this kind of damage is that the seedling falls over due to the death and collapse near soil level of part of the hypocotyl which is usually still unlignified and therefore soft at this stage. Heat damage is best known among broadleaved species, especially beech, in which the necrotic base of the hypocotyl may be stiff enough to keep the seedling upright for a long time, during which a club-shaped thickening of the stem develops due to the accumulation of assimilates above the dead zone (Fig. **6c**). Similar club-shaped swellings can—especially in conifers—be caused by the incautious use of herbicides.

Heat damage in young plants is quite often followed by infections by secondary fungi which give the impression of primary parasitism. Fungal genera which can be isolated from the dead bark include *Alternaria*, *Fusarium,* and especially *Pestalotia* (Fungi Imperfecti). *Pestalotia* species are easily recognized by their several-celled conidia which are brown except for a hyaline cell at each end. The hyaline end cells bear simple or branched setae. The most frequently occurring species [208] are:

- *Pestalotia hartigii* Tubeuf, with 3-septate conidia, 18-20 × 6 μm, on maple, birch, beech, spruce, lime, and Silver fir (Fig. **6a,b**).

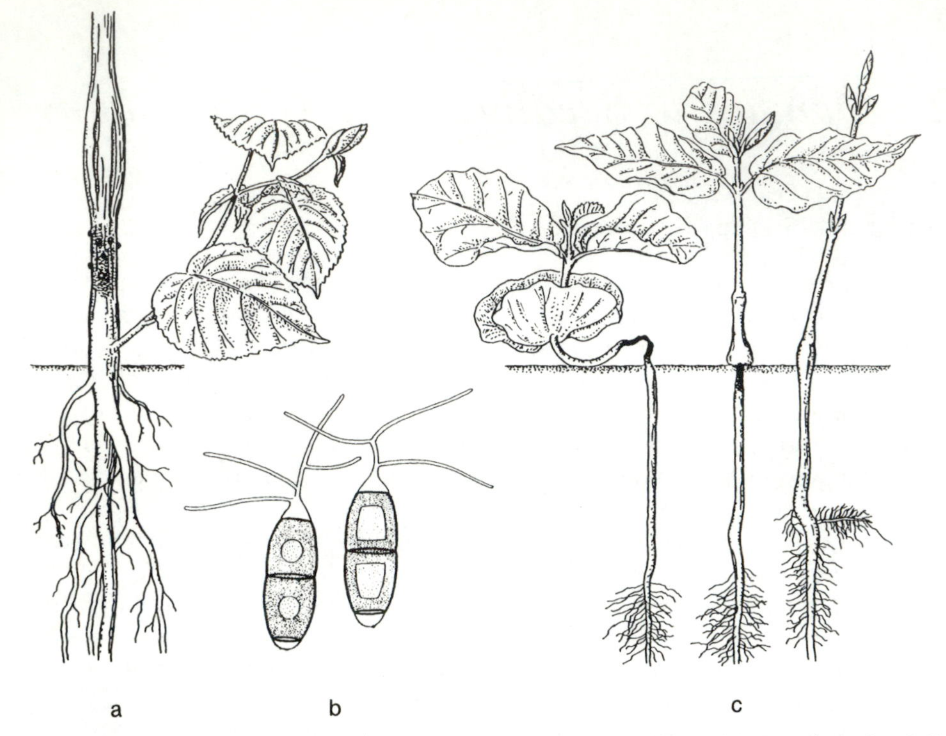

Fig. 6 Heat damage to seedlings. **a** heat damaged lime seedling with *Pestalotia hartigii*, **b** conidia; **c** various stages of heat damage to beech seedlings (c after Münch 1913)

- *P. funerea* Desm., with 4-septate spores, 21–29 × 9.5 μm (Table **II,1**); on *Chamaecyparis, Cupressus, Cryptomeria, Sequoia, Thuja,* and other conifers.

Nutritional disorders are particularly noticeable in seedlings, as their reserves of mineral nutrients are insufficient to overcome deficiencies in the soil. For example, nitrogen deficiency in conifers results in the development of relatively small pale green needles while a violet coloration is typical of a lack of phosphorus. The various deficiency symptoms can be countered by applying appropriate nutrients [15].

Smothering. Various fungal species can smother seedlings or young plants, but these vary greatly in their aggressiveness and in the degree to which their mode of action grades from the purely saprotrophic to the weakly parasitic. Their taxonomic range is as wide as their variation in aggressiveness and includes the following examples:

- *Thelephora terrestris* Erh. usually forms its fruit bodies on the ground, but these shelf-like, leather brown, and often multi-layered brackets will also develop over any nearby plant material—living or dead—which can give them extra support. In this way, young conifers can be completely enveloped and killed, with spruces

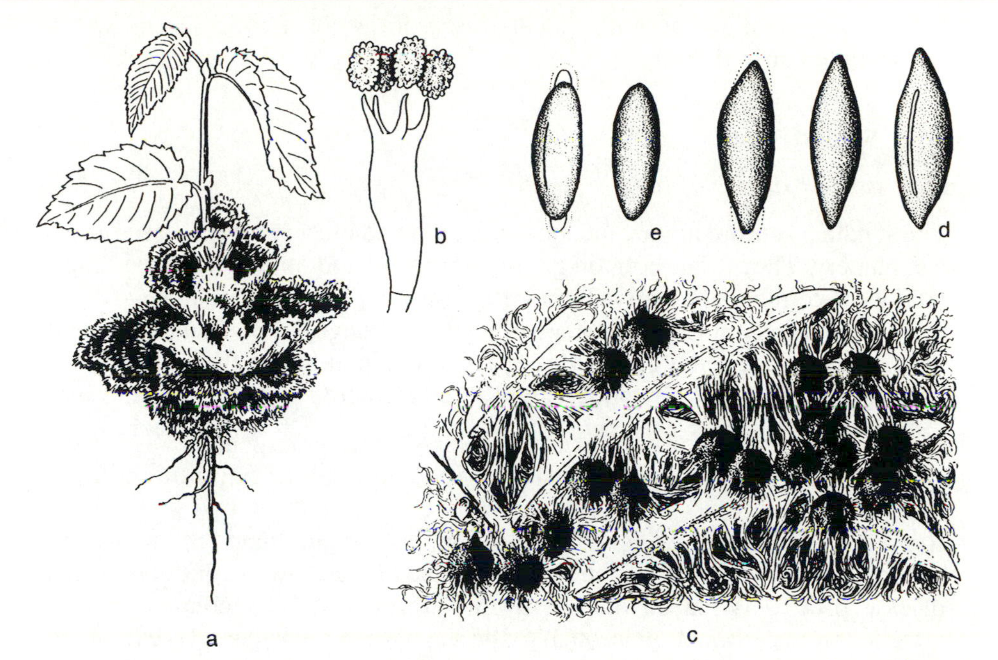

Fig. 7 'Smother fungi.' **a,b** *Thelephora terrestris*: **a** fruit bodies on birch seedling, **b** basidium with basidiospores; **c-e** *Rosellinia* species; **c** *Rosellinia minor* on spruce twig, **d** ascospores; **e** ascospores of *R. aquila* (d and e after Francis 1986)

and Silver fir being at particular risk. Among broadleaved species, birch seedlings in sandy soils are also vulnerable (Fig. **7a,b**).

- *Helicobasidium brebissonii* (Desm.) Donk (Syn. *H. purpureum* Pat.) is occasionally found at the base of stems and in the rooting zones, mainly of spruce transplants. This fungus, which occurs predominantly in the anamorph state, known as *Rhizoctonia crocorum* (Pers.) DC., first forms an epiphytic red-violet mycelium which can become parasitic, causing the affected plants to look sickly and die. It is of little significance on forest plants but it can cause considerable damage to agricultural crops by rotting roots [105].
- *Rosellinia minor* (Höhn.) Francis similarly grows epiphytically at first, covering needles or twigs in a thick brownish white hyphal mat. The affected needles later turn brown, indicating that the attack has entered a parasitic phase. Attacks occur mainly in dense stands of seedlings or transplants, where humid conditions invariably occur. Species attacked are Norway spruce and Douglas fir, less often Scots pine [77]. A reliable diagnostic feature in the later stages of attack is the development of the black, spherical fruit bodies, 0.6–0.8 mm in diameter, in which cylindrical asci with spindle-shaped, 1-celled, brown ascospores 20–25 × 6–7 μm in size, are formed (Fig. **7c,d**).
- The very similar *Rosellinia aquila* (Fr.) de Not., which likewise occurs on conifers, lives predominantly as a saprotroph. It can be distinguished from the

species described above by the smaller ascospores, 14–16 × 5–7 μm, which are rounded at both ends.

Conifer Seedling Rots

Cause: *various species of fungi*

The seedling rots are among the most common and most feared diseases in the forest nursery. They occur both on germinating seed and on first-year seedlings. Seedlings of conifers are generally more at risk than those of broadleaves.

Symptoms of seedling rots vary according to the stages of plant development at which attack occurs. Three types of attack, corresponding to these stages, can be distinguished [97], although they tend to intergrade:

- *Pre-emergence damping off* (rot of germlings before emergence)
- *Post-emergence damping off* (collapse of seedlings before cutinization of the hypocotyl)
- *Root rot of older seedlings* ('late seedling rot') The signs of disease in 'pre-emergence damping off' are mostly hidden from view, as this early-stage disease process is played out mostly underground. The seedlings may be attacked from germination through to the stage when the hypcotyl is extending and straightening up. Infection leads to a rapid decay of the affected tissues before the seedling can reach the soil surface (Fig. **8a**).

'Post-emergence damping off,' which is the type of seedling rot most often seen, occurs from two to three weeks after sowing. The first symptom, a

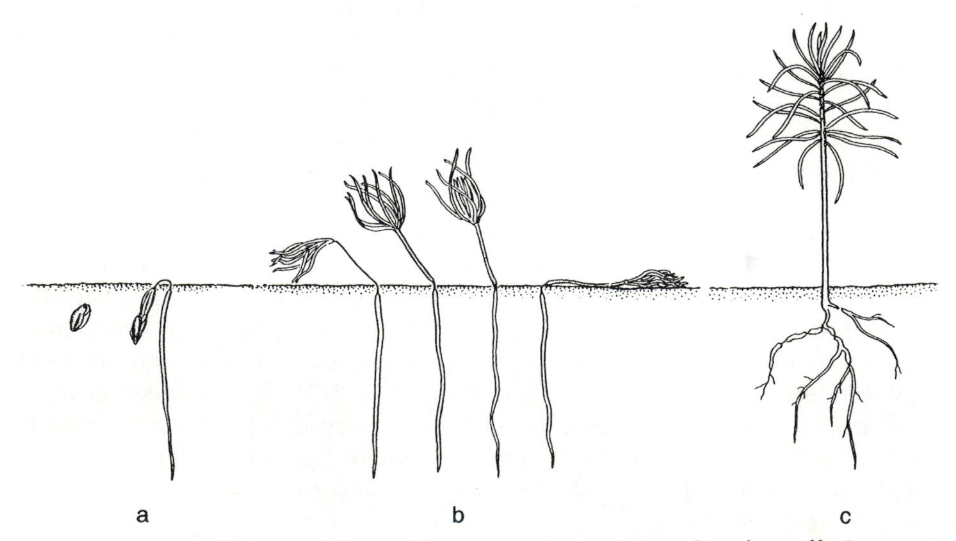

Fig. 8 Damping off in conifer seedlings. **a** pre-emergence damping off, **b,c** post-emergence damping off, **b** collapse of seedlings, **c** rootlet rot

yellowish brown discoloration of the lower part of the stem, is often not noticed; much more striking is the subsequent collapse of the seedling immediately above the soil surface which results from the destruction and shrivelling of the soft parenchyma, which is not yet cutinized (Fig. **8b**). Seedling rot can also occur eight or more weeks after germination. This 'late seedling rot' manifests itself as a partial or complete destruction of the root system. As the stem of the seedling has stiffened to some extent by this time, it can remain standing for a while, during which it turns brown. Seedlings whose root systems are only partially destroyed will, under favourable conditions, develop adventitious roots and recover. This late stage of seedling rot is on many occasions not recognized, being taken for drought or nematode damage (Fig. **8c**).

All three types of attack can be explained by the activities of soil-borne fungi. About 30 different species are now known to cause root rots in the seedlings of forest plants. Some of them are exclusively parasitic, e.g. *Phytophthora cactorum*, while others are facultative parasites, such as *Rhizoctonia solani* and some *Pythium* species, but most seedling fungi are primarily soil saprobes which can attack living plants when conditions favour this. Examples of such fungi are *Cylindrocarpon destructans* and some *Fusarium* species. The pathogenic ability of most seedling fungi rests on the production of certain enzymes or toxins which either break down the cell wall or cause physiological disturbance.

The following short descriptions are given of the most important seedling fungi, which can be selectively isolated by means of certain laboratory techniques [64]:

- *Pythium debaryanum* Hesse (Fig. **9a**), is a typical member of the Oomycetes with aseptate, much-branched mycelium and asexually produced sporangia which either function as conidia or produce zoospores. Sexually produced thick-walled oospores arise from fertilized oogonia. Other *Pythium* species which occur in neutral or alkaline soils are *P. irregulare* Buism., *P. sylvaticum* Campbell & Hendrix, and *P. ultimum* Trow.
- *Phytophthora cactorum* (Lebert & Cohn) Schöter, which also belongs to the Oomycetes, resembles *Pythium* morphologically (Fig. **10**) and in its mode of spread. Its typical hosts are broadleaved trees; less often conifers.
- *Fusarium oxysporum* Schlecht. is the most common seedling fungus of the numerous members of this genus [23], which are characterized by sickle-shaped, hyaline, multi-septate spores (Table **II,2**); in several species, 1- or to 2-celled microconidia and—mostly in culture—spherical chlamydospores occur. This fungus, which usually takes hold only after seedlings emerge, occurs in acidic substrates; it penetrates all the vascular tissue so that the seedling first wilts and later dies.—Further species: *Fusarium culmorum* (W.G. Smith) Sacc. (Table **II,3**), *F. avenaceum* (Fr.) Sacc. (Fig. **9b**). Their imperfect states are mostly *Nectria* species.
- *Cylindrocarpon destructans* (Zins.) Scholten [22], the conidial state of *Nectria*

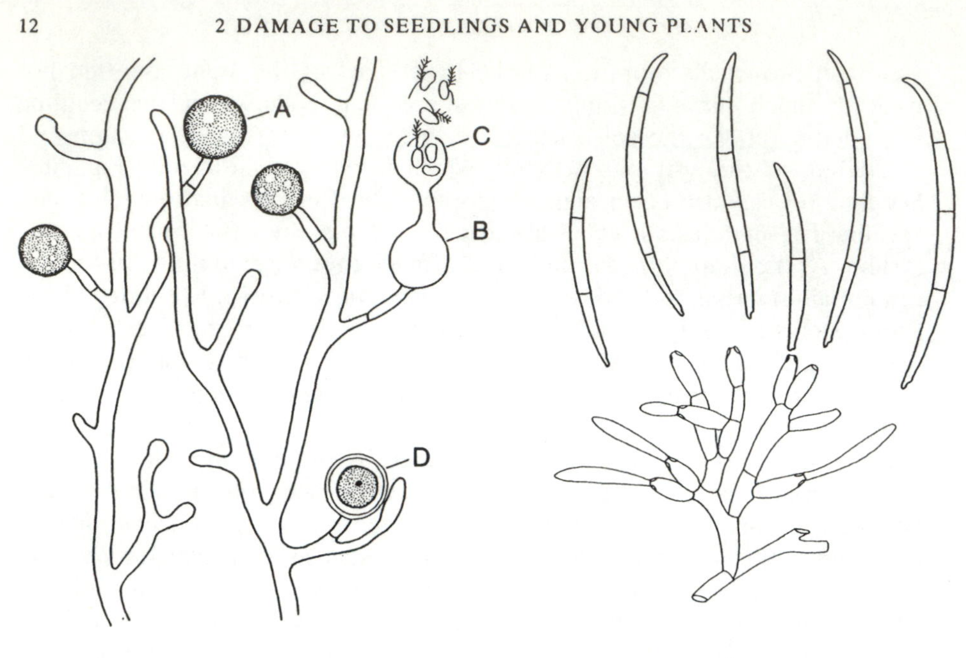

Fig. 9 Damping-off fungi. **a** *Pythium debaryanum* with sporangium (A), emptied zoosporangium (B), germination vesicle with zoospores (C) and oospore (D); **b** conidiophores with spores of *Fusarium avenaceum* (a after v. Arx 1981, b after Booth 1977)

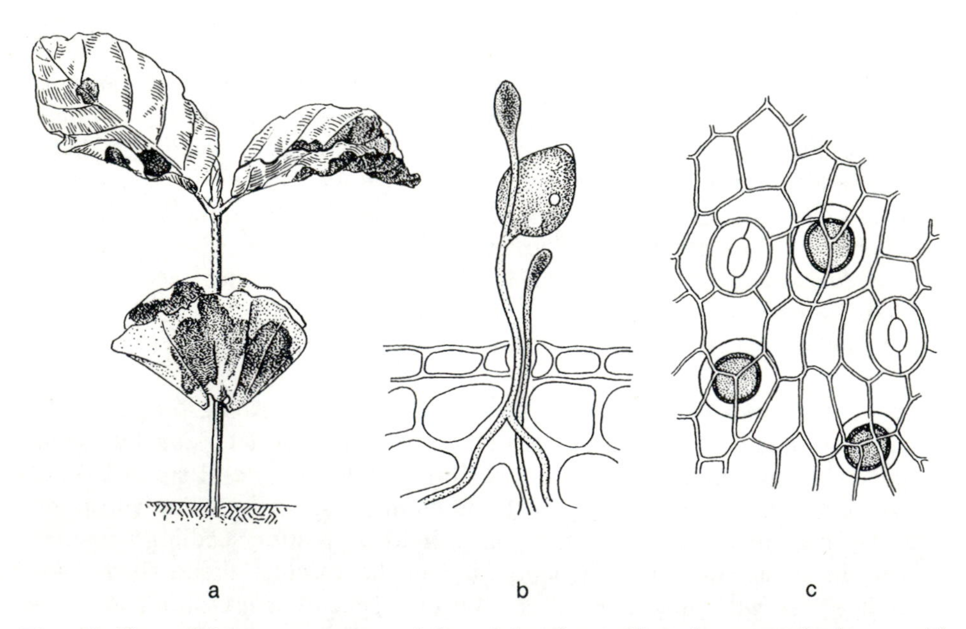

Fig. 10 *Phytophthora cactorum*. **a** infected beech seedling, **b** sporangiophores with sporangia at various stages of maturity, **c** oospores in dead cotyledon tissue (b after Hartig 1900, c after Ferdinandsen and Jørgensen 1938/39)

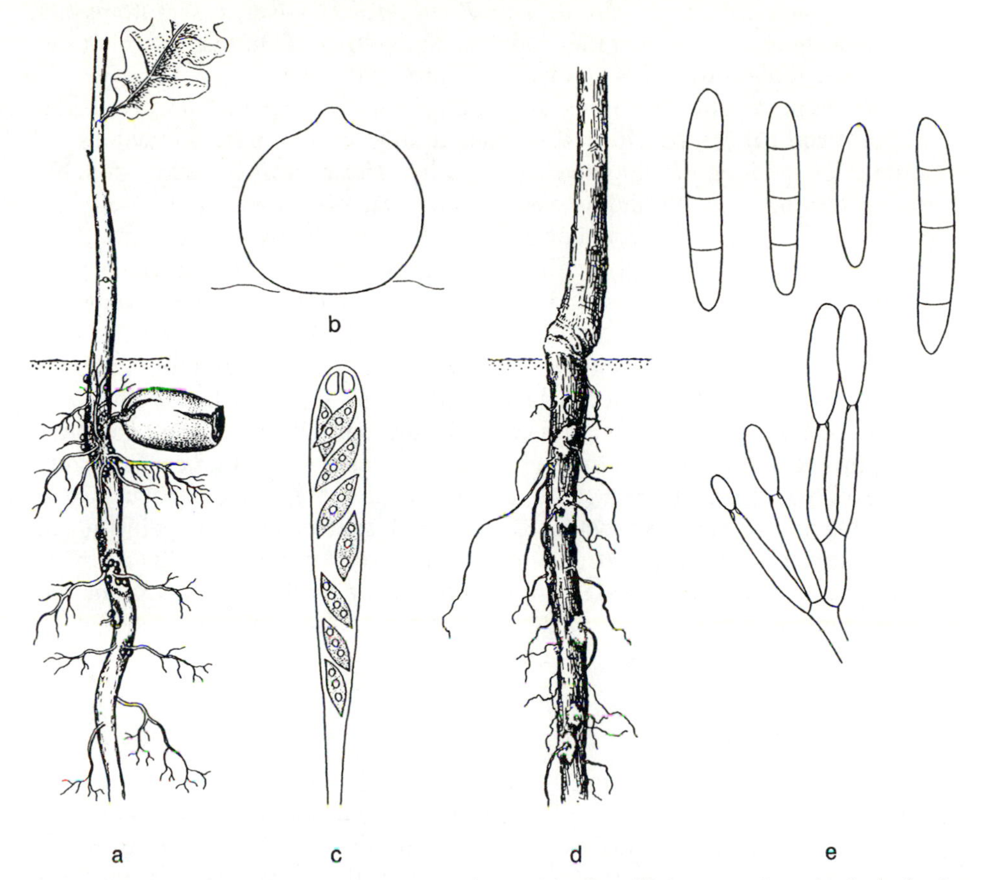

Fig. 11 Oak rootlet rots. **a** oak seedling with *Rosellinia quercina*, **b** fruit body (diagrammatic), **c** ascus with ascospores; **d** oak root with *Cylindrocarpon destructans*, **e** conidiophores with conidia (a after Hartig 1900, e after Booth 1966)

radicicola Gerlach & Nilson, is recognized from the hyaline, one or several-septate (occasionally aseptate) conidia, rounded at each end (Fig. **11e**).

- *Rhizoctonia solani* Kühn (Fig. **4d**), is the imperfect state of the basidiomycete *Thanatephorus cucumeris* (Frank) Donk. The fungus infests the root system and the root collar as a perthophyte. Its fast-growing, net-like and often superficial mycelium is septate, colourless or faintly brownish. Chlamydospores, 20–34 × 14–20 μm in size, are borne on right-angled side branches, either singly or with several forming a chain. These can also be found in the attacked tissues (a diagnostic feature). In autumn, blackish-brown sclerotia, 3–7 mm in diameter, are formed which can persist for several years. The perfect state is seldom formed (basidia with 4–6 drop-shaped basidiospores). The disease is of economic importance mainly in agricultural plants [3].
- *Macrophomina phaseolina* (Tassi) Goid., is the cause of 'charcoal root rot,' occurring on more than 300 plant species, including seedlings of

Abies, Cupressus, Picea, Pinus, and *Pseudotsuga*. Disease symptoms are the blackening of roots and the chlorosis and wilting of shoots. Distribution: in warmer temperate areas and in the tropics [58].

Various means of preventing and controlling seedling rots are available. Cultural control can be effected by maintaining optimal environmental conditions for seedling development and vitality. These must include a suitable site or rooting medium and a balanced water supply which should never be excessive. Direct control could be achieved biologically by supplementing the soil with antagonistic fungi such as *Trichoderma* species [1], but the general adoption of this technique would require more practical experience than has at present been gained. Of the physical methods of control, heat treatment of the soil can be mentioned, although this can only be considered for container systems or in situations where climate and the presence of light soils make solarization a feasible option. It is generally more practicable to use chemical plant protectants, although their effectiveness depends on numerous factors (e.g. soil type, plant species, causal agent, antagonists). If the particular type of pathogen involved is known, substances with specific activity against it can be employed. Chemical control can be carried out on the seed (by dressing it), in the soil before sowing (by soil fumigation, although environmental legislation may limit the use of such treatments), and even after emergence (by drenching or spraying).

Beech Seedling Disease

Cause: *Phytophthora cactorum* (Lebert & Cohn) Schröter

The disease manifests itself as a reddish brown, blotchy discoloration of the young cotyledons, primary leaves and stems. In wet weather, badly diseased seedlings soon collapse and rot off (in dry weather, the plants stay upright for a little longer). If the attack is confined to small parts of the cotyledons or primary leaves, the plant can recover.

In addition to the characteristic symptoms of an attack, several microscopic features are important in the diagnosis of the ‘beech seedling fungus’ which help to confirm its identity. In wet weather or if plants are kept in a moist chamber, numerous aseptate hyphae emerge through stomata and epidermal cells and give rise to hyaline sporangiophores, which bear apical or lateral sporangia [96].

If a sporangium comes into contact with water, its contents develop to produce flagellate zoospores which swarm out into the water to infect seedlings. The cycle from infection to the production of new zoospores takes only a few days, which explains the explosive spread of the fungus during wet weather. In dry weather, the sporangium functions as a conidium, being shed as a whole and aerially dispersed. In addition to wind and water, various animals can serve as vectors of the spores, including snails, insects, mice, and indeed man, by walking over the area. Thus, the ‘beech seedling fungus’ has several methods of disseminating itself to guarantee its survival. Apart from the sporangia and zoospores, which are important in the epidemiology, the fungus forms thick-walled oospores (Fig. **10**), typical of the Oomycetes to which it belongs. These have the function of resting spores, persisting in the soil for several years after the death and decay of the infected tissues within which they are formed.

Both cultural and chemical means of control are known. As the development of the fungus is encouraged by high soil moisture, shady sites, artificial shading, and waterlogged soils should be avoided. Where the disease has occurred, the infected beds should not be sown with broadleaved species for several years. A prophylactic treatment can follow by sterilizing the soil before sowing, while curative control can be effected by fungicidal spraying. If the risk of disease is high, it is necessary to check the seedbed daily and to remove and burn diseased and dead seedlings.

Other Pathogenic Species [223]:

- *Phytophthora cambivora* (Petri) Buism., is a cause of ink disease of Sweet chestnut, which is usually fatal. It causes a violet-brown discoloration of the cambium at the base of the stem and of the outer layers of sapwood. The disease occurs only in maritime climatic regions [52].
- *Phytophthora cinnamomi* Rands, is the cause of root killing in young plants of *Abies, Chamaecyparis* and yew. Needles become discoloured a dull green, later brown [228]. It can be prevented by good hygiene during plant propagation and in container plants by drenching the root ball. The fungus overwinters in the open only in mild winters. On *Quercus rubra*, the fungus causes patchy killing and blackening of the cambium at the stem base of older trees [56]. In Australia the fungus is one of the most dangerous root parasites and the cause of an epidemic tree killing disease, principally of *Eucalyptus*.
- *Phytophthora citricola* Sawada and *P. megasperma* Drechsler cause a root rot of Horse chestnut with resulting death of leaves and twigs [27].
- *Phytophthora lateralis* Tucker & Milbrath, causes a root rot of *Chamaecyparis lawsoniana* with resulting dieback. The disease occurs both in nursery stock and in older trees and is at present confined to North America.

Root Rots of Oak

Cause: *various fungal species*

Among the fungi which are involved in root damage on oak plants, the economically significant species include *Rosellinia quercina*, *Cylindrocarpon destructans*, and *Fusarium oxysporum*.

- *Rosellinia quercina* R. Hartig becomes damaging on 1- to 3-year-old oak and beech plants when they are standing in very wet soil [222]. The first symptom of disease is a yellowing and wilting of the leaves. If such plants are pulled out of the ground, the roots are found to be dead and often enveloped with a white mycelium ('white root fungus'). White fan-like mycelia are also produced by other fungi, and it is therefore necessary to use the fruit bodies for definitive diagnosis. These are perithecia which occur mostly superficially in groups on the base of the stem or deeper on the main root. They are spherical, black, and about 1 mm across. The ascospores are produced in slender asci and are fusiform, at first hyaline, later dark brown, and 21–26 × 7–9 μm in size (Fig. **11a–c**).

- *Rosellinia thelena* (Fr.) Rabenh. also lives as a root parasite on oak seedlings. It is distinguished from *R. quercina* by its long spore appendages (Table **I,1**).
- *Cylindrocarpon destructans* (Zins.) Scholten (Syn. *C. radicicola* Wollenw.) is now being increasingly found on dead roots of transplants of oak, and also on beech and other broadleaves. Plants are particularly at risk when they have been lifted and bagged, at which time their defence systems are often disturbed. Under these conditions, a fungus which at other times is a non-parasitic component of the rhizosphere of oaks and other broadleaves may be able to become pathogenic on its original host plant. It therefore follows that the best means of control is to avoid keeping the plants in bags for too long, and generally to minimize the time between lifting and planting out.

  The conidia, typically shaped for the genus, are of diagnostic value. They are produced in creamy yellow masses on the surface of affected roots. The macroconidia are hyaline, cylindrical, one- to several-celled, rounded at each end, and 25–30 × 4.5–6.5 μm in size (Fig. **11 d,e**).
- *Fusarium oxysporum* Schlecht. and other *Fusarium* species are only seen after transplants have been severely stressed by certain environmental factors such as waterlogging or drought. In these conditions, the roots of other broadleaves, such as beech and maple, are also attacked. A diagnostic character of nearly all *Fusarium* species is the sickle-shaped, multiseptate macroconidium (Fig. **9b**).

Shoot Tip Disease of Conifer Seedlings

Causes: *Strasseria geniculata* (Berk. & Broome) Höhn.
Botrytis cinerea Pers.
Sphaeropsis sapinea (Desm.) Dyko & Sutton

This disease, which occurs almost exclusively on 1–3 year-old seedlings of various conifers, has a characteristic effect on the recently flushed shoots, which are infected mainly when they start to elongate. The affected shoots become limp, curve downwards, and dry up while turning brown. The disease begins mainly in the leading shoot and then affects the laterals. After these die, new shoots develop from the lower lateral buds but these, in turn, can be attacked. The loss of several shoots can lead to the death of the whole plant. Along with pines and Silver fir, Douglas fir is particularly prone to attack.

To distinguish this disease from frost damage, the occurrence of fungi of the following species, which occur in close association with the syndrome, should be noted:

- *Strasseria geniculata* (Syn. *Allantophoma nematospora* Kleb.) seems to be the most dangerous of the three causal fungi. In seedbeds it occurs mostly

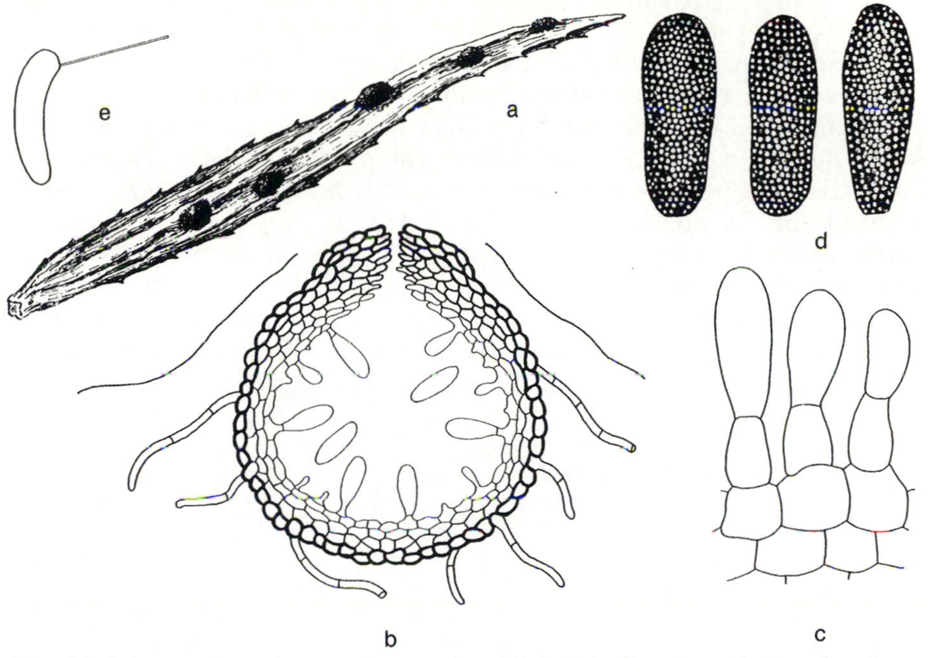

Fig. 12 *Sphaeropsis sapinea*. **a** pine needle with fruit bodies, **b** vertical section through fruit body, **c** part of the fruit body wall, **d** mature conidia; **e** spore of *Strasseria geniculata*

in patches. To identify it, the undersides of browned needles should be examined for the black pycnidia of the fungus. The cylindrical, slightly curved conidia which are formed here, are colourless, 10–13 × 3 μm in size, and furnished at the tip with a bristle about 15 μm in length (Fig. **12e**). The fungus is also one of the most common blue-stain fungi in pine timber.

- *Botrytis cinerea* is known everywhere as the cause of grey mould. It occurs every year, usually after needles have been damaged, more or less widely, and can appear on plants either parasitically or saprotrophically. Infection on germlings or 1-year seedlings is usually fatal. The fungus can immediately be recognized macroscopically by its greyish brown aerial mycelium which, in humid conditions, becomes visible on the diseased shoot tips. The spores, which are borne on di- trichotomously branched conidiophores, are ovoid and 9–12 × 6–10 μm in size (Fig. **53**).
- *Sphaeropsis sapinea* is known, in countries of the Southern hemisphere, as the cause of a shoot dieback in various conifers, especially on *Pinus nigra* and *P. radiata*. There, it occurs predominantly as a wound parasite on older trees damaged by hail or pruning. In central Europe, the fungus is seldom seen in young pine plantations but more frequently on 1–3 year-old seedlings. The hyphae, which penetrate the stomatal openings, form roundish fruit bodies on the killed needles which produce at first yellowish brown then dark brown

conidia with a warty surface; these are 30–45 × 10–16 μm in size and only rarely 2-celled (Fig. **12 a–d**).

The occurrence of all three of these fungi is encouraged by high air humidity. Other weather factors—temperature for example—are only of secondary significance. On the other hand, seedlings which have suffered from water stress, nutrient deficiency, or wounding seem to be particularly susceptible.

To prevent an outbreak of the disease, it is recommended that conifer seedbeds are not sited where the air is likely to be very humid in the early spring. Where seedlings are being raised under glass, the aerial parts of the plants can be dried more quickly if they are ventilated or if slow-running fans are employed. The control of weeds and the destruction of diseased material contribute further to the prevention of the disease. Where the danger of the disease is high, chemical control measures may be indicated.

Sirococcus Shoot Dieback of Spruce

Cause: *Sirococcus strobilinus* Preuss
Syn. *Ascochyta piniperda* Lindau

Sirococcus strobilinus is a potential cause of damage throughout the world, occurring as a pathogen predominantly in tree nurseries and gardens. Seedlings of *Picea pungens* and *Pinus contorta* are particularly at risk. *Picea sitchensis* has also proved to be susceptible. On the other hand, the fungus occurs on *Picea abies* mostly as a saprotroph with an endophytic preliminary stage. On older trees, attacks are of lesser consequence.

The beginning of the disease is seen as a brownish discoloration of the needles in the middle or at the base of the current shoot. In a mild attack it results in the curvature of the shoot tip which then hangs limply down. As the disease progresses, the shoots dry up and lose their needles; only at the tip of the down-curved, shrivelled shoot do the dead needles remain attached for some time longer. On the dead shoots and particularly on the needles at the tip of the shoot, the dark brown fruit bodies of the causal fungus develop during the course of the summer. In wet weather, spores are produced from these in the form of white drops or tendrils. The spores are 2-celled, hyaline, fusiform, and average 12 × 3 μm in size (Fig. **13**).

The dieback of shoots and spruce seedlings can also be caused by the fungus *Gremmeniella abietina*. The fungi can be differentiated by the typically sickle-shaped, multiseptate conidia of *G. abietina* (Fig. **55d**).

To prevent an outbreak of disease, it is advisable in the first instance to practise good hygiene (the removal of diseased material and avoidance of new plantings on already diseased areas). Furthermore, wet sites should be avoided. There is currently little information on the likely efficacy of fungicides but, since the infection takes place during the summer months, spraying at this time has the greatest chance of success. The most promising means of control is the selection of relatively resistant seed sources, such as have been found to exist in some species, including *Pinus contorta* [108] and Blue spruce.

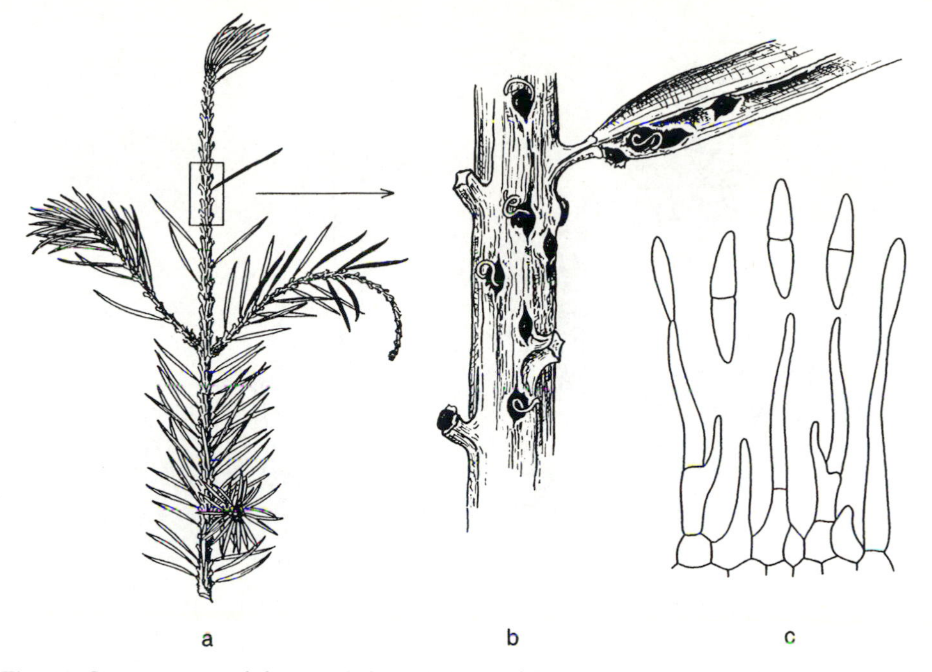

Fig. 13 *Sirococcus strobilinus*. **a** infected shoot of *Picea pungens*, **b** portion of shoot with pycnidia, **c** part of the fruit body wall with conidiophores and conidia

Meria Needle-cast of Larch

Cause: *Meria laricis* Vuill.

A yellowish or brownish discoloration and wilting of the needles of young larch in early summer indicates an attack by the imperfect fungus, *Meria laricis*. As a rule, the discoloration begins on the needles of the lower twigs and spreads more or less quickly upwards to the leading shoot. A proportion of the dead needles fall immediately, while others remain attached for some time to the twigs. The fungus does not seem to penetrate the twigs though it may, as a result of heavy defoliation, weaken the plant considerably and render it liable to attack by secondary fungi. In wet weather, conidiophores give rise at their tips to numerous 1-celled spores, 8–10 × 2.5–3 μm in size. Under the hand lens, the tufts of conidiophores appear as small white dots on the undersides of the needles. The fungus overwinters either in the needles still attached to the tree or in the fallen needles, and in the spring spores are again produced from these. Seedlings and young plants of European larch are particularly at risk whereas Japanese and hybrid larches are seldom attacked and then only mildly (Fig. **14**).

As attacks are favoured by high air humidity, larch seedbeds should be laid out on exposed, sunny sites and transplants should not be crowded. Infected beds should not be used for new plantings in the following years. Where plantings are greatly at risk, several treatments with fungicides could help to prevent excessive losses [182].

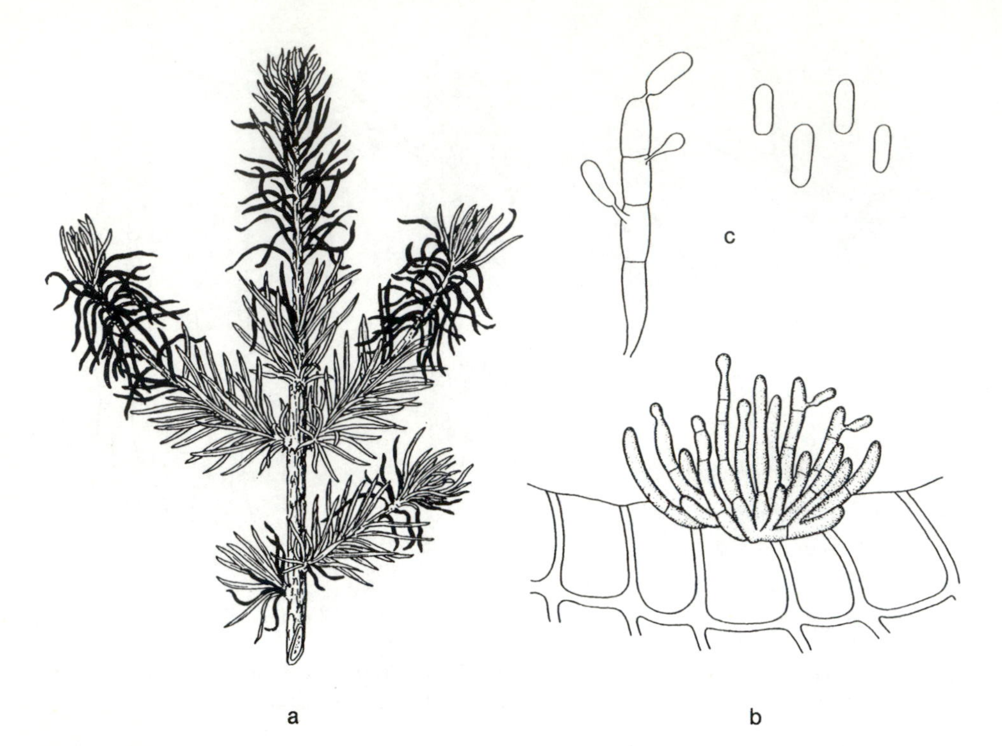

Fig. 14 *Meria laricis*. **a** symptoms on European larch, **b** conidiophores on the underside of a needle, **c** conidiophores with spores (a after Peace 1962, b after Ferdinandsen and Jørgensen 1938/39, c after Hartig 1900)

3 *Damage to Needles and Leaves*

CONIFERS: NEEDLE DAMAGE

Nonparasitic Needle Damage

Frost damage on conifers can cause various symptoms, depending on the time of year and the age of the needles. If temperatures below 0°C occur in May, when the new needles have begun to expand, the result is sudden wilting and—a day or two later—browning of most of the new flush. As a rule, only the newly forming needles are harmed by these 'late frosts,' whereas winter cold can affect older ones, especially those of the previous year's growth. This 'winter frost damage,' as the term implies, is the result of freezing injury during the dormant season. It usually occurs over much of the crown, and the affected needles typically show a marked pale or medium brown discoloration either at their tips or along their entire length. They may remain attached until some time during the following summer. This kind of damage can be seen on spruces, Silver fir, pines, Douglas fir, and also on cedars.

Winter frost/cold damage occurs when there is a change from unusually mild to unusually low temperatures in late winter [160]. The mechanism involved is clearly a premature loss of normal frost hardiness, but some pathologists believe that winter needle browning is also associated with an increase in water loss. This phenomenon, known as 'frost drought' or 'winter drought,' produces similar symptoms to freezing injury. Another complication in characterizing 'winter frost' and 'winter drought' damage is the role of various stress factors in impairing frost hardiness; these include SO_2 pollution, potassium deficiency, or dry sites. Winter browning of needles is still, therefore, a cause for controversy. True winter drought damage can be expected to occur mostly in strong sunlight—especially on young plants—when transpiration from the needles exceeds water uptake from the frozen ground. This process may be responsible for the needle browning that occurs on certain species in the mountains of central Europe; e.g. spruces, Dwarf mountain pine (*Pinus mugo* var. *pumilio*), and Cembran pine (*P. cembra*).

Deficiency diseases are brought about by the lack of certain nutrients or trace elements. Typical symptoms include a reduction in needle size or various kinds of needle discoloration which are often of diagnostic value:

- Iron/manganese deficiency (chalk chlorosis): a whitish yellow discoloration initially at the base of the youngest needles; later a complete yellowing; older needles green.
- Potassium deficiency: a faint, yellowish discoloration with the needle tips browned.
- Magnesium deficiency: yellow tips or complete yellowing of the older needles;

in Norway spruce, a striking 'new-type' symptom at higher elevations in central Europe.
- Nitrogen deficiency: a uniform pale green discoloration of all needles and often a reduction in needle size.

For confirmation of the diagnosis, chemical analysis of the needles is necessary. Further information on deficiency diseases can be found in the specialist literature [15,99].

Salt damage, which affects needles and twigs, is nowadays mainly attributed to the effects of deicing salt (NaCl), but in coastal areas it quite often occurs due to the deposition of wind-borne salt spray [63]. Damage from deicing salt is most prevalent in areas where it is heavily applied, especially along streets and footpaths. Symptoms on conifers, unlike those on broadleaved trees, are largely non-specific. In Norway spruce and Douglas fir, the needles of the current shoot turn reddish brown and later fall. In European silver fir, which is considerably less susceptible, the needles merely become olive grey to olive brown, or become spotted with yellow, even if exposure is severe. At the other end of the scale, salt damage to Serbian spruce (*Picea omorika*) is so characteristic that it has earned the German name 'Omorika-Sterben.' The symptoms of this condition are chlorotic spotting and needle browning beginning at the needle tip and developing most severely on the needles at the tip of the shoot.

Salt damage can also result from the use of fertilizers which contain Cl^-. These may include garden fertilizers; even those recommended for firs, and so it is advisable to select Cl^--free fertilizers if salt damage is to be avoided. For curative treatment of Serbian spruce, high applications of magnesium sulphate (Epsom salts) have proved effective. For most other species, the best recommendation is prevention—either by applying salt sparingly and accurately, or by using non-toxic salt substitutes.

Air pollution damage has become a special focus of attention among the abiotic disorders [200]. It can occur directly via the atmosphere when photosynthetic organs are exposed to gaseous or dissolved pollutants, or via the soil when solutions of some of these substances come into contact with roots.

In central Europe, two basic effects of atmospheric emissions can be characterized. The first occurs close to the emission source, and its symptoms have been known for more than 100 years, being attributed in the literature to 'smoke damage'. Emissions may include both particulate matter and various phytotoxic gases (e.g. hydrogen chloride, hydrogen fluoride, sulphur dioxide) which produce characteristic symptoms. For example, sulphur dioxide combines with moisture in the air to form sulphuric acid, which, in acute cases, causes characteristic needle discolorations; typically a reddish brown needle tip necrosis, as on the very susceptible European silver fir (Fig. **15c**). For an exact diagnosis of any particular case, specialist literature must be consulted [93,214].

The other types of damage now recognized do not fit the classic descriptions of smoke damage, and have been perceived as 'new phenomena' since the late 1970s, being grouped together under the term 'Waldsterben' (forest decline). They may occur in areas far from industry, and the consensus is

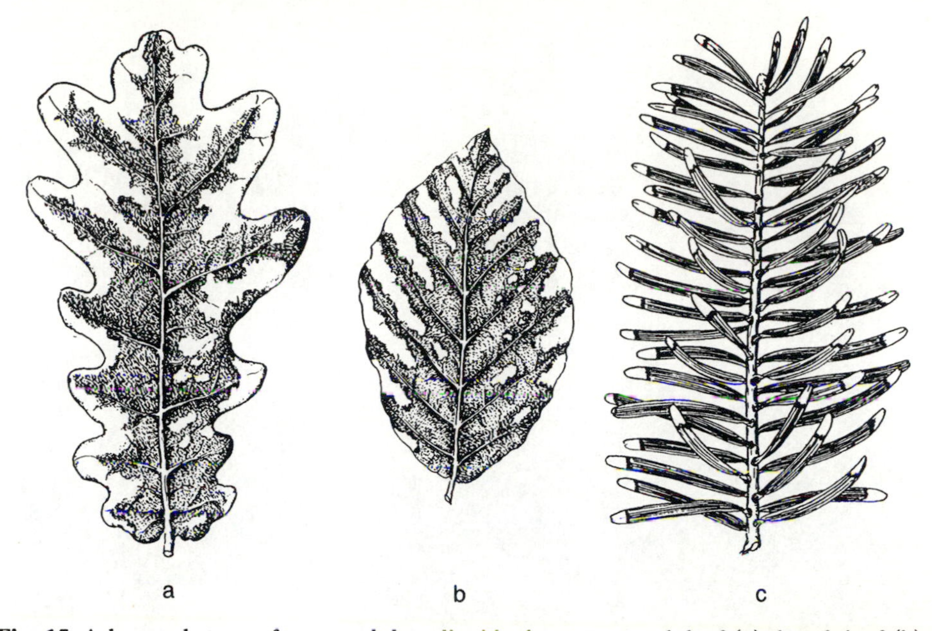

Fig. 15 Advanced stage of acute sulphur dioxide damage on oak leaf (a), beech leaf (b), and Silver fir needles (c)

that they are a complex set of disorders in which long distance transport of aerial pollutants plays a central role in combination with ordinary environmental factors such as rainfall, site and soil conditions, and silvicultural practices. The anthropogenic pollutants involved in this complex include sulphur dioxide, oxides of nitrogen, ozone, various hydrocarbons, and heavy metals [183]. They can be deposited either in wet or dry form. In the case of SO_2, one of the effects of wet deposition is to increase the acidity of the soil, which can induce aluminium or manganese toxicity in the soil.

The visible signs of 'forest damage' have been characterized in detail for Norway spruce and include needle discolorations and severe defoliation. In forms of Norway spruce which have the 'comb' type of crown, the defoliation combines with the death of buds to give a characteristic appearance; their second-order twigs become largely bare and appear to hang limply downwards, earning the term 'lametta syndrome' (Fig. **16**). Crown density is these days one of the most important criteria in forest damage surveys [93].

Rhabdocline Needle-cast of Douglas Fir

Cause: *Rhabdocline pseudotsugae* Sydow

The causal agent is sometimes known as the Swiss needle-cast fungus, despite being a native of North America and introduced to Europe as recently as 1922. It is a specific needle parasite of Douglas fir, each variety of which shows a different

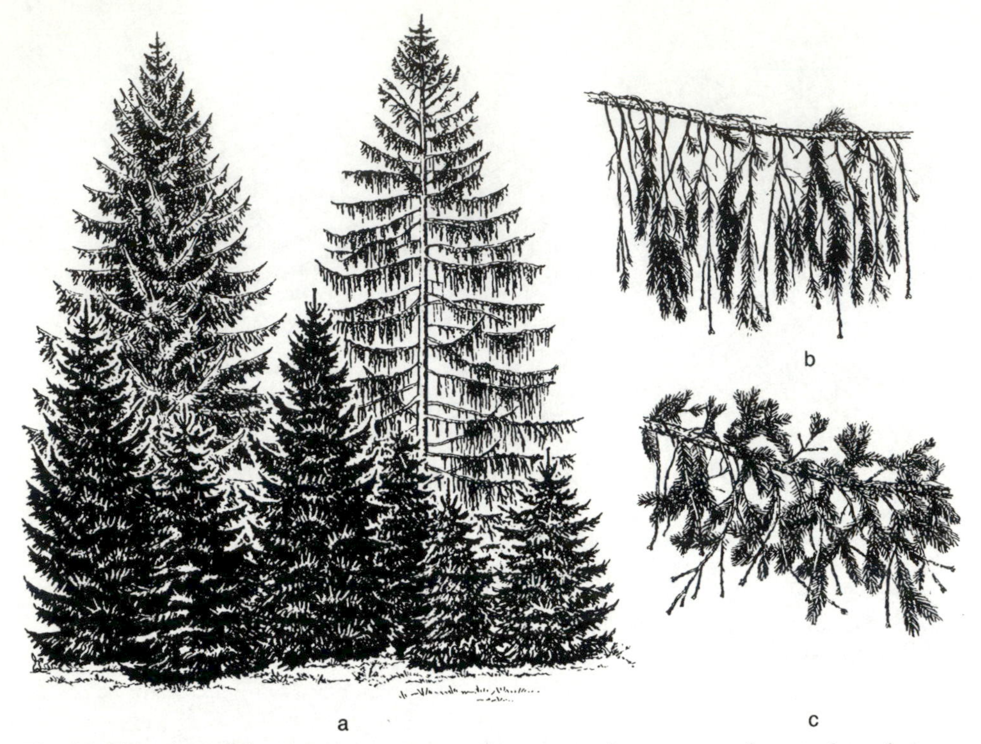

Fig. 16 Thinning of Norway spruce crowns due to complex causes. **a** damaged comb-type spruce with (right-hand tree) the 'lametta syndrome;' **b** defoliation of 1st and 2nd order twigs on a comb-type spruce, **c** the same symptom on a brush-type spruce

degree of susceptibility. The *caesia* and *glauca* forms are particularly susceptible, for which reason they are now very much less often planted in Europe than the less susceptible *viridis* form, which in turn shows variation between provenances. Trees from 2–30 years-old are attacked.

The first signs of attack are pale green flecks which in late autumn assume an orange shade. In winter, with the first frost, the flecks become a violet-brownish colour which contrasts with the green, as yet uninfected, parts of the needle, so that the needles appear marbled. During April, the fruit bodies are formed on the undersides of the needles. They ripen from May to July and then break through the epidermis to release their spores. This release takes place within the tree crowns, since the infected needles are still attached at this stage. The spores infect buds just as they are opening, and this completes the 1-year development cycle of the fungus; the old infected needles turn a uniform yellowish brown and fall in the same year, except for those that show only a few spots of infection; these are cast in the following year.

Since a whole year's needles are commonly missing as a result of this disease, it is subsequently possible to determine the year in which conditions (high air humidity in the spring or precipitation) were favourable for the development of

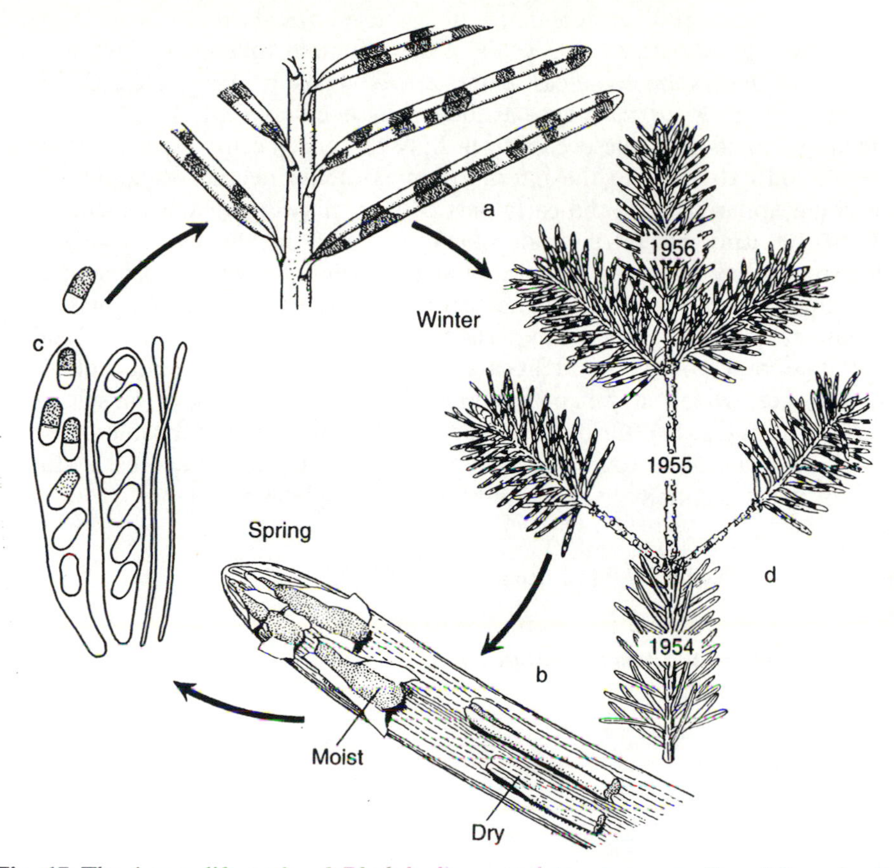

Fig. 17 The 1-year life cycle of *Rhabdocline pseudotsugae* on needles of Douglas fir. **a** development of flecks, **b** formation of fruit bodies, **c** asci and ascospores with paraphyses, **d** symptoms after the loss of a whole year's needles

the fungus (Fig. **17**).

This pathogen, which is an ascomycete, forms orange-yellow apothecia on the undersides of the needles. They are 2–4 mm long, cushion- or shoemaker's last-shaped, containing numerous clavate asci, arranged parallel with each other, with spores which are at first 1-celled and hyaline, later becoming 2-celled and light brownish in colour and measuring 18–20 × 6.5–7.5 μm. Other species and subspecies sometimes show a 2-year development cycle; these are known only from North America [147].

Phaeocryptopus Needle-cast of Douglas Fir

Cause: *Phaeocryptopus gauemannii* (Rohde) Petrak

This is another introduced pathogen, appearing for the first time in 1925, several decades after Douglas fir was brought over from North America to Europe. This pathogen is, in one respect, less damaging than *R. pseudostugae*, as it induces needle shedding only after 2 to 3 years of development following infection. As in the case of the Swiss needle-cast pathogen, Douglas fir provenances differ in their resistance to *R. pseudotsugae* as judged by needle retention [128].

Infection, which as a rule occurs from May to June, occurs via the stomatal apertures and extends into the internal tissues of the needles. Symptoms do not become apparent until the early part of the following year when the very small (50–100 μm), black fruit bodies begin to break out through the stomata. At this stage, they are characteristically arranged in lines which correspond to the locations of the stomata. In the second and third years, additional fruit bodies appear on previously green parts of the needles, giving their undersides a sooty appearance. After about 3 years, the diseased and browned needles are shed, a process which may happen sooner in a severe attack or on a highly susceptible tree (Fig. **18**). If ripe fruit bodies are examined under the microscope, sack-shaped asci will be found, each containing eight 2-celled spores, 9–10 × 3.5–6 μm in size, initially hyaline, and later turning brownish [166].

Lophodermium Needle Blight of Spruce

Cause: *Lirula macrospora* (R. Hartig) Darker
Syn. *Lophodermium macrosporum* (R. Hartig) Rehm

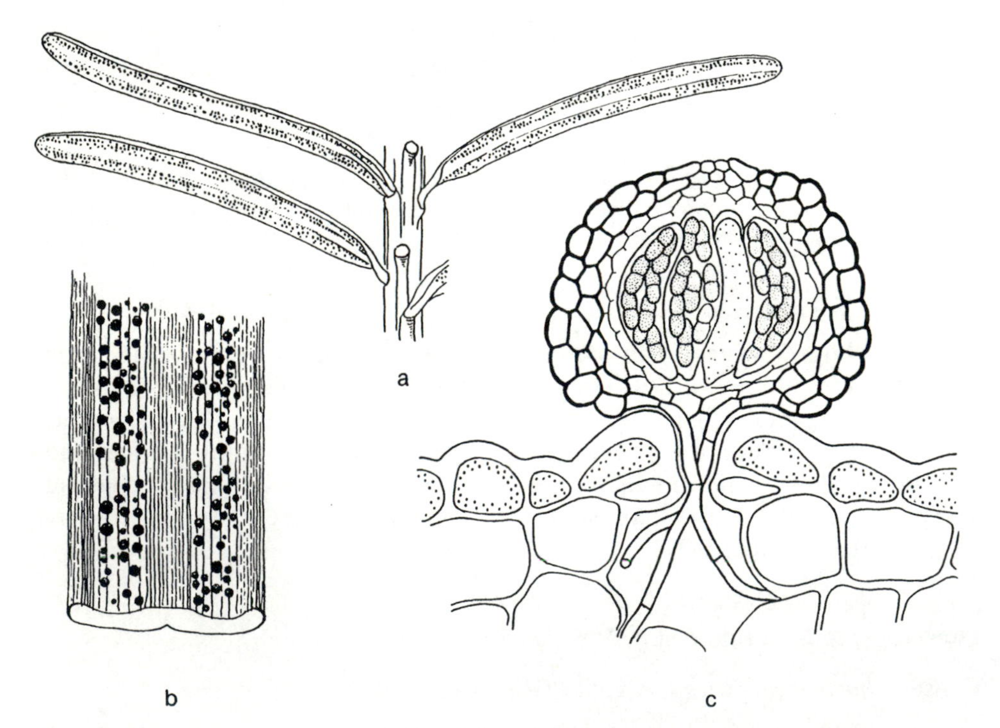

Fig. 18 *Phaeocryptopus gaeumannii*. **a** infected Douglas fir needles, **b** needle segment with fruit bodies, **c** vertical section through a fruit body

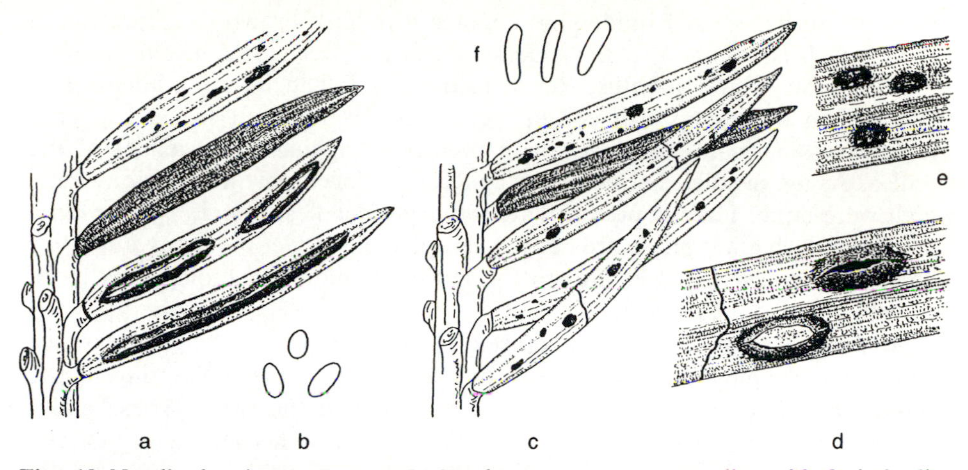

Fig. 19 Needle fungi on spruce. **a,b** *Lirula macrospora*: **a** needles with fruit bodies (hysterothecia), **b** spermatia; **c-f** *Lophodermium piceae*: **c** needles with spermogonia, **d** magnified needle segment with hysterothecia, **e** needle segment with spermogonia, **f** spermatia

This fungus, which occurs on various species of spruce, mainly attacks individual needles on the penultimate year's shoots of 10–40 year-old trees. The needles are killed during the first year of infection and then show a characteristic pale brownish to pale yellowish colour. In the following year, watery elliptical blisters appear on the surfaces of the needles, later developing black outlines. These are spermogonia, containing numerous conidia which can be seen under the microscope. They are ovoid, hyaline, 3×2 μm in size, and they function as spermatia. They are followed a few months to a year later—depending on the weather—by hysterothecia which are recognizable as black, shiny swellings, 2–8 mm long, on the undersides of the needles.

As with some other needle diseases, black phenolic substances are deposited, but these are concentrated in the form of a ring at the needle base, rather than as black transverse bands (cf. *Lophodermium piceae*). This ring occupies the abcission zone, apparently blocking the process of shedding which would normally occur after death of the needle, so that it remains attached to the twig for some time (Fig. **19a,b**).

There is some difference of opinion in the literature concerning the prevalence and significance of this fungus on spruce. It is known to cause heavy attacks and premature needle loss on wet sites, and it seems preferentially to attack needles suffering from lack of light and premature senescence. In the absence of any obvious economic losses from the disease, no experience has yet been gained in attempting to control it.

Spruce Needle Reddening

Associated fungal species: *Lophodermium piceae* (Fuckel) Höhn.
Tiarosporella parca (Berk. & Broome) Whitney
Rhizosphaera species

The term 'spruce needle reddening' denotes a reddish brown discoloration of spruce which is confined to the older needles and begins mainly in autumn. In the ensuing winter months, the dead needles fall in large numbers. This phenomenon occurs only in certain years, and is believed to be triggered by extremes of weather together with premature ageing (senescence) of the needles. As the needles die, there is a concurrent development of various fungi which were either already present latently in the needles (e.g. *Lophodermium piceae*) or which enter afterwards as saprotrophs (e.g. *Rhizosphaera kalkhoffii*). The following description relates to the more common and biologically more interesting species, *L. piceae*.

The first external sign of the presence of the fungus is the appearance of blackish spermogonia formed subepidermally on the needles during the autumn–spring period while they remain attached to the twigs. These structures contain numerous small (3.5-5.0 × 1.5 μm), cylindrical, slightly curved spermatia, which play no part in the dissemination of the fungus. At the same time, black transverse lines ('zone lines') appear on the needles and can be used as a feature to distinguish this fungus from *Lirula macrospora*. The perfect state usually begins to develop only after the needles have fallen to the ground. This is characterized by black hysterothecia which are raised, elliptic, or boat-shaped bodies, 0.8–1.4 mm in size, forming singly or in groups on the upper sides of the needles. These fruit bodies are typical of the Rhytismataceae in possessing a longitudinal slit whose opening is regulated by the uptake of moisture. The jelly-like hymenium within is composed of filiform paraphyses together with asci measuring 110–130 × 11–13 μm, in which needle-shaped ascospores (80–100 × 1.8–2.5 μm) are produced. The development cycle (Fig. **19c–f**), which can include a latent period of several years, is completed by the ripening and release of the spores in May or June [146].

Lophodermium piceae is a ubiquitous fungus in thicket-stage spruce, developing on dying needles suffering from lack of light or damaged in some other way, e.g. by frost. Its role in 'needle reddening' was long debated but its non-pathogenic character is confirmed by new findings, especially the observation that *L. piceae* can exist asymptomatically for several years in green needles without causing any visible damage to the host cells. The fungus only 'attacks' the tissues if the physiological state of the needle changes either due to endogenous senescence or to external damage. Organisms of this type, which remain for a long time in living plant tissues without causing damage, are called endophytes. In spruce, more than 50 species have to date been found to lead an endophytic life in green needles [38].

Among other endophytes which only develop after the spruce needles die is *Tiarosporella parca* [103,193]. This imperfect fungus is characterized by grey-black, subepidermal pycnidia which can easily be confused with the spermogonia of *L. piceae*; however, they contain cylindrical conidia, 34–45 × 6–7 μm in size, each with an appendage (Table **II,4**). Not all fungi occurring on brownish red discoloured needles are endophytes, some (e.g.

all the *Rhizosphaera* species on Norway spruce) start their colonization of the needles mainly after their death.

Spruce Needle Rust

Cause: *Chrysomyxa* spp.

Chrysomyxa species, which are the most important rust fungi found on spruce needles, have a similar relationship to spruce as the *Coleosporium* rusts have to pine. The symptoms of an attack are very much like those of pine needle rusts, but the needles are more seriously damaged; after flushing, yellow to pale yellow zones of discoloration appear on them and they fall prematurely. In the heteroecious species, cushion-shaped, orange-yellow, 3-mm-wide aecidia burst out of the needles and the irregularly torn whitish peridia remain on the needles after the spores have been released. The development of the uredo-, teleuto-, and basidiospores takes place on the leaves of various herbaceous or woody flowering plants. The following rust species are distinguished by their association (if any) with various dikaryotic-state hosts [80,226]:

- *Chrysomyxa abietis* (Wallr.) Unger is a microcyclic rust without an alternate host, in which the aecidio- and uredospores are lacking. Golden yellow, pustule-like teleutosori develop on 1-year needles in autumn of various spruce species. Basidiospores can infect the young needles again in spring directly. The rust is damaging because of premature defoliation, especially in Christmas tree plantations.
- *Chrysomyxa empetri* (Pers.) Schröter, which alternates between spruce (haplontic-state host) and crowberry (dikaryotic-state host), is confined to areas where *Empetrum nigrum* occurs.
- *Chrysomyxa ledi* (Alb. & Schwein.) de Bary alternates between spruce (haplontic-state host) and *Ledum palustre* (dikaryotic-state host), coincides with the range of *L. palustre,* and is confined to the northern parts of Europe.
- *Chrysomyxa rhododendri* (DC.) de Bary alternates between spruce (haplontic-state host) and *Rhododendron* species (dikaryotic-state hosts). It occurs in Europe, principally in the Alps on *R. ferrugineum* and *R. hirsutum,* and also on other *Rhododendron* species and varieties, e.g. in Great Britain, and has a haplontic-state host, mostly *Picea abies*, and occasionally also *P. sitchensis*.

Rhizosphaera Needle Browning of Spruce

Cause: *Rhizosphaera kalkhoffii* Bubák

This fungus, a member of the Fungi Imperfecti, is distributed throughout the world, living in conifer needles, especially those of the genus *Picea*. On certain species of spruce—e.g. *Picea pungens* and *P. engelmannii*—it can cause considerable and economically significant needle damage, mainly in nurseries and in Christmas tree plantations. In these ornamental spruce species, the damage

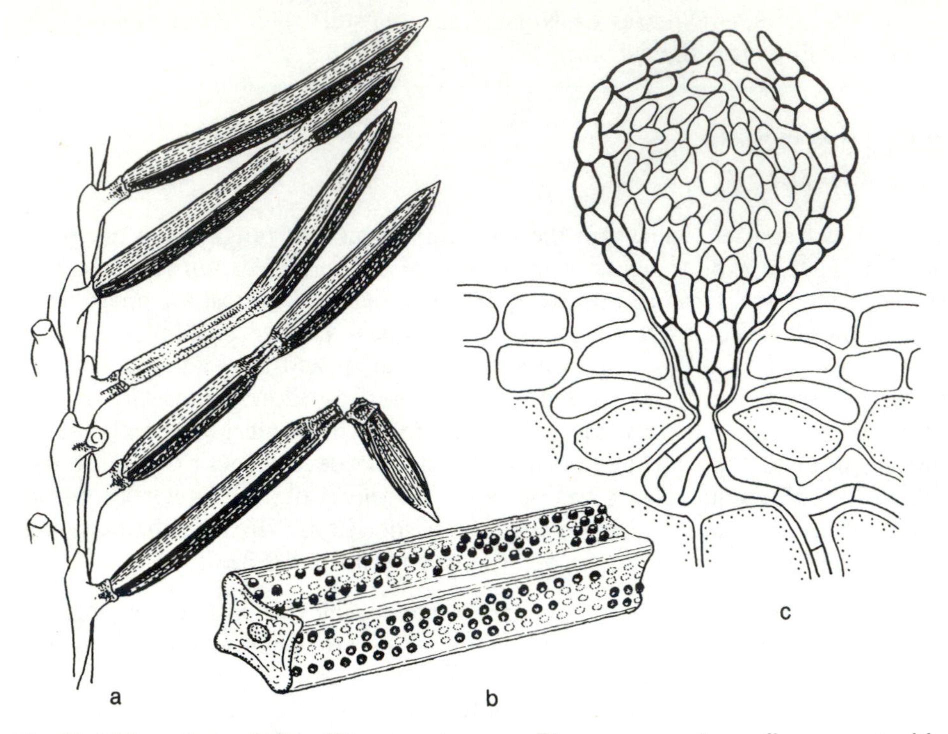

Fig. 20 *Rhizosphaera kalkhoffii*. **a** symptoms on *Picea pungens*, **b** needle segment with pycnidia, **c** vertical section through a pycnidium with spores

takes the form of discoloration, girdling, and breakage of the needles, whose bases remain in position for a long time before falling to leave a bare twig.

In the case of Norway spruce, it is doubtful whether needle reddening or browning can be caused by the fungus. Its fruit bodies certainly occur often on needles previously damaged by abiotic factors (lack of light, extremes of weather, K deficiency, or the effects of emissions) [59]; however, a direct causal relationship with the death of needles of this species has not been demonstrated. Thus, although *R. kalkhoffii* must evidently be regarded as a primary or as a weak parasite on certain spruce species, it is predominantly saprotrophic on Norway spruce and, as such, it can come into competition with *Lophodermium piceae* in certain situations.

The mycelium of *R. kalkhoffii* rapidly invades the needles after they die, and, within as little as 2 or 3 weeks, gives rise to pycnidia, which burst out through the stomata. These are black, roundish, and 80–150 μm in diameter. The conidia, which arise directly from the inner wall of the fruit body, are ovoid, hyaline, and 7–10 × 3–5 μm in size (Fig. **20**).

*Other **Rhizosphaera** Species on Spruce* [82]:

- *Rhizosphaera pini* (Corda) Maubl. is a weak parasite on various spruce, pine, and fir species. Its fruit bodies are yellowish brown and 50–65 × 125 μm in size. The conidia are ellipsoidal to ovoid, at first hyaline, later brownish, and 16–20 × 7–10 μm in size (Table **I,2**).

- *Rhizosphaera macrospora* Gourbiere & Morelet occurs on needles of *Abies alba*, and occasionally on spruce. The conidia are spherical or ovoid, 11–22 × 10–18 μm in size (Table **I,3**).

Lophodermium Needle-cast of Pine

Cause: *Lophodermium seditiosum* Minter, Staley & Millar

In forestry practice, 'pine needle-cast' is taken to mean the premature and heavy shedding of needles or dwarf shoots of pine after infection by the fungus *L. seditiosum*. Although, in central Europe, this fungus can attack various other species of pine (*P. cembra, P. mugo, P. nigra)*, it plays an economically significant role only on *P. sylvestris,* principally in plantations.

The first symptoms of this usually annual disease are tiny yellow flecks on the needles, recognizable under a hand lens from September onwards, which later become larger and browner. As each fleck is the site of an infection, their frequency can be used for the prognosis. Thus, needles which show at least 5 flecks will die and fall to the ground by the end of May at the latest. Although this 'spring defoliation' is the rule, dwarf shoots can be shed throughout the year, when—depending on the infection intensity and weather conditions—several waves of defoliation can occur (Fig. **21**).

A physiological phenomenon which ought not to be confused with the parasitic 'Pine needle-cast' is the 'autumn needle-cast' which in some pine species can suddenly set in (e.g. *Pinus cembra* and *P. mugo*). This is caused by normal senescence and as a rule only the older years' needles are involved. This process is essentially the equivalent of the autumn leaf fall of deciduous broadleaves. Various stress factors can, however, modify and intensify this process so that even the second year's needles, for example, can yellow and fall prematurely. Among those factors which trigger this premature senescence are extreme weather conditions and pollutants.

The fruit bodies of *L. seditiosum* (hysterothecia) form during the summer on the second-year needles lying on the ground. Heavy and persistent rainfall during this period (June–September) favours spore production.

The hysterothecia are black, oval to boat-shaped, and 1–1.5 mm long. They imbibe moisture in humid conditions and then open by greenish, shiny, longitudinal slits so that the thread-like, 90–130 μm-long, 1-celled ascospores can be expelled. At some time before the perfect state appears, microconidia (spermatia), 5–8 × 1 μm in size, are sometimes formed in oblong, blackish pycnidia [133]. These have no epidemiological significance, though they can be used as a diagnostic feature (Fig. **22**).

The severity of damage decreases as the age of the pine trees increases, until, by their 10th year of life, they have usually outgrown the danger. Furthermore, the survival of a plantation is endangered only if at least two consecutive wet summers have intensified the epidemic. Only in such cases should chemical protection be considered and, in any case, it is preferable to reduce the incidence

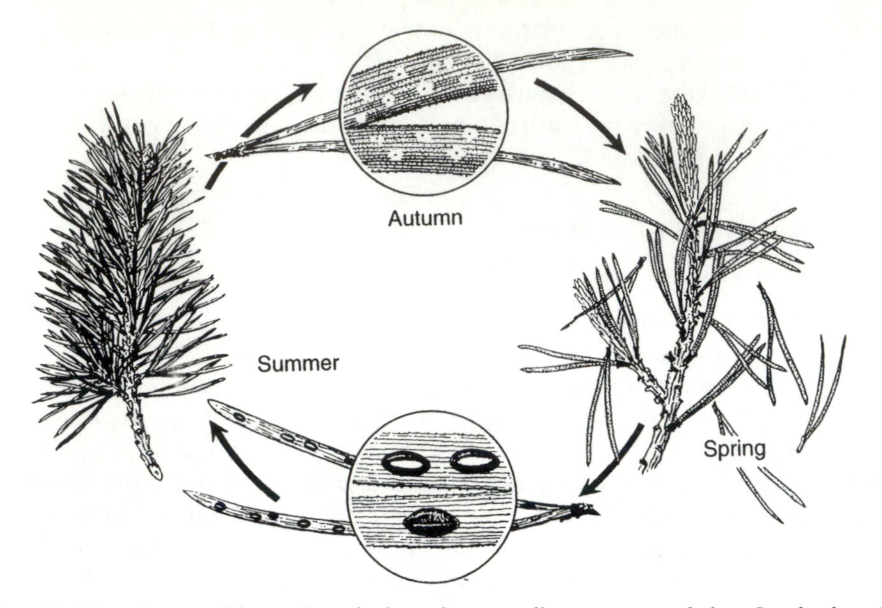

Fig. 21 The 1-year life cycle of the pine needle-cast caused by *Lophodermium seditiosum*

of the fungus prophylactically (e.g. by choosing more resistant provenances, avoiding wet sites, not planting too densely, controlling weeds, and fertilizing according to requirements). If chemical treatment proves to be the only way of achieving the desired degree of control, the spray periods should be timed according to spore-release; i.e. between the end of July and the beginning of September. The precise timing can be determined by assessing the ripeness of the fruit bodies or—more reliably—by examining the atmospheric spore density with the help of spore traps.

Other Pine Needle Fungi [69,133]:

- *Lophodermium pinastri* (Schrad.) Chev. is a saprotroph accompanying or following *L. seditiosum*, with fruit bodies (hysterothecia) that are very similar except in having reddish longitudinal slit-margins and more than 5 epidermal cells above the base of the fruit body wall. Infected needles have several black zone-lines (Fig. **22**).
- *Sclerophoma (Dothichiza) pithyophila* (Corda) Höhn., a frequent colonizer of dying or dead pine needles, is quite often secondary to insect damage e.g. after attack by the pine needle midge (*Contarinia baeri*). Fruit bodies are cushion-shaped, brownish black, with sclerotial walls, 200–300 μm in size, and uni- or multilocular with ellipsoidal to ovoid, hyaline conidia 6–8 × 3–4 μm. The fungus is an anamorph of *Sydowia polyspora* (Bref. & Tavel) E. Müller. On malt agar, it forms a slimy-looking culture with initially white, later blackish, mycelium, and buds off cells in a yeast-like manner as in *Hormonema* (Fig. **23**).

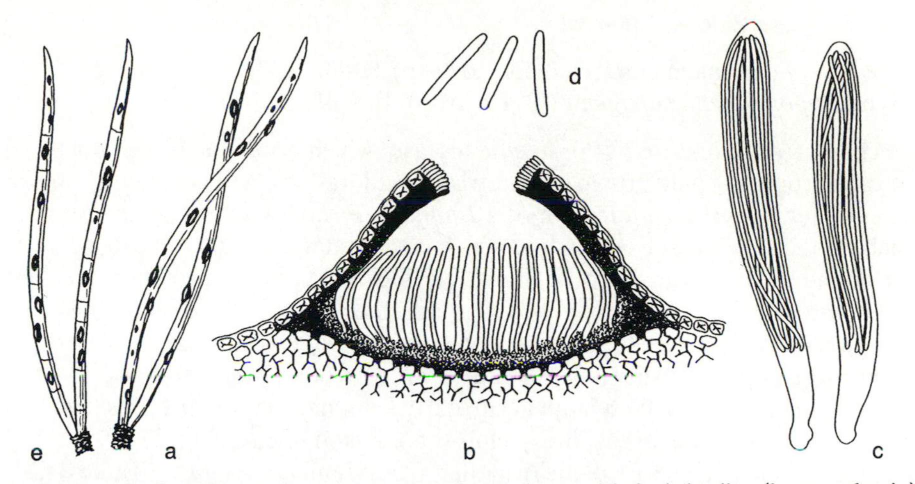

Fig. 22 *Lophodermium seditiosum*. **a** pine dwarf shoot with fruit bodies (hysterothecia), **b** cross-section through a fruit body (semi-diagrammatic), **c** asci with ascospores, **d** spermatia; **e** pine dwarf shoot with fruit bodies of *Lophodermium pinastri*

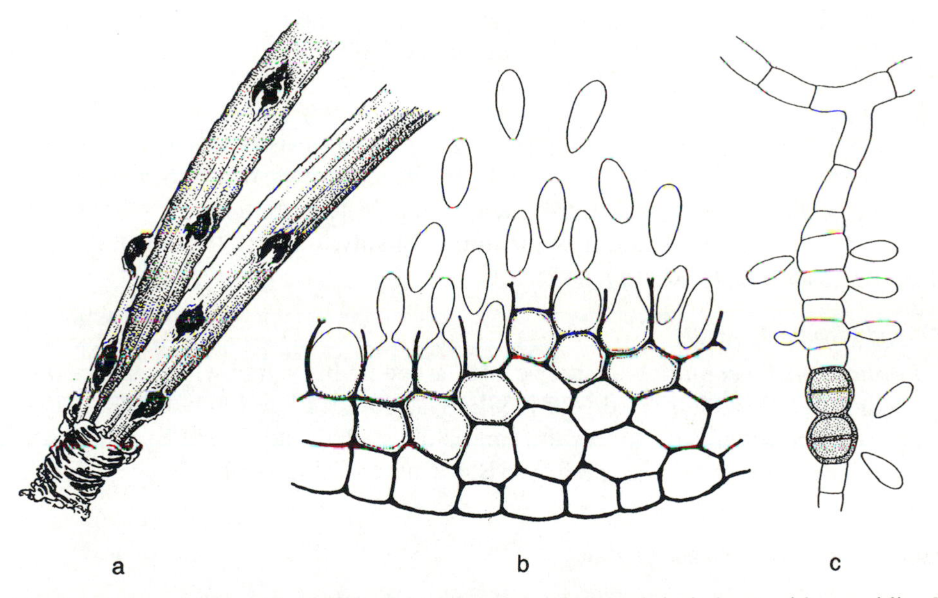

Fig. 23 *Sclerophoma pithyophila*. **a** base of a dead pine dwarf shoot with pycnidia, **b** part of the fruit body wall with ripening spores, **c** mycelium from a culture with dark resting spores and hyaline blastospores (b after v. Arx 1981)

Lophodermella Pine Needle-cast

Cause: *Lophodermella sulcigena* (E. Rostrup) Höhn.
Syn. *Hypodermella sulcigena* (E. Rostrup) Tubeuf

The characteristic feature of this needle disease, which occurs mainly in northern central Europe, is a pale brown to yellowish discoloration on the current shoots in late summer. A further indication of a *Lophodermella* attack is the needle base which remains completely green and which is separated from the apical, necrotic part of the needle by a sharp demarcation line. Infected dwarf shoots are not shed immediately but remain on the twigs until the autumn of the following year; during this time, the needles assume a brownish grey colour [134].

The most important diagnostic features are the dark hysterothecia, 5–20 × 0.3–0.4 mm in size, which appear in early summer on the previous year's infected needles. The ascospores, club-shaped and measuring 27–35 × 4–5 μm, are expelled in cool, damp weather from June to August (Table **II,5**). The 1-year development cycle of the fungus ends with the infection of the new dwarf shoots directly through the cuticle.

Various *Pinus* species are attacked, including Scots pine, Mountain pine, and Black pine. Where annual needle loss has been repeated and severe, the results can be a stunting of terminal and lateral shoots and attack by secondary needle fungi. Thus, the parts of needles killed by *Lophodermella sulcigena* are quite frequently colonized by *Hendersonia acicola* of the Fungi Imperfecti (Table **I,4**). In addition, *L. pinastri* can occur more abundantly than usual. The presence of *H. acicola* can be beneficial, as it may prevent fruit body formation by *L. sulcigena*.

Grubbing out and burning infected trees has proved to be an effective if radical means of preventing further spread of *L. sulcigena*. Fungicides, too, have been used successfully as a direct control measure against the fungus, but should be applied only in nurseries and new plantings. Finally, there is the possibility of selecting less susceptible provenances.

Related Needle Parasite:

- *Lophodermella conjuncta* Darker causes a needle browning which causes the needles to fall in their third year [135]. Hysterothecia are long-elliptical, and up to 3 mm long. Ascospores are long-club-shaped and 75–90 × 3–3.5 μm long. The parasite occurs chiefly on Black pine (Table **II,6**).

Naemacyclus Needle-cast of Pine

Cause: *Cyclaneusma minus* (Butin) DiCosmo, Peredo & Minter
Syn. *Naemacyclus minor* Butin

This fungus, which has only recently been anagrammatically renamed [61], is an ascomycete with a worldwide distribution. Under certain weather conditions it can cause a premature cast of needles between one and several years old. As a rule, infection takes place in the winter half of the year and is followed by several months of asymptomatic incubation until, in the following summer, the needles become yellowish or reddish, and soon fall to the ground. From September onwards into the winter, the fruit bodies (apothecia) are formed on the needles, mostly in groups. Unlike those of *Lophodermium seditiosum*, they are a pale cream colour and 200–660 × 250 μm in size; as they absorb moisture, each swells up in the form of a cushion, and the epidermis is pushed outwards like a pair of doors. The hyaline ascospores, eight to each ascus, are elongated to saddle-shaped, with two septa in the central portion of the spore, and are 80–90 × 2.5–3.0 μm in size (Fig. **24**).

On malt agar, the fungus forms a white, woolly mycelium in which spermogonia with 6.5–3.0 μm-long, rod-shaped microconidia develop, followed by pale yellowish apothecia with asci and ascospores [35].

Various species of pine are attacked, from transplants to thicket stage. For example, the fungus occurs frequently on *Pinus sylvestris*, although on this host it seems to be suppressed by the strongly competitive and more pathogenic *Lophodermium seditiosum*. No special control measures have, therefore, been necessary so far for *P. sylvestris*. However, on *P. mugo* [11), *P. ponderosa*, and *P. radiata*, more severe damage can occur and has occasioned the use of fungicides [118].

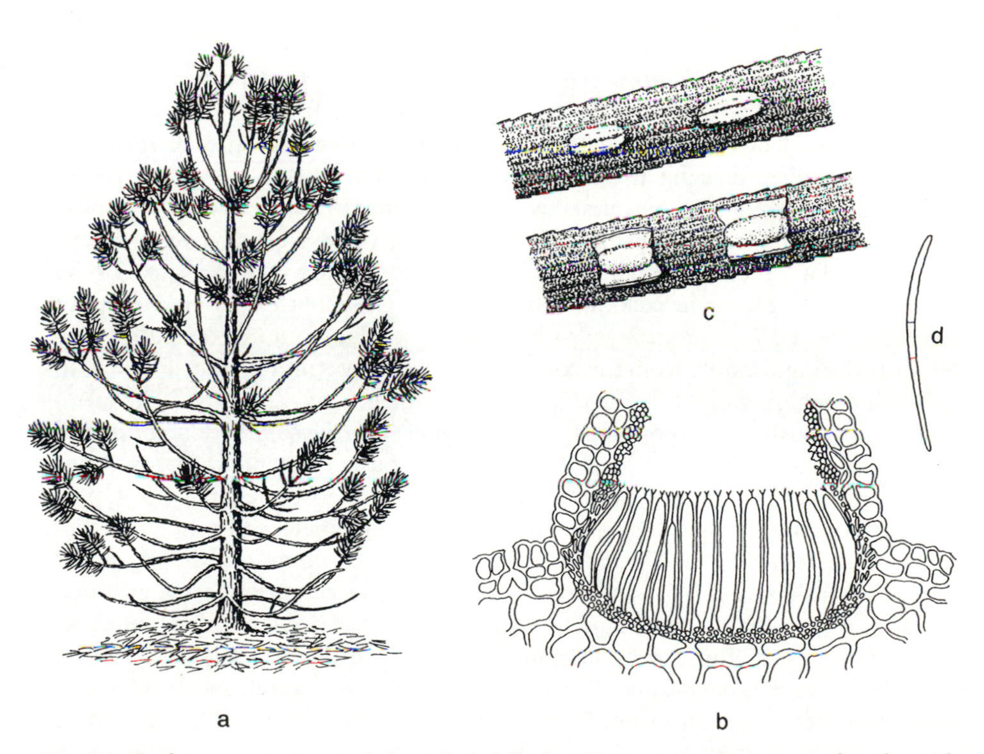

Fig. 24 *Cyclaneusma minus*. **a** infected, defoliating *Pinus mugo*, **b** cross-section through a mature fruit body, **c** needle segment with apothecia in the dry (above) and moist (below) state, **d** ascospore (a after Bazzigher 1973)

Related Species:

- *Cyclaneusma* (*Naemacyclus*) *niveum* (Pers.) DiCosmo, Peredo & Minter. Fruit bodies are bigger than those of *C. minus*. In culture, spermogonia are formed with rod- to sickle-shaped, 9–21 × 1 μm microconidia. Apothecia are not produced on malt agar. The fungus is found on *P. nigra*, *P. halepensis*, *P. pinaster*, and *P. mugo*, and is predominantly saprotrophic [35].

Dothistroma Needle Blight of Pine

Cause: *Mycosphaerella pini* E. Rostrup ap. Munk
Syn. *Scirrhia pini* Funk & Parker
Anamorph: *Dothistroma septospora* (Dorog.) Morelet
Syn. *Dothistroma pini* Hulbary

The first signs of disease—which can be confused with damage from sucking insects—are pale green flecks which later become brown. These are formed either in the middle part of the needle, causing necrotic bands, 1–2 mm broad, to form (e.g. in *P. nigra*); or at the tip of the needle, causing a spreading discoloration (e.g. in *P. mugo*). Current and one-year-old shoots are attacked. With multiple infections, the needle soon dies, though it remains attached to the twig for some time.

Confirmation of the diagnosis requires checking of the fruit bodies, of which the first to form are dark red to black conidiomata, 0.2–0.6 mm long and 0.3 mm across. They appear on the reddish brown flecks of the current and 1-year-old needles, causing the epidermis to become blistered as they ripen. Finally the epidermal blisters burst, exposing stromata in which white pycnidia are embedded. These contain numerous 2-celled, or less often 3 or 4-celled, hyaline conidia which are slightly curved and elongated, and measure 20–36 × 2.5 μm (Fig. **25**). The ascomata of the associated perfect state are quite frequently formed on 1- and 2-year-old needles of Black pine. Externally, they are hardly distinguishable from the conidiomata but they contain many loculi with asci, each containing eight 2-celled ascospores, measuring 12–14 × 3–3.5 μm.

There are various morphotypes and ecotypes of *Mycosphaerella pini* and attempts have been made to give them varietal status. The main phytopathological significance of these variants is that they include host-specific pathotypes, knowledge of which helps risk assessment for different species of pine. Thus, in Europe there is the form '*lineare*' which attacks *Pinus mugo* and *P. nigra* but which poses little risk for *P. sylvestris*. Today, all forms or varieties are grouped taxonomically under the aggregate species, *M. pini* [74]. A considerable loss of increment and killing can be attributed to this worldwide fungus, chiefly in monocultures of *P. radiata* and *P. ponderosa* in warmer countries. In Europe, the fungus is present in almost every country—although with varying frequency. To guard against further spread, all stands of Black pine should be monitored. Fungicidal sprays with materials containing copper have proved effective in controlling the fungus [153].

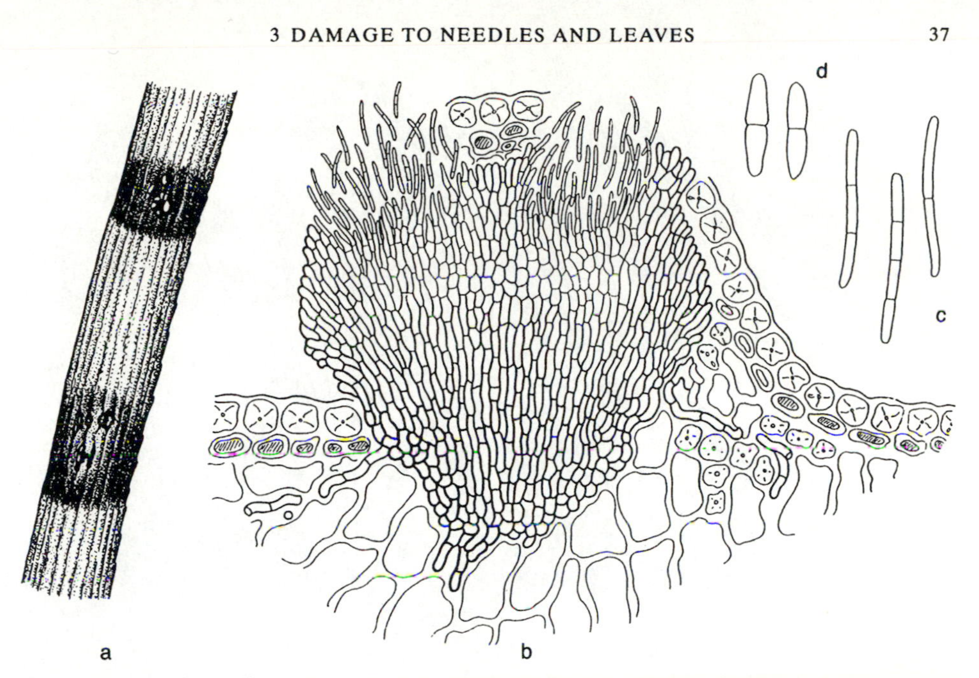

Fig. 25 *Mycosphaerella pini*. **a** needle segment with necrotic bands and fruit bodies, **b** cross section through a conidioma of the imperfect state, **c** conidia, **d** ascospores

Related Species:

– *Mycosphaerella dearnessii* Barr (Syn. *Scirrhia acicola* [Dearn.] Sigg.) is the cause of 'brown spot needle blight' of pines, which occurs mainly in North and Central America, but also in limited areas of Europe (Yugoslavia, Austria). Initial disease symptoms are numerous yellowish, later brownish, flecks with blackish borders, progressing to complete needle browning. The disease occurs mainly in conidial form, *Lecanosticta acicola* H. Sydow, with stromatic, cushion-shaped brown conidiomata, and straight or slightly curved 2–4-celled conidia which are pale to dark brown and measure 15–35 × 3–4 μm [74].

Pine Needle Rust

Cause: *Coleosporium* spp.

An attack by *Coleosporium* species on pine becomes apparent from the numerous blister-like, 1–3 mm-broad, reddish yellow aecidia which break through the epidermis of needles while they are still green. On ripening, the white, skin-like peridium tears open irregularly, and the spores are released as a yellow powder [226].

All *Coleosporium* rust fungi alternate between two different host plants, both of which are required for completion of the life cycle. All species pass through the haplontic phase on pines with the formation of aecidiospores.

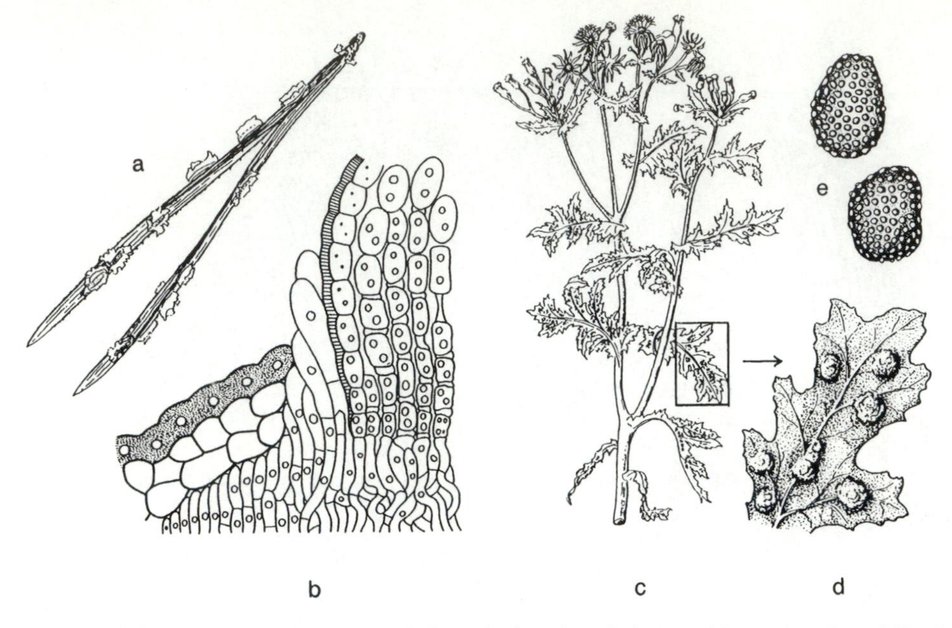

Fig. 26 *Coleosporium senecionis*. **a** infected pine dwarf shoot with ruptured aecidia, **b** part of a cross-section through an aecidium, **c** *Senecio vulgaris* infected by the rust, **d** part of a *Senecio* leaf with uredosori, **e** uredospores (c after Peace 1962)

In the dikaryotic phase (in which uredo-, teleuto-, and basidiospores are formed), they are dependent on various herbaceous plants. According to their specific association with certain plant species, the following commonly occurring *Coleosporium* species are distinguished and can be further subdivided into the following *formae speciales* [80]:

- *Coleosporium campanulae* (Pers.) Lév. alternates between pines (*Pinus* spp.) and campanula (*Campanula* spp.), or rampion (*Phyteuma* spp.).
- *C. senecionis* (Pers.) Fr. alternates between pines (*Pinus* spp.) and *Senecio* spp. (Fig. **26**).
- *C. tussilaginis* (Pers.) Lév. alternates between pines (*Pinus* spp.) and coltsfoot (*Tussilago farfara)*.

Larch Needle-cast

Cause: *Mycosphaerella laricina* (R. Hartig) Neger

This disease first shows itself in the appearance of single, or numerous, small brownish spots. The diseased needles remain attached to the twig for some time until they fall prematurely from July onwards. Usually, only the lower half of the crown is affected. Repeated attacks each year can lead to a weakening of the tree, to stunted growth, and to the death of scattered branches. European larch is the most frequently attacked species.

Mycosphaerella laricina is an ascomycete which produces its imperfect fruiting (conidial) state on the infected spots of needle tissue from June onwards. The conidial fruit bodies are very small black pustules containing cylindrical, 4-celled conidia, 30 μm in length. The ascomata of the perfect state appear during the following spring after the infected needles have fallen. They are spherical, dark brown perithecia containing ellipsoidal ascospores, 2-celled and 15–17 μm long, which are released from June onwards and cause new infections.

To prevent epidemic outbreaks of the fungus, it is advisable not to grow larch in damp places where humidity is high and the air stagnant. An established method of cultural control is to grow larch in mixture with beech (but not spruce); the autumn leaves of the beech cover the larch needles, preventing the escape of the ascospores in spring. The disease can be confused with early frost damage, premature needle loss from drought, and attacks of the Larch case bearer.

Other Larch Needle Parasites:

- *Hypodermella laricis* Tubeuf, similarly the cause of a needle-cast, occurs mainly in the Alps and their foothills. Diagnostic features are the black, elliptical hysterothecia, 0.5–0.8 mm in size and their teardrop-shaped ascospores, measuring 70–100 × 6 μm.
- *Meria laricis* Vuill. causes wilting and a yellow-brown discoloration of needles, though only on seedlings and young plants (Fig. **14**).

Herpotrichia Needle Browning of Silver Fir

Cause: *Herpotrichia parasitica* (R. Hartig) E. Rostrup
Syn. *Trichosphaeria parasitica* R. Hartig

This fungus, first described by Robert Hartig in 1884, is a typical needle parasite of firs, mainly seen on *Abies alba*, though also on other species (e.g. *A. nordmanniana, A. procera, A. veitchii*). It can also attack *Picea* species when infection pressure is high. It is characteristic of this disease that both young and old needles become brown and hang loosely from the twigs, to which they remain attached by a superficial mycelium (Fig. **27**).

Infection takes place by means of spores (conidia or ascospores) and can also occur via direct mycelial growth from an infected twig to a still healthy shoot. Under a lens, a thick, pale brown hyphal network (subiculum) can be seen covering the lines of stomata on the underside of the needle, from which hyphae penetrate both the stomata and the epidermal cells. On the twig surface, the fungus forms a mycelial network up to 300 μm thick in places, which is similarly attached to the epidermal cells beneath by haustoria.

The perithecia are, as a rule, seldom produced and then only in small numbers. They develop within the subiculum, are brownish in colour, measure 100–200 μm, and carry a number of crowded, 100–200 μm-long bristles on their apical halves. The asci are tubular and each contains 8 smoky grey spores, mostly 3-septate and 15–22 × 3.5–5.0 μm in size, which are not released until the spring. The pycnidia of the imperfect state superficially resemble the perithecia but are somewhat larger and their bristles arise singly from the pycnidial walls. They release spores from December onwards; these are cylindrical, 4.5 × 2.0 μm in size, and are borne on multiseptate conidiophores which vary in length [79].

The risk of infection is greatest in thicket-stage fir stands in damp areas. The damage in these cases is limited to a reduction in growth and to an impairment of the tree's resistance to weak parasites, which can readily take hold. Chemical treatments can reduce infection, but it is recommended in the first instance that affected stands be opened up and overhead cover removed.

This fungus can be mistaken for the ascomycete *Delphinella (Rehmiellopsis) abietis* (E. Rostrup) E. Müller which, however, occurs less often in Europe. Its black perithecia, which break through the epidermis mainly on the upper surface of the needle, contain a few club-shaped asci, each with 16–24 hyaline, 1-septate ascospores, 11–21 × 4.5–5 μm in size [57].

Silver Fir Needle Blight

Cause: *Hypodermella nervisequia* (DC.) Lagerb.

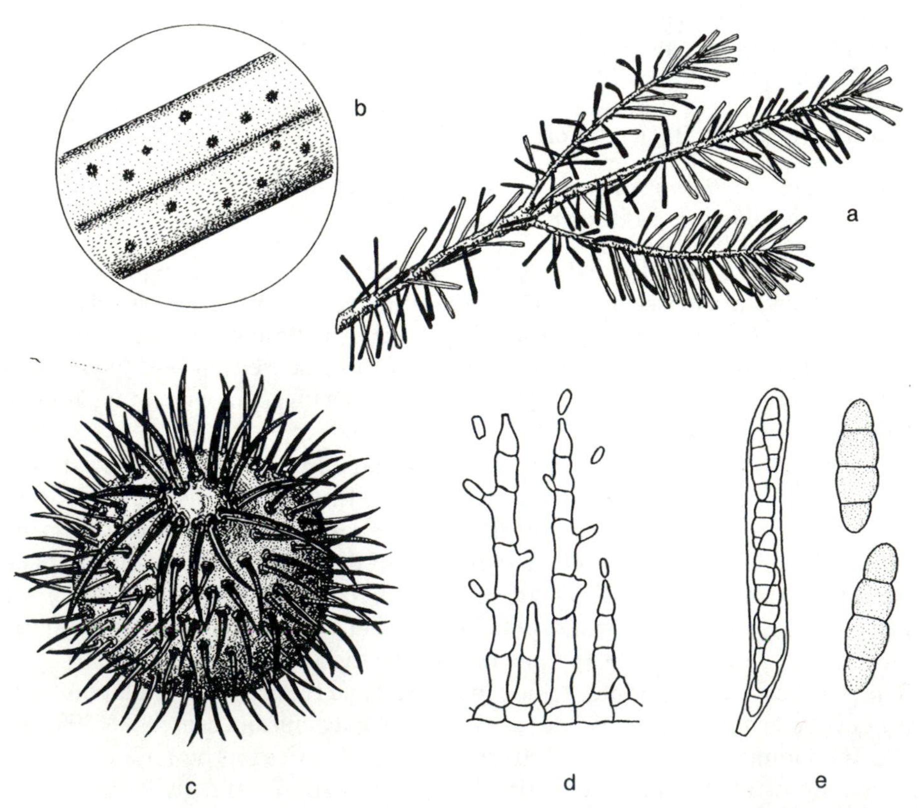

Fig. 27 *Herpotrichia parasitica*. **a** infected Silver fir branch, **b** needle segment with fruit bodies, **c** pycnidium, **d** conidiophores with conidia, **e** ascus with ascospores (c,d after Freyer 1976)

This needle-cast of *Abies alba* occurs wherever Silver fir is native or cultivated. Damage, which appears as yellowing and browning, is generally insignificant, being mainly confined to scattered needles and setting in only when these are 2 or 3 years old. The pycnidia, which are formed in the spring, can be used as a diagnostic character. These have the appearance of double, crinkled, black, elongated pustules on the upper surfaces of the yellowed needles. Later the apothecia appear in the form of black, elongated pustules on the midrib of the underside. The ascospores are club-shaped, elongated, and measure 75–90 × 3–4 μm (Fig. **29a,b**).

Silver Fir Needle Rust

Cause: *Pucciniastrum epilobii* (Pers.) Otth

Pucciniastrum epilobii is a heteroecious rust fungus (Uredinales) and so requires an alternate host, systematically unrelated to fir, for its development [226]. The cause of the disease, also known in German as 'Weißtannen-Säulenrost' (Silver fir pillar rust), begins its development on young needles, on the undersides of which white peg-like aecidia are formed in summer. When ripe, these burst open at the tip, releasing the yellow-orange coloured aecidiospores. If these spores reach the leaves of willow herb (*Epilobium* spp.), the fungus continues its development there, forming yellow uredosori and—in autumn—brownish teleutosori. In the following spring, each teleutospore germinates to form a septate basidium which, in turn, buds off basidiospores. The development cycle ends with the infection of new needles of Silver fir (Fig. **28**).

Even after repeated attacks, firs rarely suffer serious damage. Infection rarely occurs on more than a few needles, causing them to turn brown, shrivel up, and finally fall. In a heavy attack, however, the fungus can penetrate the shoot, causing brown oval necroses at the point of needle insertion. In this way, shoots become twisted and may occasionally die. More severe attacks are to be expected wherever young firs are standing among or in close proximity to stands of willow herb, the alternate host. For this reason, the most reliable method of protecting firs from attack would be to eliminate the willow herb, either mechanically or by the use of herbicides. This measure should be restricted to areas where the risk of damage is serious, so that populations of invertebrates and other wildlife dependent on willow herb are not unnecessarily harmed.

Other Needle Fungi of Silver Fir [69]:

- *Calyptospora goeppertiana* Kühn; aecidia (haplophase) occurs on needles of Silver fir with the same symptoms as for *Pucciniastrum epilobii*. Uredo- and teleutosori, and basidiospore formation (dikaryotic phase) occurs on stems of cowberry.
- *Cytospora friesii* Sacc. occurs frequently on frost-damaged or otherwise killed needles. Conidia are sausage-shaped and 4–5 × 1–1.5 μm in size (Fig. **29e,f**).
- *Rhizosphaera macrospora* Gourbiere & Morelet appears on the underside of

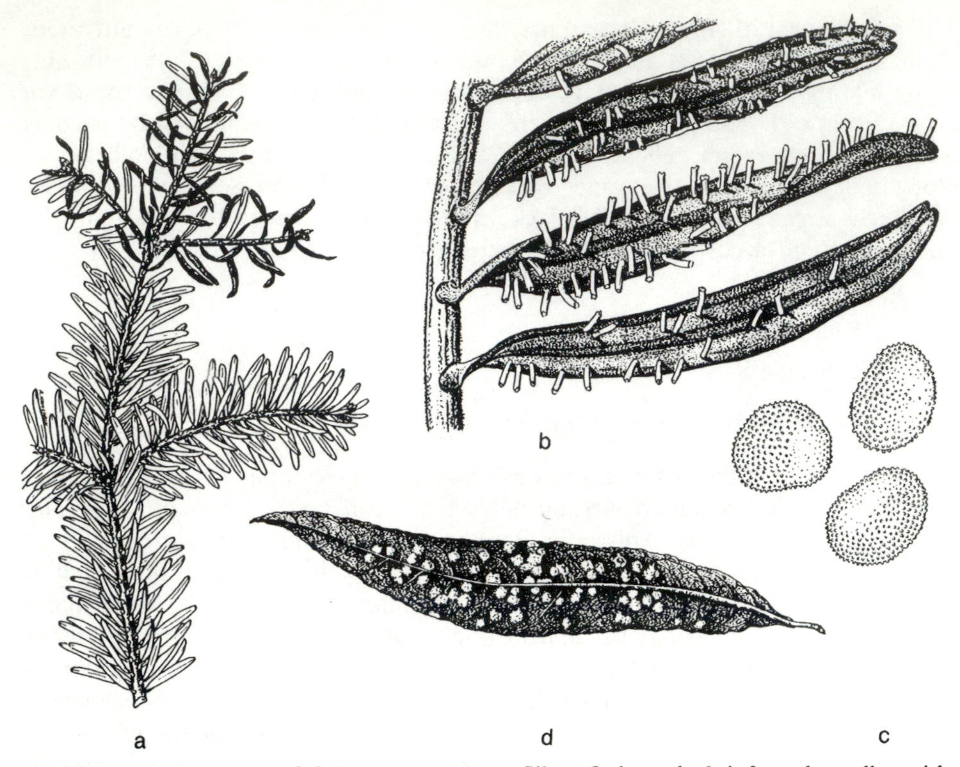

Fig. 28 *Pucciniastrum epilobii*. **a** symptoms on Silver fir branch, **b** infected needles with mature aecidia, **c** aecidiospores, **d** uredosori on the underside of a willow-herb leaf

needles of Silver fir. Conidia are spherical to ovoid, and 11–22 × 10–18 μm in size (Table **I,3**).

- *Rhizosphaera oudemansii* Maubl. occurs on the undersides of needles of various fir species. Conidia are ovoid and 9–13 × 6–9 μm in size (Fig. **29c,d**) [82]. The fungus is a weak parasite.

Black Snow Mould

Cause: *Herpotrichia juniperi* (Duby) Petrak
Syn. *Herpotrichia nigra* R. Hartig

Herpotrichia juniperi causes a needle disease which occurs widely in the high mountains of continental Europe on spruce, pine, Silver fir, and juniper. It is characteristic of the disease that the foliated branches, sometimes even entire small trees, are covered and completely enmeshed in a blackish brown mycelium. At first, the needles remain green but then become brown, die, and remain hanging on the twigs for some time until numbers of them simultaneously fall to the ground, bound together by the clinging mycelial felt. After repeated attacks, damage can extend to the bark of thin shoots so that the tips of twigs, or even

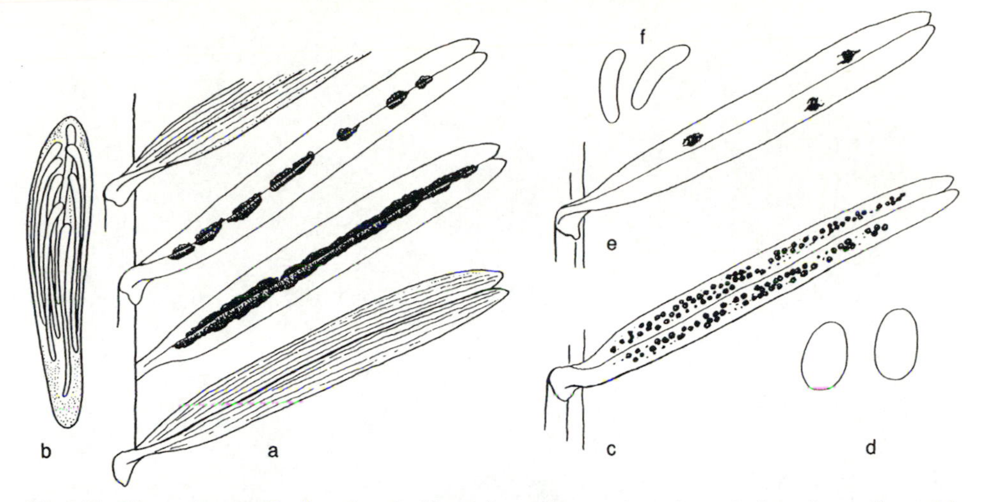

Fig. 29 Silver fir needle fungi. **a,b** *Hypodermella nervisequia*: **a** infected needles with fruit bodies (hysterothecia), **b** ascus with ascospores;
c,d *Rhizosphaera oudemansii*: **c** pycnidia on the underside of a needle, **d** conidia;
e,f *Cytospora friesii*: **e** dead needle with pycnidia, **f** conidia

whole twigs die. The disease occurs in pine stands in the Alpine Knieholz region, often in patches so that the affected areas, blackened by the fungus, give the impression at a distance of fire damage (Fig. **30**).

The 'black snow mould' develops in cavities within the winter covering of snow, where it develops at first epiphytically on the needles. In the second phase of its development, it forms haustoria which penetrate the epidermal cells of the host to take up nutrients. In the third, endoparasitic, phase the hyphae penetrate to the interior of the needle via the stomata. Only then do the attacked needles die. As the snows melt, the fungus, which is highly adapted to the microclimatic conditions beneath the snow cover, becomes dormant and is then able to survive the summer with the help of its drought-resistant mycelium.

The appearance of the mycelium is usually diagnostic: blackish brown in colour with a silky sheen and binding the needles and twigs together. Fruit bodies are not found until the second year of development. The ascospores, formed in club-shaped asci, are pale to dark brown, 22–25 × 5–7.5 μm in size and 2 to 4-celled, whereas those of a related species, *H. coulteri*, growing under similar conditions, are always 2-celled and measure 20–28 × 7–10 μm.

The White Snow Mould

Cause: *Phacidium infestans* P. Karsten *sensu lato*

Symptoms of the 'white snow mould' appear after the snows melt in the spring, and take the form of successive needle discolorations. The initial change is to a dingy yellow, soon turning to a brown or brownish red hue and finally, in summer,

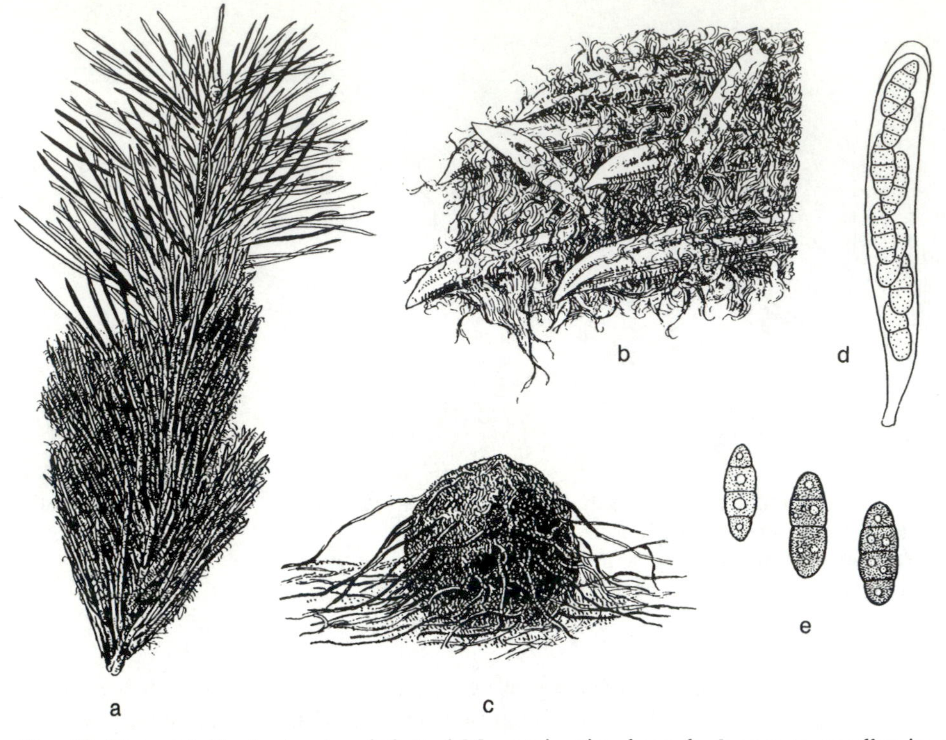

Fig. 30 *Herpotrichia juniperi*. **a** infected Mountain pine branch, **b** spruce needles in a web of mycelium, **c** fruit body, **d** ascus with ascospores, **e** mature ascospores

a pale grey. The killed needles remain attached during the next season's snow cover, and begin to fall gradually during the following summer. These symptoms are found only on conifers which are covered by snow in winter. Small trees and seedlings, which are entirely enveloped by the snow, can die even in the first year of attack; on larger trees the damage is confined to the lower branches (Fig. **31**).

The cause of the damage, *P. infestans sensu lato,* includes several varieties and geographic races which are now differentiated [171]. The form which occurs throughout Eurasia is *P. infestans* var. *infestans* and is found on species in the genera *Pinus, Abies, Juniperus,* and *Picea*. The most severely attacked is the genus *Pinus*. In Fenno-Scandia and parts of the USSR, *Pinus sylvestris* is the main host, of which the northern provenances are found to be somewhat more resistant than the southern ones [16]. In the Alps, where the fungus is represented by the southern race, *Pinus cembra* is the most severely affected. Elsewhere in northern Europe, *Pinus mugo*, *P. nigra* and *P. contorta* are occasionally attacked. Its occurrence on *Abies, Juniperus,* and *Picea* in Europe is less significant.

'White snow mould' develops exclusively under a covering of winter snow where—in the snow cavities—it meets with the necessary high air humidity.

There too, it is able to take advantage of the prevailing temperatures, for it can grow relatively well at 0°C. (In this respect, *Phacidium infestans* can be compared to *Herpotrichia juniperi* which is similarly adapted to the peculiar environmental conditions beneath a covering of winter snow.) Beneath the snow, the fungus grows from needle to needle, covering them with a greenish white mycelium which dries up and disappears once the snow begins to melt.

To diagnose the 'white snow mould' with certainty, it is necessary to find the fruit bodies (or check its identity in the laboratory), although fruiting is sometimes absent on species of *Abies, Juniperus,* and *Picea*. In the unripe state, the primordia of the apothecia are visible as small, roundish, dark spots scattered along the needles. In August when they are ripe, the papillate swellings in the epidermis tear open, forming irregular epidermal flaps so that the reddish grey hymenial disc, about 1 mm across, becomes visible.

The asci are club-shaped, 70–150 × 12–21 μm in size, and each contains 2,4, or (normally) 8 ascospores; these are hyaline, 1-celled, ovoid-elongate, slightly flattened on one side, and measure 15–26 × 5–9 μm.

For control of the disease, various treatments are recommended: cutting large clearings; separating re-afforestation areas by strips of some other tree species; or favouring more resistant species or provenances. Fungicidal control is usually economic only in nurseries [73].

Keithia Disease of *Thuja*

Cause: *Didymascella thujina* (E. Durand) Maire
Syn. *Keithia thujina* E. Durand

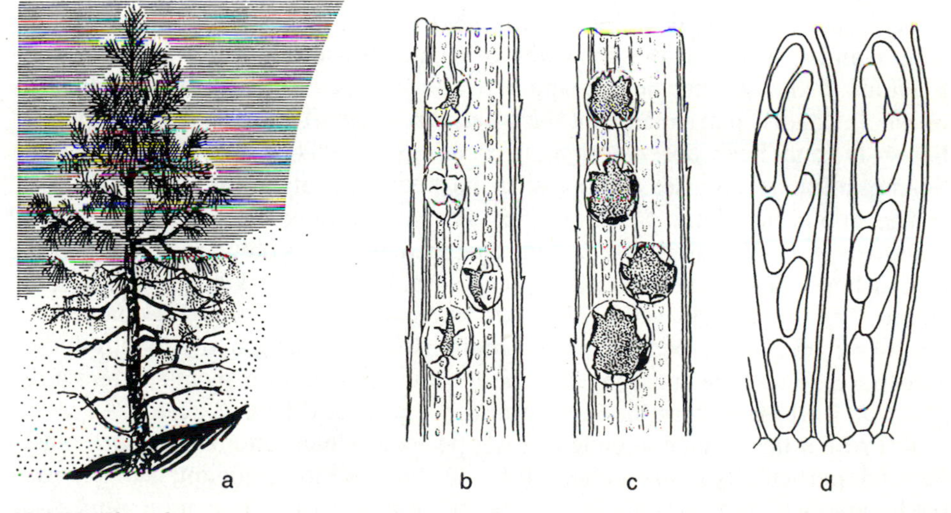

Fig. 31 *Phacidium infestans*. **a** symptoms on a young pine tree, with needles under the covering of snow killed, **b,c** needle segment with apothecia in the dry state (left) and moist state (right), **d** asci with ascospores and paraphyses

Fig. 32 Needle fungi of the Cupressaceae. **a-c** *Didymascella thujina*: **a** symptoms on *Thuja plicata*, **b** single infected scale leaves with apothecia, **c** paraphyses and asci with immature (left) and mature (right) ascospores; **d,e** *Phomopsis juniperovora*: **d** portion of shoot tip of *Juniperus chinensis* with fruit bodies (acervuli), **e** conidia

This disease of *Thuja plicata* and *T. occidentalis* characteristically begins during winter and spring with the browning of single, scattered, scale leaves on the previous year's shoots. Severe and repeated attacks can result in the death of whole shoots, particularly on the lower branches. The loss of whole plants has so far only been observed among seedlings under 4 years old in nurseries. Susceptibility to the disease diminishes markedly with increasing age [150].

The apothecia of the fungus, which start developing in May on the dead scale leaves, can be used as a further diagnostic feature; they are cushion-like, roundish to oval-elongate in outline, 0.5–2 mm long and dark olive-brownish in colour. They are initially yellowish brown when they burst through the epidermis. Their asci are club-shaped and each contains two uniseptate, ovoid to ellipsoidal ascospores. These are at first hyaline, later brownish to dark brown, and measure 21–25 × 13–17 μm, with the septum in the apical third (Fig. **32 a–c**).

To prevent this often epidemic disease, perfectly clean, uninfected stock must be used, particularly in the nursery. It has also proved to be advantageous to use fresh nursery beds each time; i.e. the continual raising of plants on the same site should be avoided. Fungicidal control by spraying during the growing season is of limited value because of the long period of spore production [19].

Other Needle Fungi on the Cupressaceae:

- *Didymascella tetraspora* (Phill. & Keith) Maire, is like *D. thujina*, but the asci are 4-spored. Ascospores are 21–24 × 13–16 μm in size. The fungus occurs on living needles of *Juniperus communis* [57].
- *Pestalotia funerea* Desm. colonizes both single scale leaves and portions of this host. It is a weak parasite on *Thuja* and other Cupressaceae. Conidia are 21–29 × 7–9.5 μm in size (Table **II,1**).
- *Lophodermium juniperinum* (Fr.) de Not. is a widespread colonizer of older and suppressed needles of *Juniperus* species [51]. Hysterothecia are solitary, black, elliptic-boat-shaped, and 0.5–1.9 mm long. Ascospores are vermiform, and 70–90 × 2–3 μm in size (Table **II,8**).
- *Chloroscypha sabinae* (Fuckel) Dennis occurs on still living and dead needles of *Juniperus communis* and *J. sabina*. Apothecia are erumpent, dark olive green to black, gelatinous when moist, and horny when dry. Ascospores are hyaline, and 17–21 × 7–8 μm in size (Table **I,5**).

BROADLEAVED TREES: LEAF DAMAGE

Nonparasitic Leaf Damage

Frost damage on leaves of deciduous trees is mainly the result of 'late frosts,' when the newly flushed leaves are subjected to temperatures below 0°C. The symptoms may be partial necrosis, mostly confined to the leaf margin or tip, or browning of the whole leaf. Where leaves have not completely unfolded, the necrosis can arise in the tissue between the lateral veins, extending to the midrib in severe cases. Such leaves appear tattered after they expand, a condition which is perhaps best known in Horse chestnut.

Damage from wind can arise in the spring as a result of the effect of warm and dry winds (e.g. the Föhn wind in Alpine regions). Desiccation of the marginal parenchyma results in a brown leaf edge ('marginal scorch'), or the browning of the intercostal areas; the vascular bundles mostly remain unchanged in colour.

Heat damage is recognizable from the yellowing of leaves and premature leaf fall ('heat defoliation,' 'summer defoliation'). The damage occurs mainly on leaves in the interior of the crown when they become exposed to the sun. The most susceptible genera include lime, robinia, poplar, plane, and elm. The cause is excessive heating of the leaves, combined with a diminishing supply of water from the drying soil. Similar prerequisites are assumed to apply when the leaves roll or fold longitudinally upwards.

Pollutant damage to leaves produces discolorations and growth disturbances which are characteristic of whichever harmful substances and tree species are involved. Symptoms have been documented for damage by various toxic gases, including ozone, hydrogen chloride, hydrogen fluoride and sulphur dioxide. For example SO_2, in cases of acute damage, causes characteristic marginal leaf necroses, tiny circular spots scattered over the leaf blade, or browning of the interveinal areas

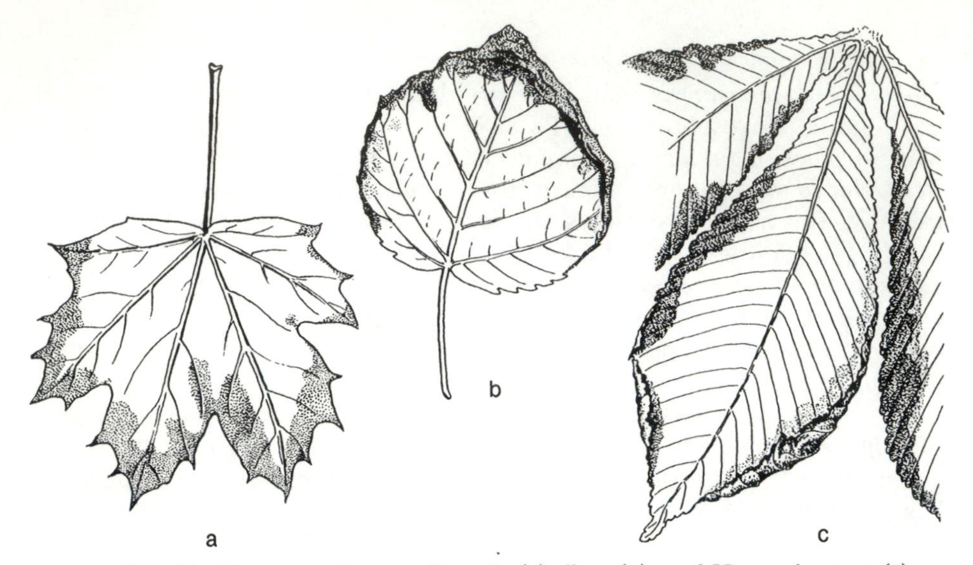

Fig. 33 Chloride damage on leaves of maple (**a**), lime (**b**), and Horse chestnut (**c**)

(Fig. **15 a,b**). In general, broadleaves suffer less from emissions than conifers. Types of tree that are only slightly sensitive include oak, elm, robinia, aspen, and ginkgo.

For the precise diagnosis of particular symptoms, specialist literature should be consulted [93,99], and foliar analyses carried out.

Salt damage is mostly a problem of urban areas and of stands along road and motorways, being very closely associated with the application of deicing salt (NaCl) for the clearance of snow and ice. Damage is also frequent in coastal districts due to deposition from salt spray, and occasionally occurs far inland when spray is swept up by storms. Deicing salt that dissolves into the soil water is taken up by the roots and is then translocated to green tissues where it results in marginal leaf scorch (Fig. **33**) and stunted leaf growth. Here, the critical Cl^- concentration is 10 mg/g dry weight (= 1% of the dry substance) of a leaf. Such damage may be confined to particular parts of the crown. Sycamore (*Acer pseudoplatanus*), hornbeam, beech, rowan, and the shallow-rooted Horse chestnut have proved to be particularly sensitive [63]. Oaks, wych elm, and White willow are among the most salt-tolerant central or west European species while, among exotic genera, *Robinia*, *Ailanthus*, *Gleditsia,* and *Sophora* have proved to be relatively salt-resistant.

Nutrient deficiency damage can involve a shortage of one or more essential nutrients. The symptoms include various kinds of discoloration, necroses, or leaf deformities, depending on the plant species, the season at which damage occurs, and on the identity of the nutrient that is lacking. Diagnosis requires a series of thorough differential tests which, along with symptomatology, include chemical analysis and pot experiments under controlled conditions [12,99].

The foliar symptoms of the more common nutrient deficiencies include the following examples:

- *Nitrogen deficiency*; leaves generally a uniform pale green and smaller than normal
- *Magnesium deficiency*; yellowish green to yellow interveinal chlorosis with subsequent browning
- *Potassium deficiency*; chlorosis developing from leaf margins with subsequent browning; interveinal areas often puckered upwards
- *Iron deficiency*; youngest leaves lemon yellow, otherwise yellowish white interveinal chlorosis contrasting sharply with the green veins ('lime-induced chlorosis').

Leaf Viruses

Trees react to infection by plant-pathogenic viruses (phytophages) chiefly with changes in leaf colour and morphology. These changes result from metabolic disturbances in infected host cells, where biosynthetic pathways are diverted to enable viral replication. A virus has no such resources of its own, being simply a particle of high molecular weight, consisting of a nucleic acid core (RNA or DNA) surrounded by a protein coat. Most viruses are either rod-shaped/filiform with a spiral structure (helical symmetry), or spherical with polyhedral symmetry. In a very few cases, they exhibit a complex structure with an additional lipoprotein coating. The size range of filiform types is from 10–20 nm diameter and 100–2000 nm in length, while spherical forms range from 20–300 nm in diameter; they can thus be seen only with the electron microscope. They are detected and identified in trees principally with the help of serological tests (reactions of tissue extracts containing viruses with virus-specific antibodies produced from mammals), and by their transfer to sensitive host plants (indicator plants).

In nature, viruses are disseminated by vectors (insects, gall mites, nematodes, fungi), and by means of seeds and pollen. In the soil, they can also be transferred through mechanical wounds on roots or via root grafts. Man plays a leading role in spreading them by using virus-infected propagation material (seed, cuttings, scions).

The only practicable means of controlling viruses in forest and parkland trees are prophylactic: selecting healthy plants; favouring resistant species and clones; using virus-free propagation material; and grubbing out diseased trees to eliminate sources of potential transmission by vectors.

Viral infection in forest trees can have economic significance, either directly or indirectly. Direct effects may include losses of increment, and degenerative phenomena which can lead to the death of trees. The indirect effects involve damage by other parasites or by environmental stresses (weather, air pollutants, etc.), to which the virus-infected tissues are predisposed through premature aging processes (senescence).

Viruses have been detected in some conifer species but have not been shown to cause any disease symptoms, apart from growth reductions. The symptoms of damage mentioned above occur in broadleaves but, as a rule, they are not diagnostic proof of viral infection; this requires use of the methods mentioned above.

The following list includes the more striking and common leaf viruses [50,99,144]:

Maple virosis: chlorotic band and line patterns, mosaic-like flecking or mottle on *Acer pseudoplatanus, A. platanoides, A. negundo*; virus species not yet identified.

Oak virosis: chlorotic mottle or flecking on slightly deformed and usually undersized leaves; tobamovirus (tobamovirus group) with rigid rods (300 nm long), mechanically transmissible; or potexvirus (potexvirus group) with slightly flexuous threads (470–500 nm long), mechanically transmissible; on *Quercus robur, Q. petraea*: chlorotic ring flecking; thought to be a virus in the nepovirus group (cf. birch), on *Q. petraea*.

Ash mosaic: chlorotic mottle, star-shaped or ring-shaped spots, leaf dwarfing; on Black poplar (*Populus nigra*) and *euramericana* hybrids; mosaic and leaf deformation on white poplar (*P. alba*); stunted growth and premature death; virus (carlavirus group) with flexuous threads (600–800 nm long); no vector known (Fig. **34a**).

Ring-fleck of rowan: chlorotic ring and line patterns on leaves of *Sorbus aucuparia*.

Beech virosis: chlorotic (in part mosaic-like) flecking and mottling, leaf dwarfing, shoot mottle.

Elm mottle: chlorotic mottle, ring or line patterns on slightly deformed leaves; (ilarvirus group) with spherical particles about 30 nm diameter, seed transmissible; on *Ulmus minor, U. glabra* (Fig. **34b**).

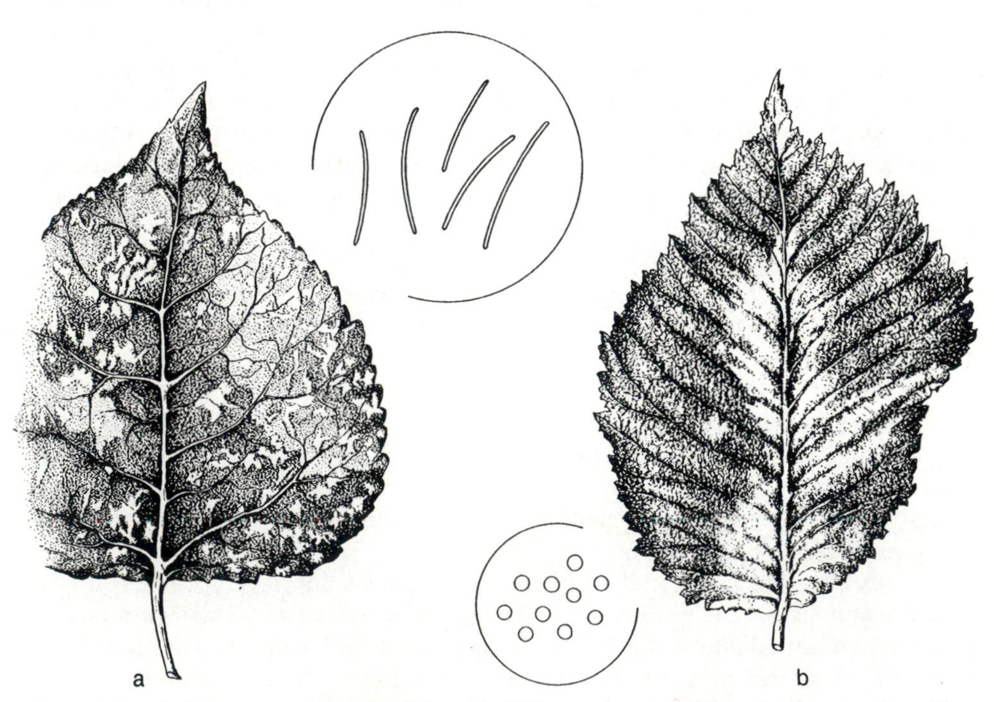

Fig. 34 Leaf viruses. **a** poplar mosaic with filiform virus particles, **b** elm mottle with spherical virus particles (after Schmelzer 1977)

Virus decline of birch: chlorotic and necrotic flecking, leaf dwarfing and leaf roll; gradual death of shoots and branches or of whole trees; cherry leaf roll virus (nepovirus group) with polyhedral particles (28 nm), seed and pollen transmissible; all *Betula* species.

Giant Leaf-blotch of Sycamore

Cause: *Pleuroceras pseudoplatani* (Tubeuf) Monod
Syn. *Ophiognomonia pseudoplatani* (Tubeuf) Barrett & Pearce
Anamorph: *Asteroma pseudoplatani* Butin & Wulf

This disease, which occurs only on sycamore, is characterized by relatively large (2–5 cm) brownish blotches, which initially have finger-like, diffuse borders and which occur in small numbers on any one leaf. Typically, the underside of the leaf shows blackish vein necroses. At an advanced stage, the blotches become brownish grey and develop smooth margins. Affected leaves may be deformed and can also become a patchy yellow, then fall prematurely because they are easily detached from the leaf trace. Depending on the weather and the age of the tree, the severity of attack varies from infections of single leaves to blotching of almost all the leaves on the tree (Fig. **35**).

From July, small, and therefore easily overlooked, acervuli are laid down on the blackish discoloured veins on the underside of the leaf. These are slightly protuberant cushions, mostly elliptic and measuring 50–350 × 50–200 μm. They contain numerous drop-shaped conidia, 6–7 × 2–3 μm in size, which are extruded from flask-shaped conidiogenous cells (Fig. **35c**). The fruit bodies of the associated perfect state appear in the spring on the fallen leaves. They are dark spherical perithecia with long ostioles, containing needle-like, 2-celled ascospores, 45–65 × 0.5–1.5 μm in size, which are responsible for the new infections in May.

From various reports in the literature and from personal observations, the causal agent of giant leaf blotch of sycamore seems to be widely distributed, but attacks do not occur with equal severity every year. They are observed mainly where the fungus can overwinter on fallen leaves that are left lying until the spring. This indicates that the removal of the leaf litter is the easiest and most effective method of control and can, at the same time, prevent spread to new sites.

Powdery Mildew of Maple

Cause: *Uncinula* species

In central Europe, two different host-specific mildew species occur on maple leaves, which are described in the following:

- *Uncinula tulasnei* Fuckel forms roundish spots on upper leaf surfaces of *Acer platanoides*, occasionally over the entire blade. These consist of a thick, persistent, shining white hyphal web. From midsummer, this ectotrophic mycelium forms conidiophores with cuboid to roundish microconidia, 7–11 × 7–9 μm

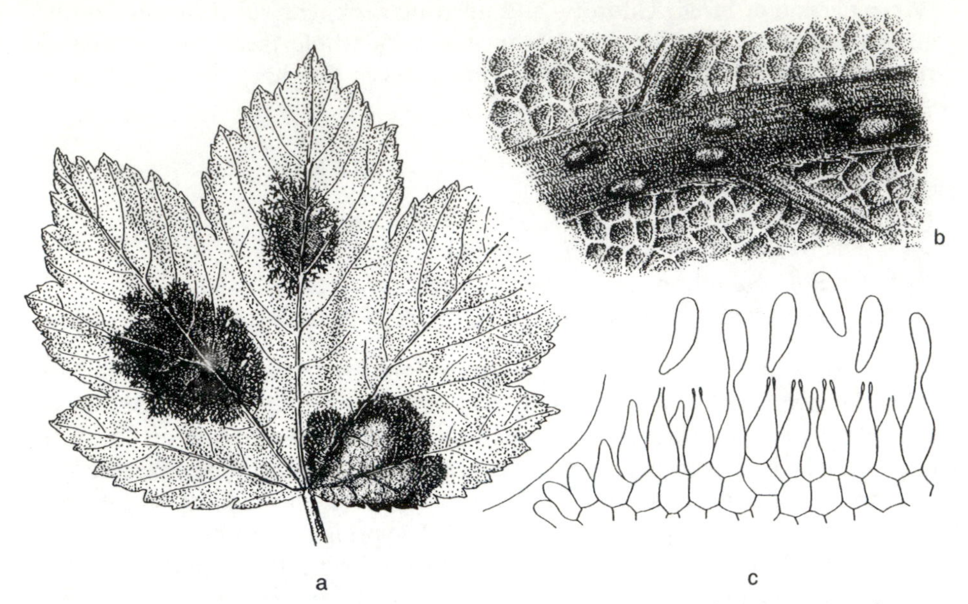

Fig. 35 *Pleuroceras pseudoplatani*. **a** symptoms on sycamore leaf, **b** spermogonia on a vein on the underside of the leaf, **c** cross-section (detail) through a spermogonium with conidiogenous cells and spermatia

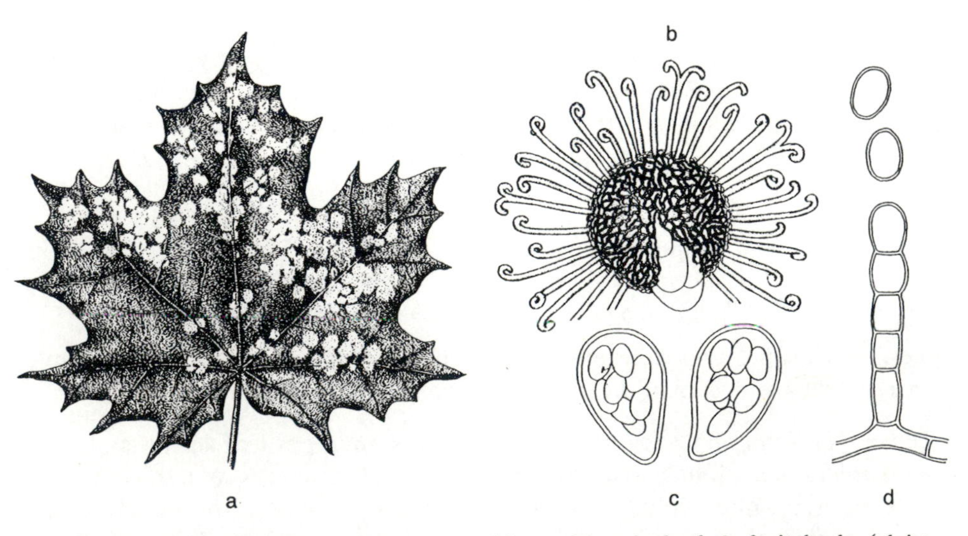

Fig. 36 *Uncinula tulasnei*. **a** symptoms on Norway maple leaf, **b** fruit body (cleistothecium) with appendages, **c** asci with ascospores, **d** conidiophore with conidia

in size. Less often, macroconidia are found, which are more ellipsoidal in shape, with almost parallel sidewalls and 19–25 × 13–18 μm in size. In autumn, cleistothecia begin to form. These are at first spherical, later lenticularly flattened, about 150 μm long, and with 50 to 70 hyaline appendages which, typically for this species, have their ends bent or rolled in spirally. Few of these appendages, unlike those of *U. bicornis*, are dichotomously branched. The fruit bodies burst open in the spring and contain 8 to 15 sac-shaped, 8-spored asci (Fig. **36**). In gardens, mildew generally spoils only the appearance of the plant; in nurseries, it can result in growth reduction.

- *Uncinula bicornis* (Wallr.) Lév. forms a cohesive, easily removable, fine coating on both leaf surfaces. The ellipsoidal to angular conidia, which are formed in chains, reach 26–36 × 16–20 μm in size, much larger than those of *U. tulasnei*. The cleistothecia are found on the underside of the leaf and have typical dichotomously branched appendages with hooked ends. The principal hosts for this species of powdery mildew are *Acer pseudoplatanus* and *A. campestre.*

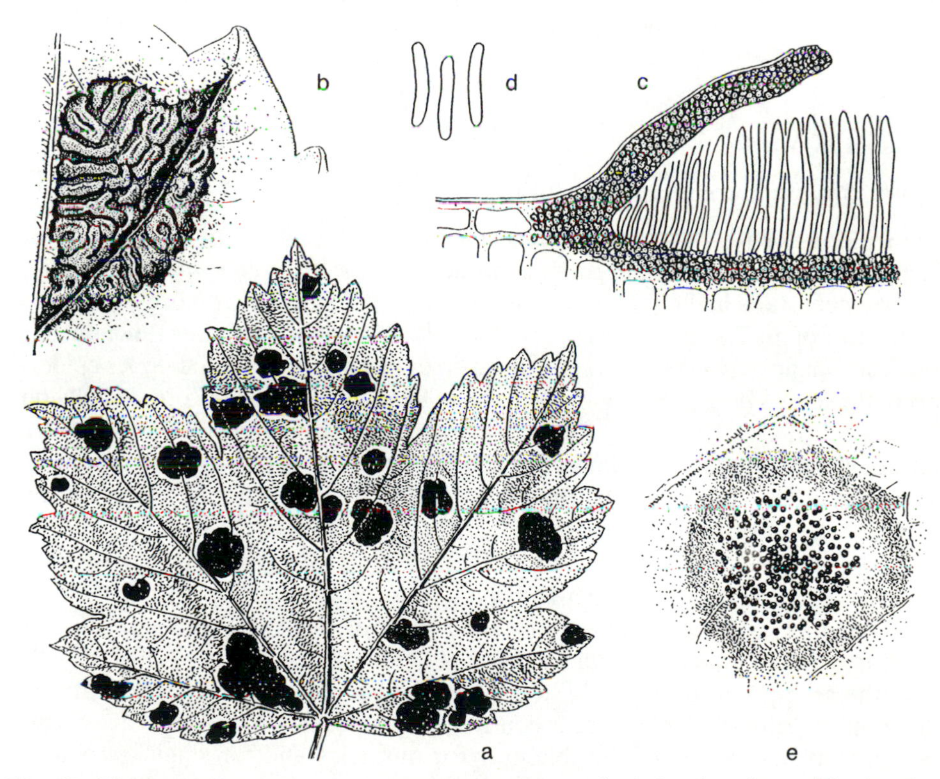

Fig. 37 *Rhytisma acerinum*. **a** symptoms on sycamore leaf, **b** sclerotium with ripening apothecia, **c** cross-section through an ascoma with asci and paraphyses, **d** conidia; **e** leaf spot with sclerotia of *Rhytisma punctatum*

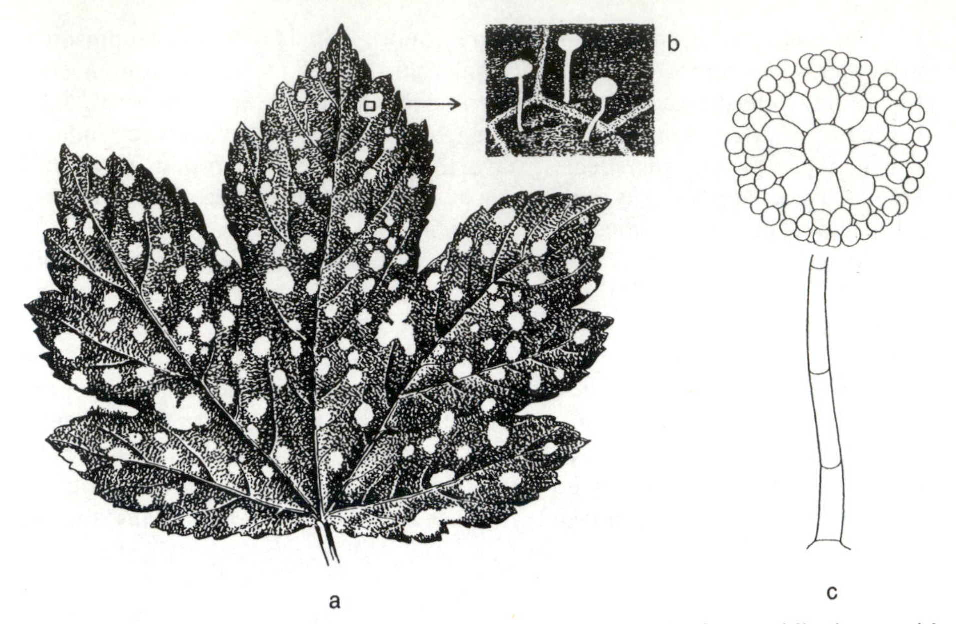

Fig. 38 *Cristulariella depraedans*. **a** symptoms on a sycamore leaf, **b** conidiophores with spore heads on the underside of a leaf (detail), **c** conidiophore with spore head

Tar Spot of Maple

Cause: *Rhytisma acerinum* (Pers.) Fr.

This striking disease is distinguished by black spots, 1–2 cm in diameter, which occur during the summer on the upper leaf surfaces of various maple species. Disease development starts in the spring when the germ tubes of the ascospores penetrate to the interior of the leaves via the upper epidermis. Soon afterwards, the fungus forms tough stromatic networks (sclerotia) which emerge from beneath the upper epidermis of the leaf as black, shining, tar-like spots. Shallow fruit bodies form initially on the leaf spots, giving rise to rod-shaped microconidia which function as spermatia during dikaryotization. This imperfect state of the fungus is known as *Melasmia acerina* Lév. The fruit bodies of the perfect state do not ripen until the following spring on the leaves lying on the ground. During this process, the covering layer of the stroma arches irregularly upwards so that the spots assume a wrinkled aspect—accounting for the alternative German name 'Ahornrunzelschorf' (wrinkled maple scab)—so that the ascospores can be expelled. If the airborne spores reach new maple leaves, the 1-year development cycle of the fungus begins anew (Fig. **37a–d**).

As the ascospores are the only stage in the fungal life-cycle capable of causing infection, tar spot of maple is easily controlled by removing the fallen leaves. This measure may obviously be feasible only for nurseries and for single garden or park trees.

Rhytisma acerinum occurs in several physiological races, each of which is able to infect only specific species of maple [140]. The severity of an attack is strongly affected by the inoculum density and the proximity of the infection source, but not by the SO_2 concentration in the air [122].

If the blackened areas consist of clusters of non-coalescent, pinhead-sized sclerotia, the fungus is the less commonly occurring *R. punctatum* (Pers.) Fr.; this occurs only on sycamore (Fig. **37e**).

Cristulariella Leaf Spot of Maple

Cause: *Cristulariella depraedans* (Cooke) Höhn.

This disease manifests itself in the form of 0.2–1 cm, greyish white, roundish spots, distributed more or less evenly and usually in large numbers on the leaf blades of *Acer pseudoplatanus*. Attack is most prevalent on the leaves of twigs which hang down low on younger trees. In severe cases leaves may undergo premature abcission of leaves while they are still green.

It is possible to confuse this with damage by the leaf mite, *Dasyneura vitrina* ('window-gall'). Leafhopper damage also produces whitish spots and these can also look very similar when occurring in groups. They have a separate name in German: 'Weisspunktkrankheit' (white spot disease).

The causal agent, *C. depraedans*, is a member of the Fungi Imperfecti. Its delicate, stalked spore heads are formed singly on the undersides of the leaves; they are roundish to lens-shaped, 0.1–0.2 mm in diameter, and composed of numerous, small, roundish cells (Fig. **38**). During the summer, the whole spore head, which functions as a conidium, is cast off and conveyed to other leaves in the rain and wind. On germination, long germ tubes emerge from the peripheral end cells, penetrate the leaf parenchyma, and bring about infection. Overwintering is achieved with the help of black, vermiform sclerotia which form on the fallen leaves in autumn.

Epidemic attacks by this generally uncommon fungus are linked to persistently very wet weather. Should control become necessary—e.g. in nurseries—the recommendation would be to remove the leaves that are lying on the ground and to cultivate the ground regularly.

Further Leaf Parasites of Maple [208]:

- *Didymosporina aceris* (Lib.) Höhn. causes roundish or irregular, ochraceous spots 1–3 cm in diameter. Acervuli are hypophyllous, brownish. Conidia are pale brown, obovoid or subcylindrical, truncate at the base, 1-septate below the middle, and 7–9.5 × 3–4 μm in size. The parasite occurs mostly on Norway maple.
- *Diplodina acerina* (Pass.) Sutton gives rise to circular or irregularly shaped, greyish yellow, 0.5–1 cm spots with dark margins, often extending from dead leaf galls. Fruit bodies are hypophyllous, brownish, with brownish, spindle-shaped, 2-celled conidia, 12–16 × 3–3.5 μm in size (Table **I,7**). It occurs mostly on sycamore.

- *Discula campestris* (Pass.) Arx induces medium-sized, irregular, brown spots. Fruit bodies are hypophyllous, amber coloured, with ellipsoidal or cylindrical, hyaline conidia, 6–9 × 2–3 μm in size (Table **I,6**). The disease affects various maple species [218].
- *Kabatiella (Aureobasidium) apocrypta* (Ell. & Ev.) Arx forms necrotic spots, mostly circular, often with a dead gall in the centre. If the petiole is attacked, premature leaf fall may occur (Fig. **41c,d**). It occurs on various maple species [218].
- *Phloeospora aceris* (Lib.) Sacc. gives rise to very small, non-coalescent, greenish brown spots which, when emptied of spores, dry out from the centre and become ash grey. Conidia are cylindrical, 4-celled, constricted at the septa, 20–30 × 4–5 μm in size. Attacks occur from July onwards on green leaves of *Acer pseudoplatanus* and *A. campestre* (Table **II,9**).
- *Phyllosticta aceris* Sacc. induces roundish to angular, ochre, 5 mm spots on older leaves. Pycnidia are pale and 100–170 μm in diameter. Conidia are ovoid to long-ellipsoidal, 1- and 2-celled, and 6–8 × 3.5 μm in size (Table **I,8**).
- *Phyllosticta minima* (Berk. & Curtis) Underw. & Earle gives rise to brown spots up to 1 cm across. Pycnidia are black and roundish. Conidia are roundish or ovoid in outline, 12–16 × 9–11 μm in size, and occasionally mixed with the dumbbell-shaped spores of the *Leptodothiorella* state (**Table I,9**).

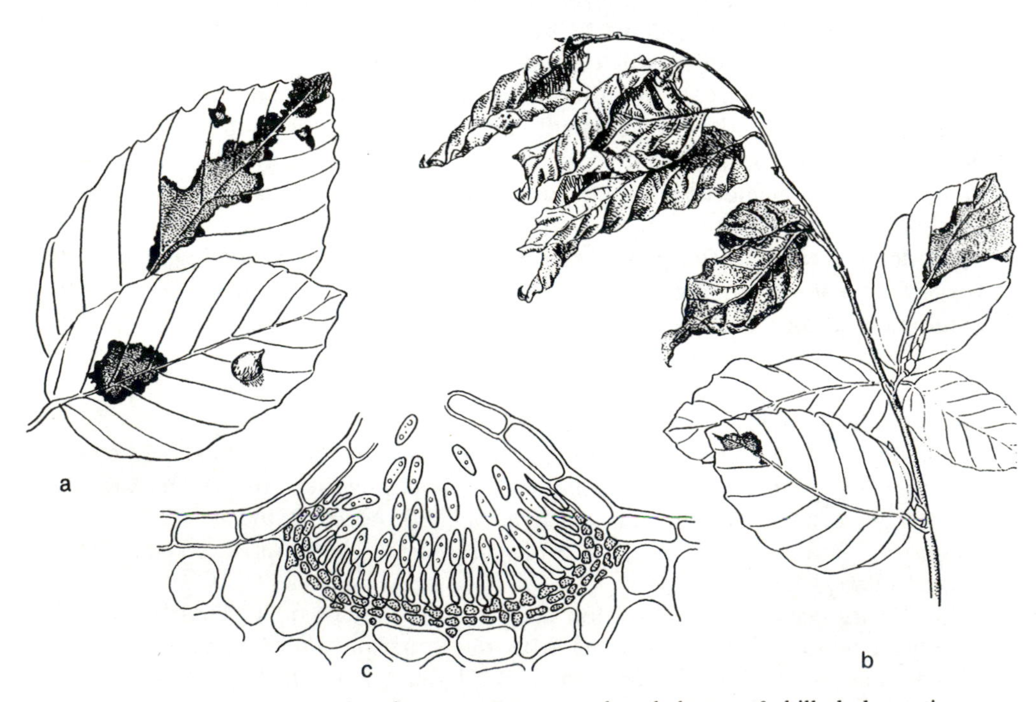

Fig. 39 *Apiognomonia errabunda*. **a** symptoms on beech leaves, **b** killed shoot tip, **c** cross-section through a fruit body of the imperfect fruiting state

Birch Leaf Rust

Cause: *Melampsoridium betulinum* (Pers.) Kleb.

This heteroecious rust fungus begins its development in the spring on the needles of *Larix decidua*, initially forming spermogonia and later orange aecidia. Infection of birch leaves by the aecidiospores leads to the development of yellow, pustule-like uredosori; often over the entire leaf undersurface. In autumn, development continues with the production of teleutospores. The life cycle of the rust is completed with the formation of the next generation of basidiospores which are budded off from the basidia in the spring [226]. By overwintering in the birch buds, the rust can also survive without passing into the haplontic-state host (*Larix*) and thus omit the formation of the spermogonia and aecidia (microcyclic development).

A severe attack can result in premature leaf fall. Serious damage can occur in this way, principally in nurseries. Additionally, attacks of rust can lead to increased frost-susceptibility and to a greater liability to attack by secondary pathogens (e.g. *Melanconium betulinum*).

Other Leaf Fungi on Birch:

- *Discula betulina* (Westend.) Arx causes the formation of small blackish spots on the upper leaf surface, with premature yellowing. Acervuli are hypophyllous, 100–180 μm in size, with cylindrical to ellipsoidal, often curved conidia, 10–16 × 2.5–3.5 μm in size (Table **I,10**).
- *Phyllactinia guttata* (Wallr.) Lév. Mycelium hypophyllous as an ephemeral coating or as discrete white spots. Conidia are born singly on conidiophores. Cleistothecia are spherical to flattened, mostly less than 200 μm, and bear 6 to 15 stiff, straight appendages with basal swellings. The fungus appears on the undersides of the leaves and also on other woody plants, e.g. *Carpinus*, *Corylus,* and *Fagus* (Fig. **40a–d**).

Beech Leaf Anthracnose

Cause: *Apiognomonia errabunda* (Roberge) Höhn.
Anamorph: *Discula* sp.

This disease is characterized by discrete, irregular brown necroses, distributed in patches on the leaf blade. In years when the fungus is abundant, young shoots up to 30 cm long can be attacked (Fig. **39**). Severe, extensive leaf browning can give the impression that there is a serious threat to the tree. The significance of such symptoms should not, however, be overestimated, as experience has shown that even severely attacked trees flush again normally without any noticeable after-effects. On rare occasions, the fungus can also form necrotic spots on the cotyledons of seedlings; these are irregularly circular.

Similar necroses are caused in the spring by the larva of the beech leaf miner, but these can be distinguished by the presence of the mined area which consists of a pale brown patch linked by a narrow tunnel to the midrib of the leaf.

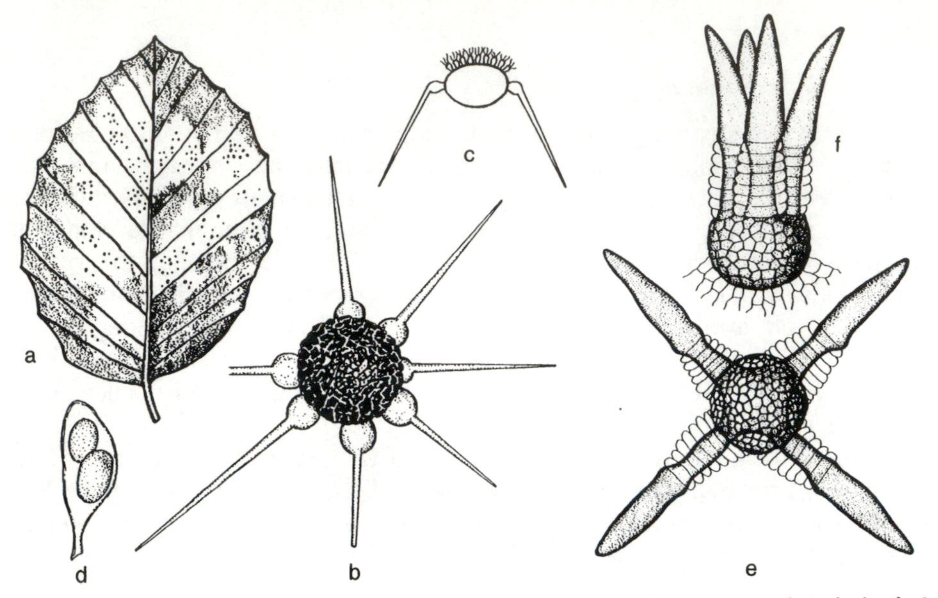

Fig. 40 Beech leaf fungi. **a-d** *Phyllactinia guttata*: **a** symptoms on a beech leaf, **b** cleistothecium (plan view), **c** cleistothecium (side view), **d** ascus with 2 ascospores; **e,f** *Valdensia heterodoxa*: **e** young conidia (plan view), **f** older, germinated conidia (side view) (a after Von Tubeuf 1895, b after Viennot-Bourgin 1949)

In summer, the tiny conidiomata of the fungus form in the dead tissues of the leaves and cortex. These contain large numbers of 1-celled, long-ovoid to ellipsoidal conidia, 9–14 × 4–6 μm in size. In the literature, this conidial state has been assigned to various form-genera [138,218]. These have now been superseded by the revised taxon: *Discula umbrinella* (Berk. & Br.) Sutton [208], although this name should be reserved for a similar fungus which occurs on oak. In view of the strong host specificity of the beech form, the name *Gloeosporium fagi* (Desm. & Roberge) Westend. [8] would be more appropriate.

From a biological standpoint, it is noteworthy that the fungus occurs in almost all green beech leaves as an endophyte without producing disease symptoms [194]. The fungus only switches to a pathogenic phase if certain gall-forming insects develop on the same part of the leaf. The main species acting as triggers are the pouch-gall midges, *Mikiola fagi,* and *Hartigiola annulipes*. The galls themselves die at an early stage of the process. It is interesting to regard this as a kind of symbiosis where the fungus is incorporated into the tree's defence system.

Other Beech Leaf Fungi [69]:

- *Mycosphaerella punctiformis* (Fr.) Schröter is an ascomycete with black, spherical, 60–120 μm-diameter fruit bodies occurring in clusters, immersed in the leaf tissue. Asci are sack-shaped with 2-celled ascospores, 8–13 × 3–4 μm. This is a common decay fungus of fallen leaves.
- *Phyllactinia guttata* (Wallr.) Lév. is a powdery mildew producing a whitish, unbroken, thick mycelial covering on the upper and lower surfaces of the

leaves. Cleistothecia are spherical to flattened, less than 200 μm in diameter, with 6–15 stilt-like appendages. The fungus is conspicuous, especially in autumn on fallen leaves (Fig. **40 a–d**).

- *Valdensia heterodoxa* Peyr. is a polyphagous parasite on various herbaceous and woody plants, occurring mainly on *Vaccinium myrtillus* (bilberry), and occasionally on *Fagus sylvatica* [10] on which brown blotches of various sizes develop, sometimes leading to the complete browning and wilt of the leaves. It also occurs on oak with the formation of light coloured flecks. The conidia of the hyphomycete arise singly on the undersides on the leaves of young plants, at first flat, like starfish, with 4–5 brownish, horn-like appendages encased in several transparent layers. They are later raised, cylindrical, 200–300 μm high, with appendages in bundles, and smaller basal cells which germinate and are responsible for new infections (Fig. **40 e,f**).

Oak Leaf Browning

Cause: *Apiognomonia quercina* (Kleb.) Höhn.
  Anamorph: *Discula* sp.

Typical symptoms of this disease are irregularly shaped, light brown necroses,

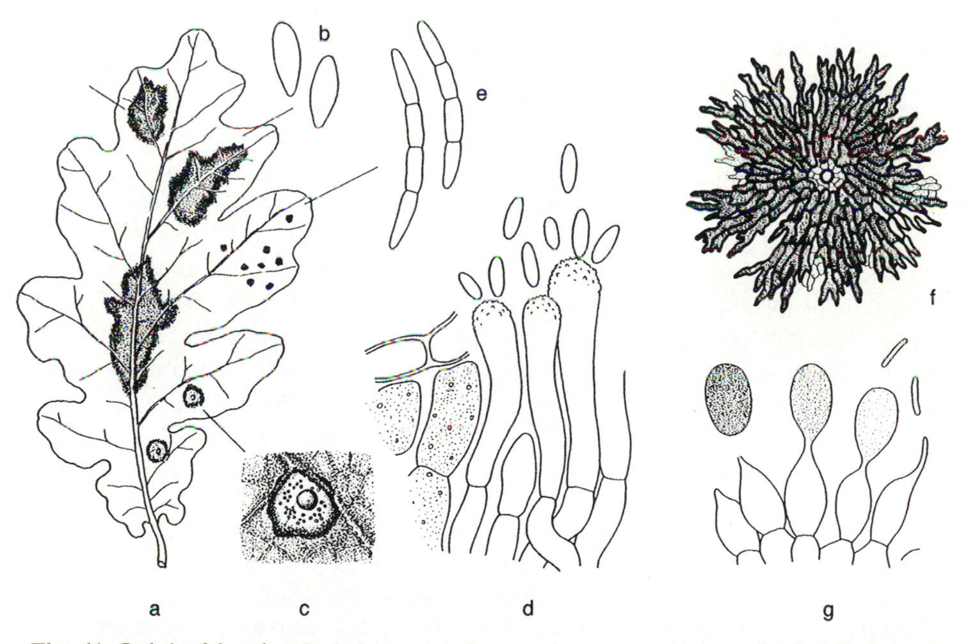

Fig. 41 Oak leaf fungi. **a,b** *Apiognomonia quercina*: **a** symptoms on leaf of Pedunculate oak, **b** conidia; **c,d** *Kabatiella apocrypta*: **c** symptoms with dead gall, **d** cross-section (detail) through a fruit body; **e** conidia of *Septoria quercicola* with symptoms; **f,g** *Tubakia dryina*: **f** plan view of an acervulus, **g** conidiogenous cells with macro- and microconidia at various stages of maturity

0.5–2 cm in diameter, appearing mostly on the upper part of the leaf blade and less often at its base or even on the petiole. Gall-forming insects can trigger leaf spot formation by stimulating the pathogenic development of the fungus from a previously established endophytic phase. During this process the insect dies [40]. There is thus a parallel with *Apiognomonia errabunda*.

The conidiomata on the undersides of the leaves provide a reliable diagnostic feature; they exude golden yellow drops containing ellipsoidal macroconidia, 9–13 × 4–6 μm in size. In culture the fungus produces smaller, elliptical microconidia measuring 6–7 × 2.5–3.5 μm, although these occur less often in nature. In the literature, the most frequent names for the conidial form are the combinations *Discula quercina* (Westend.) Arx [218] and *Gloeosporium quercinum* Westend. [8]. The fruit bodies of the associated perfect state develop only on leaves lying on the ground. Release of their ripe spores, from May onwards, results in infection of the newly flushed leaves (Fig. **41 a,b**).

It is possible to confuse this damage with feeding damage caused by an oak leaf aphid, *Phylloxera coccinea* v. Heyd. The underside of the leaves should be examined for the imagines or the remains of skins [99].

Other Oak Leaf Parasites:

- *Cryptocline cinerescens* (Bubák) Arx gives rise to medium sized, roundish, brown spots. Acervuli are hypophyllous and brownish. Conidia are hyaline to faintly brownish and 10–16 × 8–9 μm in size (Table **I,11**).
- *Kabatiella (Aureobasidium) apocrypta* (Ell. & Ev.) Arx is the cause of small, roundish, pale brown leaf necroses, mostly associated with feeding sites of psyllid larvae. Acervuli are 40–70 μm in size. Conidiogenous cells are club-shaped, resembling basidia, bearing 1-celled, ellipsoidal conidia at their apices (Fig. **41 c,d**).
- *Microsphaera alphitoides* Grif. & Maubl. is an ectoparasite of young oak; forms a powdery coating (see 'Oak mildew' below).
- *Mycosphaerella punctiformis* (Pers.) Starb. is a primary colonizer of dying leaves, which occurs occasionally in late summer on spotted leaves in the *Asteromella*-spermatial form. Spermatia are rod-shaped, and 4 × 1.5 μm.
- *Septoria quercicola* (Desm.) Sacc. is the cause of reddish brown, usually numerous, and evenly distributed leaf spots, 0.5–1.5 mm in size. Pycnidia are hypophyllous, dark brown. Conidia are hyaline, curved, 3 septate, often constricted at the septa, and 34–44 × 3–4 μm in size (Fig. **41e**).
- *Taphrina caerulescens* (Desm. & Mont.) Tul. produces pale green to yellowish or reddish leaf spots, about 1 cm across and upwardly puckered. Abaxially, these bear asci of variable size which contain numerous blastospores. The parasite occurs on various oak species and is not common anywhere in Europe. It is economically significant only in North America [137].
- *Tubakia (Actinopelte) dryina* (Sacc.) Sutton is the cause of brownish leaf spots, variable in size. It also occurs as a symptomless endophyte. Acervuli are hypophyllous, yellowish to blackish grey, with radially arranged, often incomplete scutellum. Macroconidia measure 13–15 × 6–8 μm and microconidia measure 5–7 × 1.5 μm (Fig. **41 f,g**).

Oak Mildew

Cause: *Microsphaera alphitoides* Grif. & Maubl.

The true powdery mildews occupy a peculiar position among the parasitic leaf fungi. They are highly host-specific and with few exceptions live as ectoparasites. The mycelium, which is predominantly superficial, forms specialized feeding hyphae (haustoria) which penetrate the host cells through the outer epidermal wall. These affected cells remain alive for a prolonged period following penetration.

Microsphaera alphitoides, which did not occur in central or northwestern Europe epidemically before 1907, is the most striking of the powdery mildews which occur on forest and parkland trees in those regions. It is also the most important economically. An attack is manifested by superficial white fungal patches, variable in size and often running together, on both the upper and lower leaf surfaces. At an advanced stage, the infected leaves look as if they have been dusted with flour. Severely attacked leaves curl up and can die. The shoot tips also can become misshapen and twisted, though this can also be due in part to *Botrytis cinerea*.

Both Sessile oak and Pedunculate oak are susceptible; the former rather less so, although the severity of attack on either species can vary considerably between various provenances and indeed between individual sibling trees. The North American oak species are in general resistant or less susceptible to powdery mildew.

Environmental factors, including atmospheric pollutants, may influence the severity of an outbreak of oak mildew. Thus in areas with a high air SO_2 content, mildew attacks have been observed to be less severe than in areas of 'clean' air [121].

Microscopic examination of the white patches reveals numerous, colourless, rod-shaped conidiophores which bud off chains of barrel-shaped conidia, 30–36 × 19–23 μm in size. Fresh infections can be established by these spores alighting on leaves 3 weeks old or less, giving rise to a new generation of conidia within as little as 3 days, given suitable weather conditions. Where spore production is continuous—encouraged by low air humidity and strong sunshine—a huge infection pressure builds up, impinging mainly on the Lammas shoots.

In late summer, spherical, closed cleistothecia may develop in clusters. These initially appear yellow, later reddish, finally blackish, and reach 0.1 mm in size. Each of them is furnished with several appendages which, as with other mildew species, are distinctive in form, having spreading dichotomous branches at their tips. Inside the cleistothecia, 8–15 asci are formed which imbibe moisture and swell up in the spring, bursting the wall of the fruit bodies to liberate the ascospores (Fig. **42**).

The fungus overwinters as mycelium beneath the scales of buds destined to develop into apical shoots during spring. The fruit bodies serve as another means

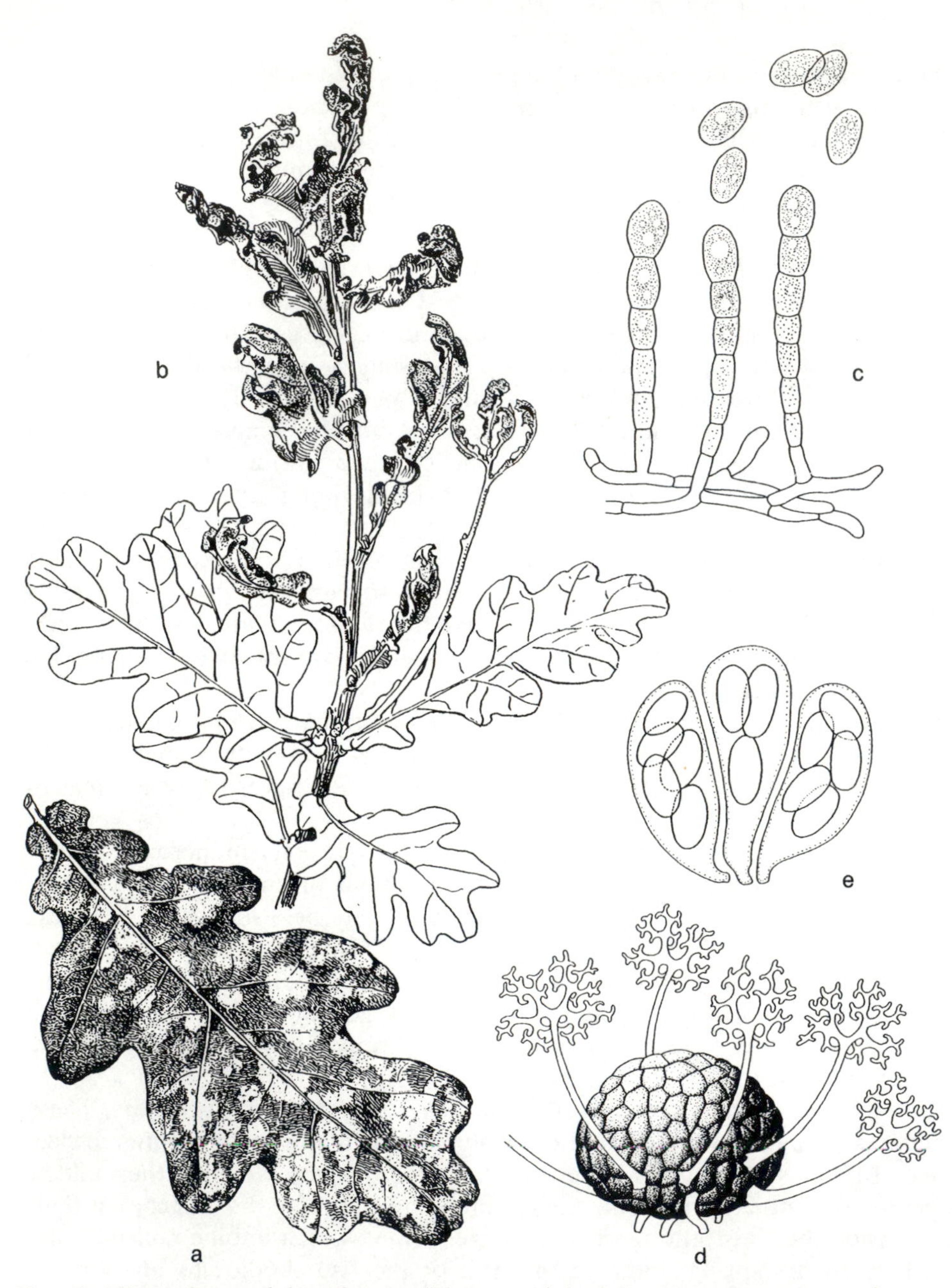

Fig. 42 *Microsphaera alphitoides*. **a** mildew on oak leaf, **b** symptoms on young shoot, **c** conidiophores with conidia, **d** fruit body (cleistothecium) with appendages, **e** asci with ascospores

of overwintering—although they do not develop every year—and in May their ascospores are released from the leaf litter to infect the new season's leaves.

Chemical control—if used at all—should be confined to seedlings in nurseries and to young transplants, even though mature trees can be conspicuously attacked in years when epidemics are severe. Fungicides which have proved effective include sulphur-based preparations and systemic compounds.

Taphrina Gall of Alder

Cause: *Taphrina tosquinetii* (Westend.) Magnus

The symptoms of this disease, which can be seen on Common alder as well as on various exotic alder species, are blister-like swellings and mussel-like distortions of the green leaves together with hypertrophy of the infected shoots. Where outbreaks recur annually, the use of fungicides may become necessary, but should be confined to nurseries.

This member of the Taphrinales is a highly specialized, host-specific parasite which, by producing growth substances, can interfere with the metabolic processes and morphogenesis of the host plant. Its mycelium occurs beneath the leaf cuticle and also colonizes the buds in autumn as the overwintering stage. Spread of the fungus is achieved by means of ascospores and yeast-like blastospores which are budded off from the ascospores while still in the ascus. As the symptoms are not always unmistakable, the leaves should be checked for the densely packed asci, mostly on the undersides, which in shape resemble those of *Taphrina populina* (cf. Fig. **45**).

Other Leaf Fungi and Mycoplasmas of Alder:

- *Asteroma alneum* (Pers.) Sutton causes brownish, roundish leaf spots. Acervuli appear on both leaf surfaces and are greenish black. Conidia are fusiform, straight or slightly curved, and 7–11 × 1.5–2.5 μm in size (Table **I,12**). Perfect state occurs (*Gnomoniella tubaeformis*) in the following year on fallen leaves [138].
- *Melampsoridium hiratsukanum* Ito is a heteroecious rust fungus forming its spermogonia and aecidia on *Larix,* and uredosori and teleutosori on various species of *Alnus.* The uredosori form small, orange-yellow pustules on the undersides of the leaves. In some areas—where the haplontic-state host is absent—the fungus can also persist on alder alone [226].
- *Microsphaera penicillata* (Wallr.) Lév. Mycelium are cobwebby and hypophyllous. Cleistothecia are spherical with radially arranged appendages 3–6 times dichotomously branched at their ends. The fungus occurs on all European and some exotic *Alnus* species [17].
- *Monostichella alni* (Ell. & Ev.) Arx produces roundish brown blotches, 0.2–2 cm in size. Conidiomata are between the epidermis and cuticle with conidia ovoid to ellipsoidal and 13–19 × 6–9 μm in size.
- *Phyllactinia guttata* (Wallr.) Lév. Mycelium is inconspicuous, hypophyllous, causing blotches several centimetres across. Cleistothecia are at first spherical

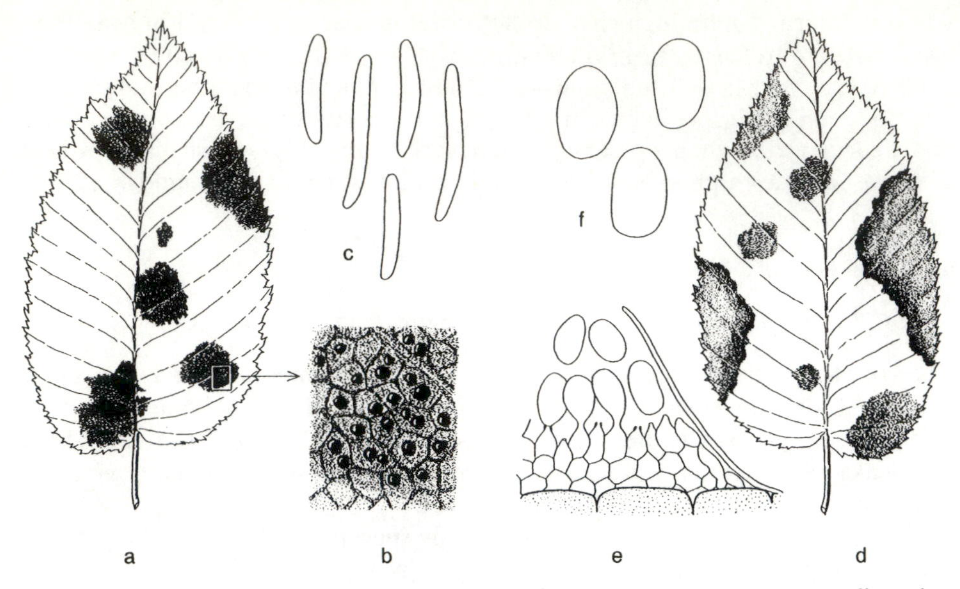

Fig. 43 Hornbeam leaf fungi. **a** leaf spots caused by *Asteroma carpini*, **b** acervuli on the underside of a leaf, **c** conidia; **d** marginal leaf necroses caused by *Monostichella robergei*, **e** cross-section of fruit body (detail), **f** conidia

and yellowish, later flattened and blackish, about 200 μm in diameter with 6–14 stilt-shaped appendages attached to the base. It occurs on many woody plants, and is regarded as an aggregate species (Fig. **40a–d**).

- *Mycoplasma-like organisms* (MLO) cause a yellowish green discoloration of the leaves and marginal leaf necrosis, and also dwarfing of leaves. An attack leads to a generally unhealthy appearance in the tree [186].

Gnomoniella and Asteroma Leaf Browning of Hornbeam

Cause: *Gnomoniella carpinea* (Fr.) Monod
Anamorph: *Monostichella robergei* (Desm.) Höhn.
or: *Asteroma carpini* (Lib.) Sutton
Syn. *Cylindrosporella carpini* (Lib.) Höhn.

There are at least two fungi on hornbeam that can cause the appearance of sometimes large leaf spots. Similar lesions can also result from the influence of abiotic factors (e.g. drought, deicing salt).

The first signs of infection by *G. carpinea* are a dense sprinkling of pale brown lesions on the upper surface of the leaf, involving at first only the outer layers of cells. As numerous spots coalesce, larger areas of discoloration develop and can cover the whole leaf; these are diffuse, brown, and often show a silvery sheen. The imperfect fruiting state, *M. robergei*, forms acervuli in the necrotic leaf tissues. Initially yellow-brown and later almost black, these are rounded

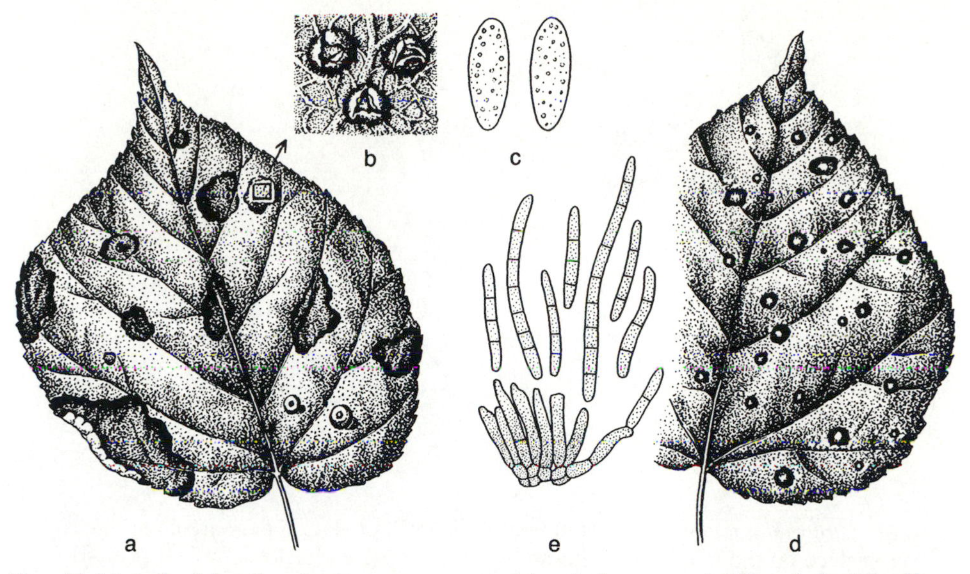

Fig. 44 Lime leaf fungi. **a** leaf necroses caused by *Apiognomonia tiliae*, **b** fruit bodies of the imperfect state (detail), **c** conidia; **d** leaf spots caused by *Cercospora microsora*, **e** conidiophores with conidia (e after Ellis 1976)

or occasionally angular, and have a small-celled conidiogenous basal layer. The hyaline conidia are ovoid to broadly cylindrical, and measure 12–17 × 7–9 μm (Fig. **43d–f**). The associated perfect state occurs in the following spring on the leaves lying on the ground.

In natural stands of hornbeam, this fungus is relatively common though unimportant because it is a weak parasite which usually attacks only ageing leaves. It can, however, cause widespread leaf browning and premature defoliation, given certain predisposing conditions: e.g. after hornbeam hedges have been cut hard back.

Asteroma (Cylindrosporella) carpini causes similar spots on attached leaves of hornbeam, but these do not form until the autumn. They reach 0.5–2 cm in size and are at first blackish-grey, later grey-brown. The acervuli, thickly scattered on the undersides of the leaves, appear as tiny, grey-brown spots and contain rod-shaped conidia, 9–13 × 2 μm in size (Fig. **43a–c**). Reliable information concerning the occurrence of an associated perfect state is not yet available.

Apiognomonia Leaf Browning of Lime

Cause: *Apiognomonia tiliae* (Rehm) Höhn.
Anamorph: *Discula* sp.

This, like the other leaf browning diseases mentioned above, is characterized by irregularly shaped and sometimes large, dark-edged necroses. A definite diagnosis can be provided only by the conidiomata, which are formed on both

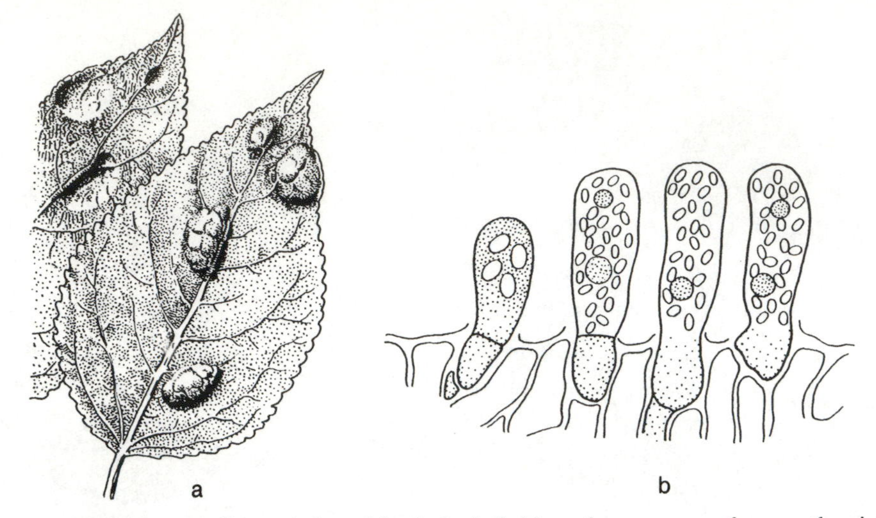

Fig. 45 *Taphrina populina*. **a** infected leaf of a hybrid poplar, upper surface on the right, lower surface on the left, **b** asci with ascospores on the underside of a leaf (detail)

sides of the leaf and whose conidia resemble those of *Apiognomonia veneta* in shape and size. The development of *A. tiliae* is largely dependent on gall-forming mites or insects. Common examples of these are the gall mite *Eriophyes leiosoma* and the gall fly *Didymomyia reaumuriana* (Fig. **44a–c**).

Other Leaf Parasites of Lime:

- *Cercospora microsora* Sacc. attacks leaves and petioles and occasionally causes total defoliation. Leaf spots are spread over the whole leaf blade and are 1–3 mm in size, roundish, with a black margin. Conidiomata are hypophyllous, brownish, with vermiform, multiseptate, pale olive-brown conidia, 35–90 × 3–4 μm in size (Fig. **44d,e**).
- *Didymosphaeria petrakiana* Sacc., a weak parasite, producing brownish-black spots, 1-3 cm across, near the end of the growing season; the spots are roundish, with diffuse finger-like marginal extensions, as with *Pleuroceras pseudoplatani* (cf. Fig. 35). During autumn *Asteromella*- like conidiomata form on the underside of the leaves; perithecia on fallen leaves form early in the following spring, containing two-celled, pale brown ascospores which measure 13-16 × 5-6 μm.

Poplar Leaf Blister

Cause: *Taphrina populina* Fr.

This fungus causes upwardly domed contortions of the leaves of various poplar species. Underneath, these are covered in a yellow bloom, whose striking colour

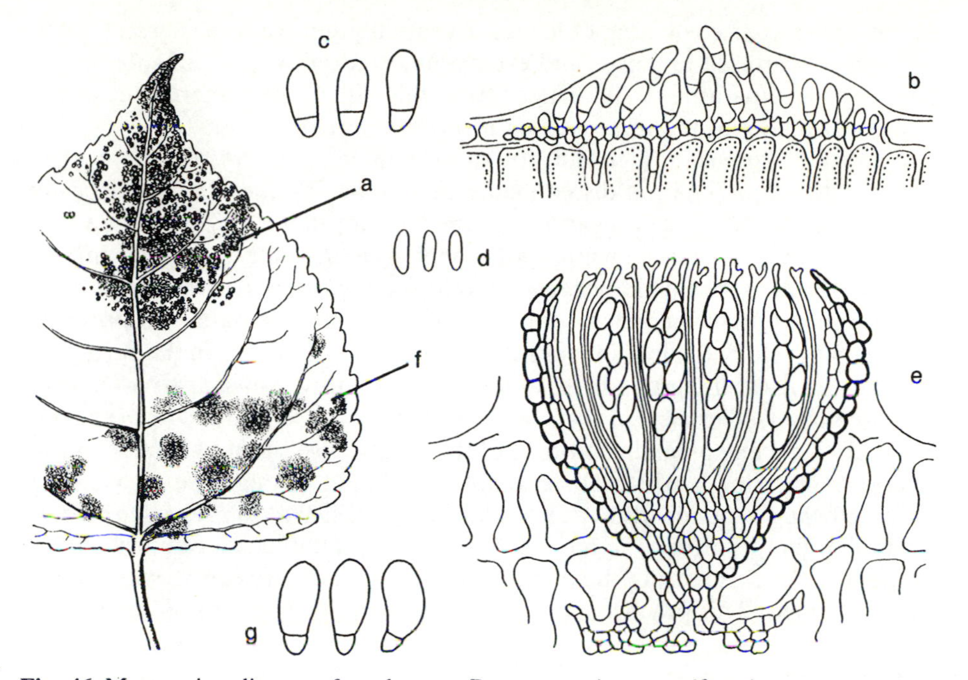

Fig. 46 Marssonina disease of poplar. **a-e** *Drepanopeziza punctiformis*: **a** symptoms on a Black poplar leaf, **b** cross-section through a fruit body of the imperfect state, *Marssonina brunnea*, **c** macroconidia, **d** microconidia, **e** cross-section through an apothecium of the perfect state; **f** symptoms of infection by *M. populi-nigrae*, **g** conidia

is due to the spore contents of the asci which are densely packed between the cuticle and the epidermis. These spores are released to the outside when ripe. True fruit bodies are lacking in all members of the Taphrinales. The asci are broad-cylindrical, apically truncate, and measure 50–112 × 15–40 μm. At first each contains four ascospores, but these bud off numerous secondary cells which eventually fill the ascus. The spores are very resistant to external influences and are the means by which the fungus overwinters. No significant destruction of leaf tissue is caused by this conspicuous fungus (Fig. **45**).

Marssonina Leaf-spot of Poplar

Cause: *Drepanopeziza punctiformis* Gremmen
 Anamorph: *Marssonina brunnea* (Ell. & Ev.) Magnus

This disease is restricted to *Populus nigra*, *P. deltoides,* and their hybrids. It appears in summer in the form of brownish black, thickly distributed leaf spots which may run together to form larger blotches. By late summer, the affected leaves show yellowing and wilting and are then shed prematurely. If heavy attacks occur in several successive years, shoot development is inhibited. The lateral buds on the branches fail to flush, or flush only weakly so that the tree has only a sparse covering of leaves. Eventually such outbreaks can lead to the death of scattered branches and even young trees as a result of subsequent attacks by weak parasites. The danger from this disease lies, therefore, in the chronic weakening of the tree, making it prone to other diseases.

Marssonina brunnea, which did not cause epidemics in western or central Europe until about 1960, is disseminated in summer by the characteristically shaped macroconidia which measure 14–16 × 6 μm; these develop inside thin acervuli, each of which has a white spot (opening) in its centre. Another imperfect state, the microconidial form, also develops quite often; in some years even more commonly than the macroconidial form. The perfect state—conforming to the classic cycle of the leaf-inhabiting ascomycetes—develops in the spring on the leaves lying on the ground. It is represented by tiny, top-shaped apothecia which are immersed in the leaf tissue and which contain asci and ascospores (Fig. **46a–e**). To avoid an epidemic outbreak of the fungus, use should be made of the resistant poplar clones which are available. It should, though, be noted here that, by growing a monoculture of a poplar merely resistant to *Marssonina* (e.g. 'Robusta'), the occurrence of another disease (e.g. Dothichiza canker) can be encouraged. For this reason, the planting of a number of different poplar clones is recommended. This practice can obviate the need for chemical protectants.

Related Species:

- *Marssonina castagnei* (Desm. & Mont.) Magnus occurs on aspen and grey and white poplar. Conidia measure 19–21 × 5–9 μm [48].

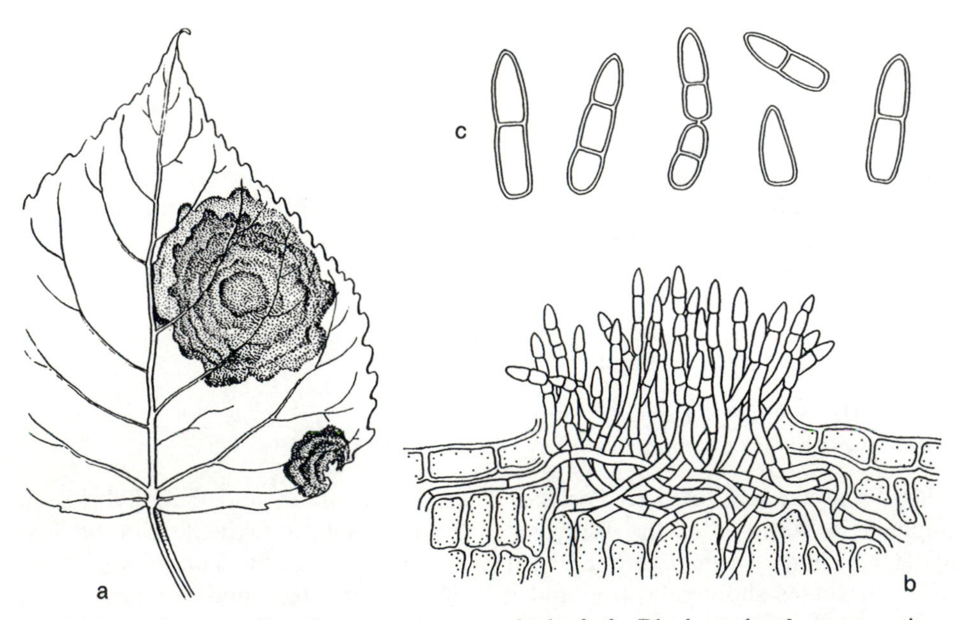

Fig. 47 *Septotinia populiperda*. **a** symptoms on the leaf of a Black poplar, **b** cross-section through a fruit body of the imperfect state, **c** conidia

- *Marssonina populi-nigrae* Kleb. occurs on Black poplar and its hybrids and on Balsam poplar. Conidia measure 20–22 × 8–11 μm (Fig. **46f,g**).

Septotinia Leaf Blotch of Poplar

Cause: *Septotinia populiperda* Waterman & Cash ex Sutton
Anamorph: *Septotis populiperda* (Moesz & Smár.) Waterman & Cash

This disease, which occurs only on Black poplar hybrids, is characterized by roundish, brown leaf blotches, 1–3 cm in diameter and often containing several, dark, circular zones within the necrotic areas. Affected leaves soon wilt and fall prematurely (Fig. **47**). Susceptibility is clone-dependent [65].

A distinctive diagnostic feature are the white acervuli of the imperfect state which are formed on the upper and lower leaf surfaces. They contain hyaline, 1- to 4-celled conidia, 20–30 × 5–6 μm in size. The perfect state, which develops on the fallen leaves in the spring, is characterized by brownish, stalked apothecia, 1–3 mm across, which resemble those of *Ciboria batschiana* (Fig. **5**). Leaf infection requires epidermal wounding, such as may often be caused by hail or leaf-chewing insects. In creating such infection courts, certain *Phyllodecta* species can also play a part in spreading the fungal spores. In addition to this, successful infection is dependent on a long period of high atmospheric humidity. Fungal invasion of the leaf tissue is finally halted by defensive host reactions, including the production of barriers composed of tannins.

Other Causes of Poplar Leaf Diseases [33]:

- *Melampsora* species produce striking, orange-yellow patches of spores on the undersides of leaves (see 'Poplar leaf rust', p. 70).
- *Phyllosticta osteospora* Sacc. causes irregularly shaped, pale brown, 0.5–1.5 cm spots. Pycnidia are spherical, black, with hyaline conidia, 3.5–6.5 μm in size, and swollen at both ends (Table **I,13**). It occurs on leaves of Balsam poplar and Black hybrid poplars.
- *Phyllosticta populorum* Sacc. forms 3–10 mm, greyish white, roundish spots with blackish brown margins. Pycnidia are spherical with elliptical conidia, 6–7 × 3–4 μm in size (Table **I,14**). It occurs on leaves of Balsam poplar.
- *Pollaccia elegans* Servazzi, the imperfect state of *Venturia populina* (Vull.) Fabr. cause a leaf disease and shoot tip dieback on Black and Balsam poplars. Conidia measure 32–38 × 11 μm (Table **II,10**).
- *Pollaccia radiosa* (Lib.) Bald. & Cif., imperfect state of *Venturia macularis* (Fr.) E. Müller & Arx, causes necrotic leaf spots and also kills young shoots. It occurs only on White poplar and Aspen (see 'Shoot tip disease of Aspen,' p. 89).
- *Septoria populi* Desm. forms small (1–3mm), roundish, blackish brown necroses with greyish centres. Pycnidia are roundish with sausage-shaped, curved, 2-celled conidia, 30–45 × 3 μm in size. It occurs on leaves of Balsam and Black hybrid poplars (Table **II,11**).
- *Uncinula adunca* (Wallr.) Lév., poplar mildew forms a greyish white, extensive

mycelial coating on both leaf surfaces. Conidia are ellipsoidal and appear in short chains. Cleistothecia are dark, 100–150 μm in diameter, with numerous hooked appendages. It also occurs on leaves of willow.

Poplar Leaf Rust

Cause: *Melampsora* spp. [80,226]

A rust attack on poplar becomes noticeable as a more or less thick, orange-yellow coating on the undersides of the leaves. It is caused by various species of *Melampsora* which can be differentiated both on the basis of their haplontic-state hosts and morphologically according to the form and size of the uredospores and paraphyses [156]. Of the more common European species, the following can be mentioned:

- *Melampsora allii-populina* Kleb. occurs on leaves of *Populus nigra* and its hybrids, *P. trichocarpa,* and *P. balsamifera*; alternate hosts, *Allium* and *Arum* species.
- *Melampsora larici-populina* Kleb. occurs on leaves of *Populus nigra* and its hybrids; alternate hosts, *Larix* species.
- *Melampsora larici-tremulae* Kleb. occurs on leaves of *Populus alba, P. tremula,* and *P. canescens;* alternate hosts, *Larix* species.
- *Melampsora magnusiana* Wagner occurs on leaves of *Populus alba, P. tremula,* and *P. canescens*; alternate hosts, *Chelidonium majus* and *Corydalis* species.
- *Melampsora rostrupii* Wagner occurs on leaves of *Populus alba, P. tremula,* and *P. canescens*; alternate host, *Mercurialis biennis*.

Most of these species are heteroecious, producing their uredosori and teleutosori on poplars and their spermogonia and aecidia on various alternate hosts. To the forester, only the states which are formed on poplars are of any practical interest. An explosive distribution of uredospores can lead in summer to an attack which results in the shrivelling and premature falling of the leaves. A general loss of host vigour is quite often associated with this primary damage, favouring other diseases, such as Dothichiza canker.

The problem of poplar rust diseases has today been largely solved by the breeding of more or less resistant clones. Where, in spite of this, rust-susceptible clones are planted, a severe attack can be avoided by fungicidal applications during the growing season. Early-season attacks can be prevented by excluding the particular alternate host from the neighbourhood. However, this precaution may fail, since some *Melampsora* species can re-infect poplars without the intervention of the alternate host when favourable conditions (i.e. mild winters), allow overwintering of the uredospores.

Anthracnose of Plane

Cause: *Apiognomonia veneta* (Sacc. & Speg.) Höhn.
Anamorph: *Discula* sp.

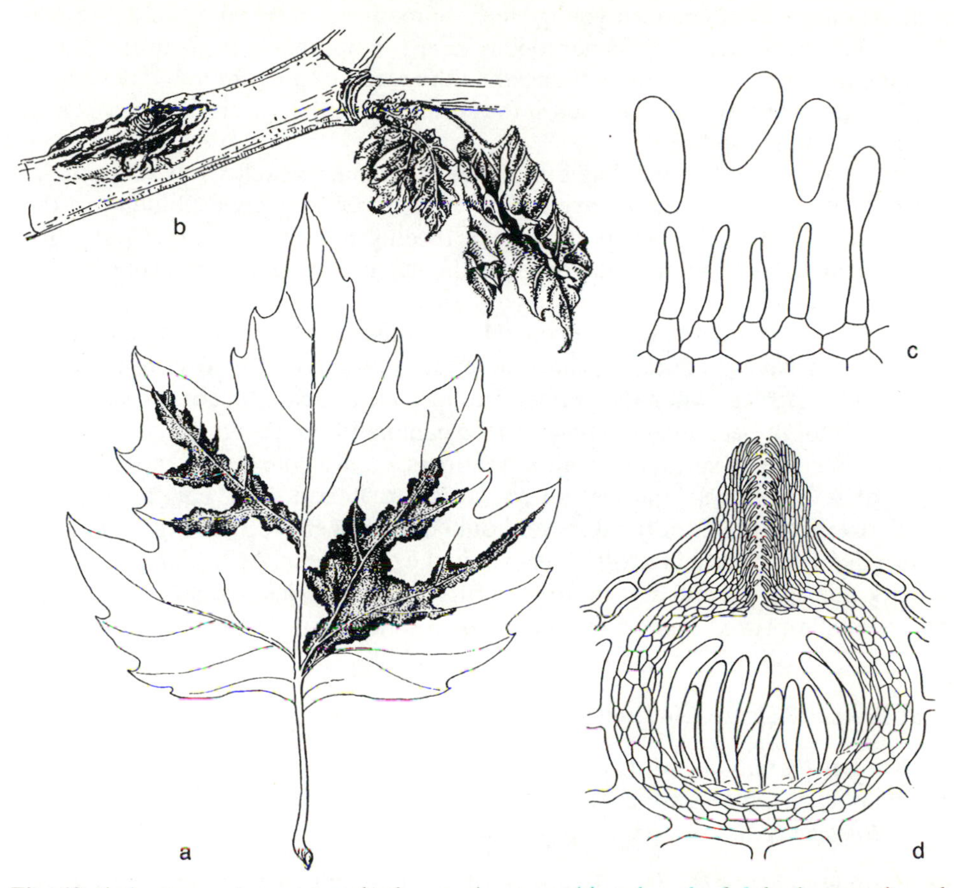

Fig. 48 *Apiognomonia veneta*. **a** leaf necrosis on an older plane leaf, **b** bark necrosis and wilting young leaves; **c** conidiophores with spores of the imperfect state, **d** cross-section through a perithecium

This fungus is one of the most common and most widely distributed causes of disease in plane trees. In Britain and central Europe, damage occurs mainly on London plane (*Platanus hybrida*) planted in streets and parks, although other species can be attacked (*P. occidentalis* more and *P. orientalis* less severely) after wet, cool spring months [207].

Of the various symptoms, partial browning of the leaves is the most distinctive. This manifests itself as relatively large, jagged necroses which mostly run along the veins. The necroses may spread into the petioles, causing the leaves to fall prematurely, and the fungus may then grow into the shoots, becoming dormant during the rest of the growing season. In mild weather during the tree's dormant period, these lesions may be reactivated, leading either to the death of bark around the buds, which then fail to flush, or to complete girdling and hence dieback of the twigs. Older trees are particularly affected by this bark necrosis. The fungus can also attack bark on the main stem when, for example, flush pruning cuts are inflicted on young trees. Sometimes, following the leaf blight stage, the fungus overwinters harmlessly near the bud bases but can then grow into the flushing shoots, killing them when they are up to 30 cm long. This shoot blight is favoured by cold weather in the spring, when host resistance is reduced by slow growth (Fig. **48**).

The form of fructification that is produced in summer is exclusively the conidial state which has also been referred to as *Gloeosporium platani* Oudem. in the literature [8,218]. It also occurs along the leaf veins, initially under the epidermis, in the form of dark conidiomata, in which the hyaline conidia, measuring 9–13 × 4–6 μm, mature.

The main recommendation for control of the disease is the preventive removal of the fallen leaves in autumn, since these are the source of ascospore infection in the following year when the perfect fruiting state appears. In some cases it can be helpful to cut back infected twigs, as the conidia formed on them can give rise to new infections in spring. Chemical treatment is advisable only if, after several years of severe attack, the tree's survival is threatened. The significance of a single severe outbreak of the disease should not be overrated, as experience has shown that trees so affected often recover in the same year. As for the prevention of bark damage, careful and thorough pruning in nurseries, including that even of the smallest twigs, can be more successful than the use of fungicides.

Leaf Blotch of Horse Chestnut

Cause: *Guignardia aesculi* (Peck) Stew.
Anamorph: *Leptodothiorella aesculicola* (Sacc.) Siwan. and
Phyllosticta sphaeropsoidea Ell. & Ev.

Of the few diseases to affect Horse chestnut, Guignardia leaf blotch is among the most striking and seems to have become ever more widespread since its first appearance in Europe in 1950. In July or August the leaves of affected trees become blotched brown, cup upwards, and mostly fall prematurely. The necroses, which reach several centimetres in size, are irregular in shape and often have bright yellow or pale brownish margins [149].

As drought or salt can also sometimes cause leaf browning, albeit extending from the leaf margin, an examination should be made for the two types of imperfect fruit body which, even with the naked eye, are recognizable as black specks in the necrotic tissue. The *Phyllosticta* form, which is usually produced first, is distinguished by, among other things, ovoid, hyaline conidia, 13–20 × 10–14 μm in size. In contrast, the *Leptodothiorella* form possesses rod-shaped conidia, 4–9 × 1-2 μm in size. The fungus overwinters in the leaves lying on the ground where, in spring, the fruit bodies of the perfect state are formed. They produce ascospores which are assumed to infect the new leaves in spring (Fig. **49**).

Of the species of *Aesculus* used as parkland and street trees in Europe, *A. hippocastanum* and *A. pavia* are equally susceptible, while the less common shrub-like *A. parviflora* is not attacked [142]. In older trees, the damage is mainly

Fig. 49 *Guignardia aesculi*. **a** Horse chestnut leaf with symptoms of an attack, **b** pycnidia of the imperfect state on the underside of a leaf (detail), **c** cross-section through a pycnidium of the *Phyllosticta* imperfect state, **d** conidia of *Phyllosticta*, **e** conidia of the *Leptodothiorella* imperfect state

aesthetic; in nurseries it can lead to quite severe losses, especially if attacks occur year after year.

In non-woodland situations where control is desired for aesthetic reasons, the removal of the fallen autumn leaves is recommended as an 'environmentally friendly' preventive measure. In addition, the fungus can be suppressed with chemicals, although only during the period between bud burst and full leaf expansion. This method should preferably used only in nurseries and even then only if young plants have already suffered heavy attacks in previous years.

Marginal leaf necrosis, stunted growth, and defoliation can also be induced by various *Phytophthora* species which can attack the stem base and the branches

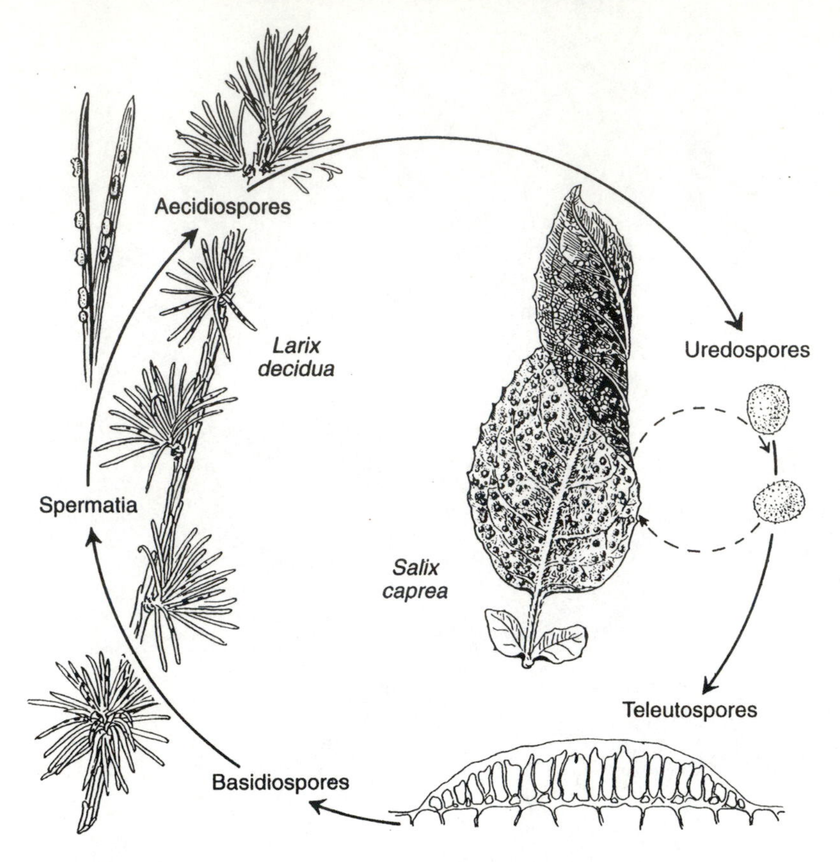

Fig. 50 Willow leaf rust. The 1-year life cycle of *Melampsora salicina* on a leaf of *Salix caprea* and needles of *Larix decidua* (from Butin 1960)

as well as the roots. The cause of such symptoms can be diagnosed with certainty only by isolation of the causal organism [27].

Willow Leaf Rust

Cause: *Melampsora* spp. [80,226]

Willow leaf rust—like its counterpart on poplar—can be recognized from the yellow to orange-yellow uredosori on the undersides of the leaves, often at quite some distance from the observer. In a heavy attack, the leaves curl and dry up. The economic significance of willow leaf rust is mainly limited to specialized cropping systems, especially biomass production, in which the yield losses from the disease can be reduced by avoiding the use of particularly susceptible species and cultivars.

Several species and forms of heteroecious rust fungi are recognized on willows, but these can be identified with certainty only if the alternate host is known.

Examples of haplontic-state hosts include Silver fir, larch, or *Euonymus*, or herbaceous plants from various families. If the systematic relationships cannot be given precisely, it is necessary to hark back to the old collective name, *Melampsora salicina* Lév. Similarly, the term 'willow leaf rust' should be understood as a collective name for a disease which is caused by various fungal species.

Other Leaf Parasites of Willow [34,69,173]:

- *Marssonina salicicola* (Bres.) Magnus single or clustered, irregularly roundish, brown spots on the upper leaf surfaces. Conidia are unequally two-celled in stromatic acervuli (see p. 91 and Fig. **61a–d**)
- *Monostichella salicis* (Westend.) Arx causes small leaf spots and premature leaf fall in *Salix alba, S. americana,* and *S. fragilis.* It spreads in summer by means of ellipsoidal, 10–17 × 5–8 μm conidia (Table **I,15**. In spring, the fruit bodies of the perfect state (*Drepanopeziza salicis*) form on fallen leaves.
- *Pollaccia saliciperda* (All. & Tubeuf) Arx causes brown leaf spots. More serious damage occurs if shoot tips and bark are infected (see p. 93).
- *Rhytisma salicinum* (Pers.) Fr., causes black, tar-like spots up to 2 cm across. In summer, cylindrical conidia, 5–6 μm long, develop. Apothecia develop in the following spring on fallen leaves.
- *Uncinula adunca* (Wallr.) Lév., willow powdery mildew, forms a whitish covering on both sides of the leaves. Conidia appear in short chains; cleistothecia, with numerous, long, hooked appendages, similar to powdery mildew of maple (Fig. **36**)

4 *Damage to Buds, Shoots, and Branches*

Frost Damage to Unhardened Shoots

Shoots tend to be damaged most severely by frost when it occurs in spring ('late frost'), since the tissues are soft at this time and more susceptible to thermal shock than in autumn or winter. However, frosts at the end of the growing season ('early frosts') can also be damaging to certain species that have a continuous flushing habit and set buds late in the summer (e.g. poplars) or not at all (e.g. eucalypts). Severe frost damage causes shoots to wilt and droop, becoming brown and desiccated after a few days. If the damage is sublethal or confined to one side of the shoot, subsequent growth may be stunted, curved, or otherwise distorted, especially in the case of conifers (Fig. **51**). In most broadleaved species, the loss of the shoot tips stimulates compensatory growth either from axillary buds (e.g. oak) or from dormant buds (e.g. beech), but this new growth usually bears undersized chlorotic leaves. Conifers are less able than broadleaves to replace lost shoots; an exception to this is larch, which can refoliate entirely.

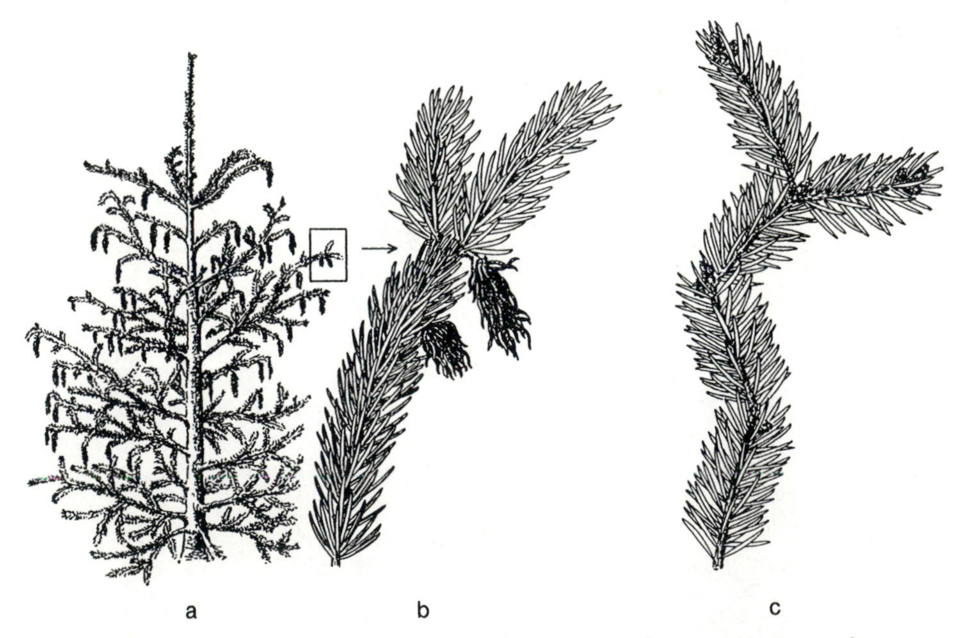

Fig. 51 Spring frost damage on conifers. **a** symptoms on a young spruce tree, **b** spruce branch with killed side shoots; **c** zigzag growth on Douglas fir after repeated loss of the terminal shoot

Tree species differ in the susceptibility of their young shoots to late frosts. Among the broadleaves native to Europe, beech, lime, ash, and oak (especially sessile oak) are relatively susceptible. Aspen and birch are notably frost hardy, since their leaves do not freeze until the temperature has dropped to −6°C. Species introduced from warmer climates are particularly at risk from frost if either they flush early or are inherently very frost tender. Different species are at risk depending on the climate of the region concerned; in central Europe, for example, these include Sweet chestnut, plane, walnut, and ornamentals like the Tree of heaven and Indian bean tree. Among the conifers, the European silver fir and certain provenances of Douglas fir, Corsican pine, and Sitka spruce are particularly at risk from frost, more so than Norway spruce. In contrast, Scots pine seems to be completely unaffected by late frosts. Some tree species, such as oak or elm, escape the danger of freezing despite being very sensitive to frost because they flush relatively late.

Cucurbitaria Bud Blight of Spruce

Cause: *Gemmamyces piceae* (Borthw.) Cassagrande
Syn. *Cucurbitaria piceae* Borthw.
Anamorph: *Megaloseptoria mirabilis* Naumov

This disease affects *Picea pungens* and certain other spruces, including *P. abies*, though not *P. sitchensis*, and is also seen occasionally on *Abies alba*. It leads to the death of buds and abnormal shoot development (Fig. **52 a–d**). Buds may swell in spring, but they flush incompletely and crookedly, twisting into a spiral. Loss of the terminal bud stimulates premature development of laterals, so that the normal length ratio of first and second-order shoots is disturbed. Repeated death of buds may occur over several successive years, producing distorted and twisted branches or short, claw-like shoots. The affected shoots gradually lose their needles and eventually die. From June onwards, the dead buds acquire a black, warty, crust-like coating consisting of the basal stroma of the fungus and its pinhead-sized fruit bodies. Microscopic examination in summer generally reveals the conidial form, with its filiform, multiseptate, hyaline spores, 150–300 × 6–8 μm in size. Less often the perfect state appears, forming fruit bodies which are similarly spherical and dark coloured but contain numerous asci, each with eight muriform, brownish ascospores, measuring 36–50 × 13–15 μm [47,69].

Current evidence indicates that the fungus is most prevalent in regions where there are frequent spells of wet weather, sometimes occurring in epidemic outbreaks and causing severe losses in Christmas tree plantations. The recommended control is to remove infected material by pruning in the first instance; fungicidal sprays are also of some value if applied several times during the year.

Other Causes of Bud Death:

A different syndrome can easily be confused with Cucurbitaria bud blight. This takes the form of arrested bud development and has an abiotic cause. The

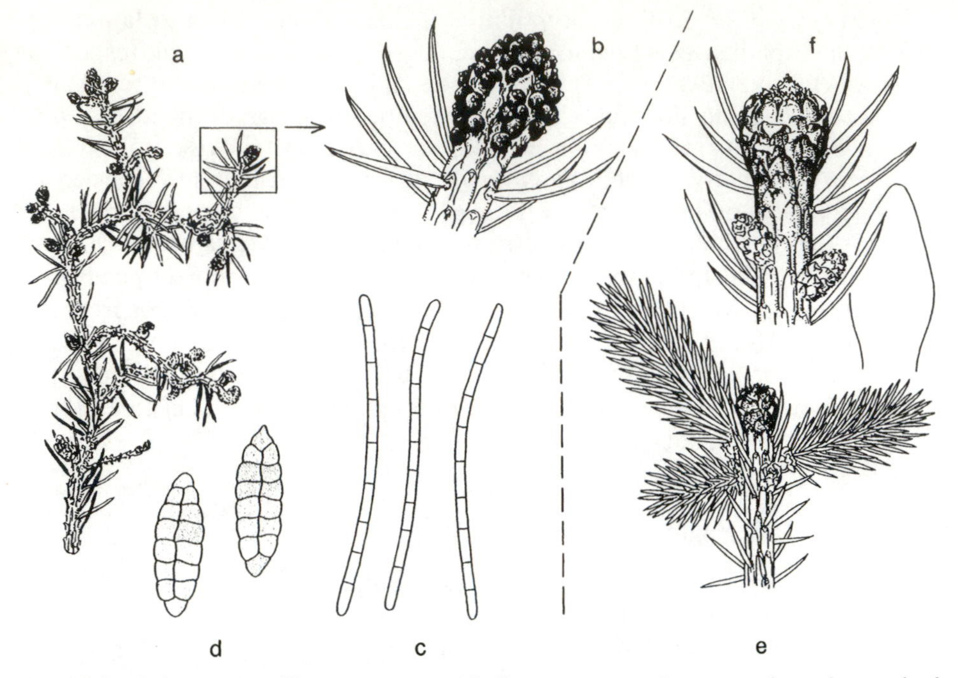

Fig. 52 Bud damage to *Picea pungens*. **a-d** *Gemmamyces piceae*: **a** a branch attacked over a number of years, **b** a killed bud covered with fruit bodies, **c** conidia, **d** ascospores; **e,f** bud failure from abiotic causes: **e** appearance of a terminal bud which has failed to flush, **f** swollen, aborted bud (healthy bud in silhouette)

symptoms, which occur most commonly on *Picea pungens*, appear in spring when the affected buds—mainly the terminal ones—fail to flush, before becoming brown and desiccated in early summer. These buds can usually be distinguished from normal ones during the previous autumn, since they are abnormally swollen and cylindrical in shape, instead of conically pointed (Fig. **52 e,f**). The reason for their failure to flush is evidently early frost damage, although this only harms very enlarged buds.

The death of the terminal bud results in impaired development of the leading shoot, which is a particular problem in Christmas tree plantations. For this reason, trees which are found to bear enlarged terminal buds in autumn should be selected for sale in time for the following Christmas, before obvious damage occurs. Good form can be restored in damaged plants up to 4 years old by tying 1-year-old lateral shoots into the vertical position, so as to restore normal height growth as quickly as possible. In the long term, there is the possibility of using provenances less susceptible to frost.

Grey Mould

Cause: *Botryotinia fuckeliana* (de Bary) Whetzel
Anamorph: *Botrytis cinerea* Pers.

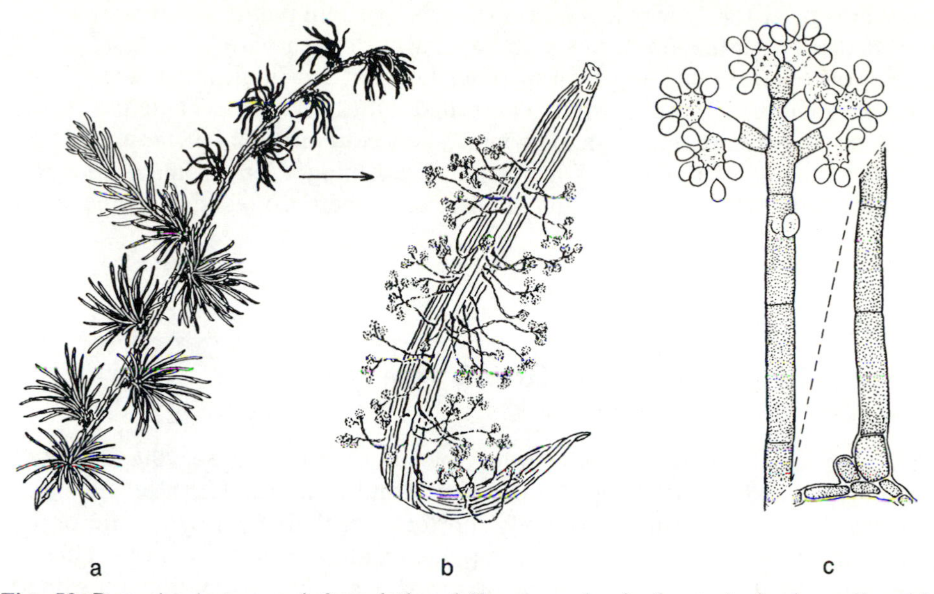

Fig. 53 *Botrytis cinerea*. **a** infected tip of European larch shoot, **b** dead needle with conidiophores and spore heads, **c** conidiophores with spores

The grey mould fungus, which is generally known under its anamorphic name, is a facultative parasite with a very wide host range. It occurs throughout the world, principally on dead plant material, though under certain conditions it is able to attack living plants. The damage is mainly confined to young tissues, causing the collapse of seedlings or the death of buds and young shoots on older trees. Affected shoots typically hang down limply and then turn brown and dry up. In this form, the damage resembles that caused by late frost but an important difference is that it occurs only on scattered shoots, not generally over the lower twigs or the entire tree. A further diagnostic feature of the diseased tissues appears at a later stage; this is the grey-brown aerial mycelium, usually abundant with silvery brown branched conidiophores, bearing hyaline or pale brownish, ovoid spores which measure 9–12 × 6–10 μm. This is followed in autumn by the appearance of small, roundish, dark-coloured resting structures, sclerotia, by which the fungus overwinters.

Although grey mould can damage broadleaves, it occurs predominantly on conifers of both Eurasian and North American origin, particularly Douglas fir, European silver fir, Norway spruce, southern provenances of Sitka spruce, larch, *Sequoia sempervirens,* and *Cupressus* spp. It is also common on eucalypts grown in maritime climates. As the fungus can infect only young tissues, damage occurs mainly on needles and young shoots which have not yet hardened. In some species, including larch, *B. cinerea* can pass via the dwarf shoot into the bark of the previous year's long shoot and cause its death (Fig. **53**). Experience has shown that the disease occurs only in conditions of high air humidity and relatively low temperatures. Botrytis damage can therefore be expected if the weather is persistently cool and moist, especially where plants are overcrowded. In broadleaves, infection seems to be encouraged by a variety of factors that injure soft tissues, including frost or other kinds of mechanical damage.

Grey mould can be successfully controlled with fungicides. Their use in the forest or in gardens may not, of course, be economic, and in any case, an attack on shoots of older trees is not life-threatening. Control may, however, be essential in the nursery and greenhouse, e.g. when propagating by means of cuttings.

Sphaeropsis Shoot-killing of Pine

Cause: *Sphaeropsis sapinea* (Fr.) Dyko & Sutton
Syn. *Diplodia pinea* (Desm.) Kickx

This fungus, which occurs on various species of pine, causes the death of currently expanding shoots of *Pinus sylvestris* and *P. nigra*. The affected shoot tips turn brown and remain distinctly shorter than their healthy counterparts. The shoot death is accompanied by copious exudation of resin, noticeable as numerous dried drops on the bark of the shoots. The formation of wound periderm soon halts the spread of the necrosis, restricting it to the current year's shoots. As a rule, twig development is restored by the activation of dormant buds, but this sometimes produces bushy growth. Attacks occur on trees from 10–40 years old and occasionally on seedlings (Fig. **54**).

The fruit bodies of the fungus are spherical, dark brown pycnidia which appear in the killed bark or at the base of dead needles, often in clusters. These fruit

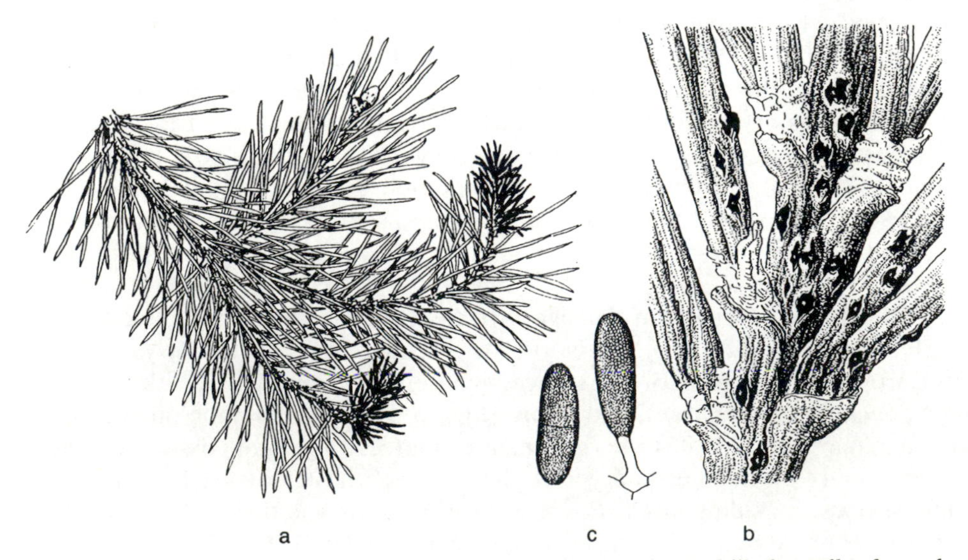

Fig. 54 *Sphaeropsis sapinea*. **a** symptoms on pine, **b** pycnidia on killed needles from the shoot tip, **c** conidia

bodies contain conidia which are at first hyaline and 1-celled, later dark brown and in part 2-celled; they are ellipsoidal to long-ovoid and become warty with age, the warts arising on the endospore, that is to say on the inner surface of the spore wall; they measure 25–36 × 13–16 μm.

This pathogen is known all over the world, but it tends to be more aggressive in warmer climates where, in contrast to its role in central or northwestern Europe, it can kill the mature bark of branches and stems. However, such attack depends on predisposing factors among which are nutrient deficiencies, soil dryness and particular site conditions (low-lying land). In these warmer climates, tree species greatly at risk include *Pinus radiata, P. nigra,* and *P. ponderosa*. As a saprotroph, *S. sapinea* grows on fallen dead branches and cones and also on logs of pine, where it can cause considerable loss of timber value by producing an intense grey-blue discoloration of the wood ('blue stain').

Pine Twisting Rust

Cause: *Melampsora pinitorqua* E. Rostrup

This pathogenic fungus, also known by the collective name of *Melampsora populnea*, is responsible for the curvature of shoots and growth disturbances in various pine species. The disease first becomes apparent on the mid-zones of expanding shoots where cushion-shaped aecidia develop on patches of yellowed bark. The damage interrupts the expansion of the shoot on one side, but the opposite side continues to develop unimpeded so that the shoot axis becomes curved. Vertical growth is, however, resumed due to the negative geotropism of the shoot tip, and this produces an S-shaped bend which usually persists for some years [17]. The damage leads to reduced height growth [131] and possibly a permanent stem deformity. The chief host in Europe is *Pinus sylvestris*. In addition, *P. mugo, P. nigra,* and *P. strobus* can be damaged. Attacks occur over the age range of 1 to 10 years [226].

In cool, moist summer months, rust-damaged shoots can be secondarily infected by *Botrytis cinerea*. This results in death of the affected shoots, recognizable from the browning of all the dwarf shoots (needle fascicles).

Similar shoot curvature can be induced by the pine shoot moth (*Rhyacionia buoliana*). In this case the lowest, sharp bend of the 'post-horn' deformity is located exactly at the junction between two annual shoot increments. In the case of the pine twisting rust, however, the first bend is in the middle or lower third of the shoot.

Melampsora pinitorqua is a heteroecious rust fungus, alternating between two different host plants. The pine is the haplontic-state host on which the pycnio- and aecidiospores of the fungus are produced. The dikaryotic state develops on the leaves of White and Grey poplars and aspen, forming uredospores in summer, teleutospores in autumn and basidiospores in the following year. On the pine, the fungus persists for no more than one year, but on poplars it can overwinter in the buds and in bark, producing basidiospores each year which can cause repeated new infections of any pines in the vicinity.

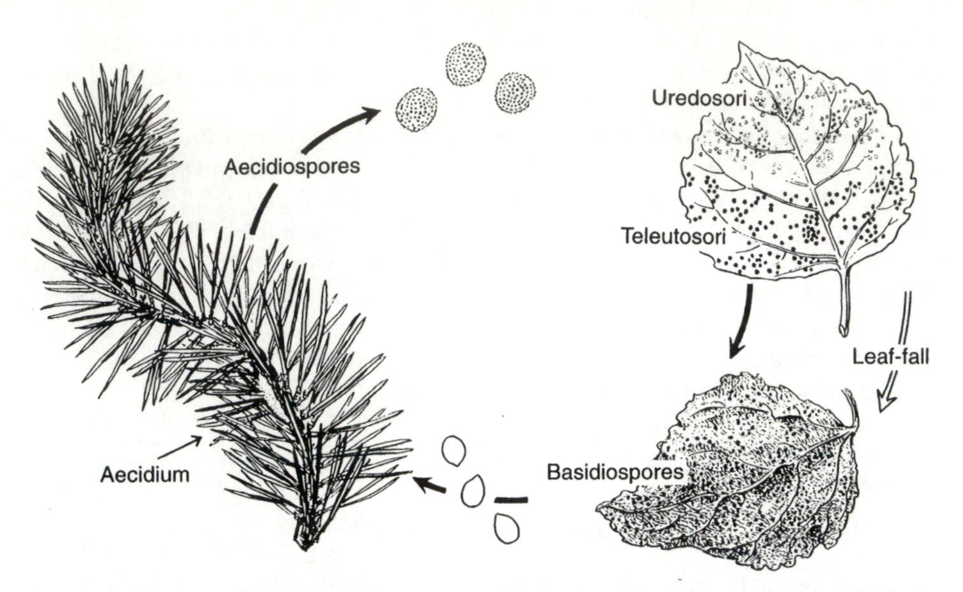

Fig. 55 *Melampsora pinitorqua*. The 1-year life cycle with symptoms respectively on Scots pine (left) and aspen (right)

The most effective way of preventing pine twisting rust damage in a plantation is to remove all aspens and Grey and White poplars within a radius of 500 metres, although this may not be desirable in areas where such trees have special ecological value. In tree nurseries, growing the complementary host plants close to each other should be avoided. As differences in susceptibility have been observed in various Scots pine provenances, there is also scope for making resistant selections.

Brunchorstia Dieback of Conifers

Cause: *Gremmeniella abietina* (Lagerb.) Morelet
Syn. *Ascocalyx abietina* (Lagerb.) Schläpfer-Bernhard
Scleroderris lagerbergii Gremmen
Anamorph: *Brunchorstia pinea* (Karsten) Höhn.

Brunchorstia dieback is one of the most serious of tree diseases. In recent decades it has led to the destruction of large numbers of conifer stands in central Europe and also in North America and eastern Asia. In its typical form, the syndrome is characterized by a death of shoots accompanied by bark damage, the extent of which depends on the tree species and location of the stand. To date, almost 50 tree species from 7 different genera of the Pinaceae have been listed as hosts. In Europe, *Pinus nigra* and its varieties such as *maritima* (Corsican pine) are noted for their susceptibility and can suffer much damage, especially in the thicket stage. Recently, damage seems to have worsened on Norway spruce [7]. In Europe, other host plants are *Pinus cembra, P. contorta, P. mugo, P. strobus, P. sylvestris,* and *Abies alba* [66], with variations in susceptibility between these species and amongst their provenances [169].

The cause of Brunchorstia dieback, *Gremmeniella abietina*, is an ascomycete fungus with a very checkered taxonomic history [139]. Two varieties with differing imperfect fruiting forms and host plants are now recognized in Europe. The form designated *Brunchorstia pinea* var. *cembrae* Morelet occurs chiefly on *Pinus cembra* in the Alps, whereas *Brunchorstia pinea* var. *pinea* occurs at lower elevations principally on *Pinus nigra* and other species of pine, less often on *Picea abies*.

The symptoms of the disease vary both with its stage and the tree species. *Pinus nigra* deserves special mention here because of its importance in central Europe and the major use of its variety *maritima* in Great Britain. On *P. nigra* the first symptoms of shoot tip infection show during the six cooler months of the year with browning at the base of the bud (visible in cross-section) and resin exudation from the buds. Then, early in the new year, the needles turn brown from the tips of the twigs downwards and fall prematurely. Similar symptoms occur if infection takes place below the shoot tip; this is initiated via the stomata of the scales. Replacement shoots, derived from dormant buds below the dead branches, produce a bushy growth pattern. In the case of chronic disease in which the replacement shoots are also attacked, the whole tree can eventually die. However, trees can recover even from severe attacks when epidemic spread, favoured by a wet summer, is arrested by a following dry summer.

The fruit bodies of the imperfect state form on the killed buds and extension shoots, often in the first year. These are pycnidia and contain hyaline conidia, curved like sickles. In var. *pinea* the conidia are 3–4 celled and 24–48 × 2.5–3.5 μm in size; in var. *cembrae* they are 6–8 celled and up to 73 μm long. The fruit bodies of the associated perfect state are brownish-black apothecia, 0.5–1.2 mm across, whose asci contain eight 3-4-celled, hyaline ascospores 14–20 × 3.3–5.0 μm in size. Whereas apothecia are common in the variety *cembrae*, in the variety *pinea* they are seldom formed and then only in the second year of infection (Fig. **56**).

Damage from *G. abietina* can be prevented or reduced using silvicultural methods in the first instance. These include the avoidance of underplanting and of cool, wet sites, timely thinning to prevent the development of dense stands, and—during reafforestation—the use of less susceptible provenances. The use of fungicides (from June to September) is economic only in nurseries.

Related Species:

- *Ascocalyx (Encoeliopsis) laricina* (Ettlinger) Schläpfer-Bernhard is the cause of a shoot dieback of European larch, particularly damaging in high elevation afforestation, and often associated with frost. Pycnidia are of the conidial state *Brunchorstia laricina* Ettlinger on dead, 2-year-old long shoots with

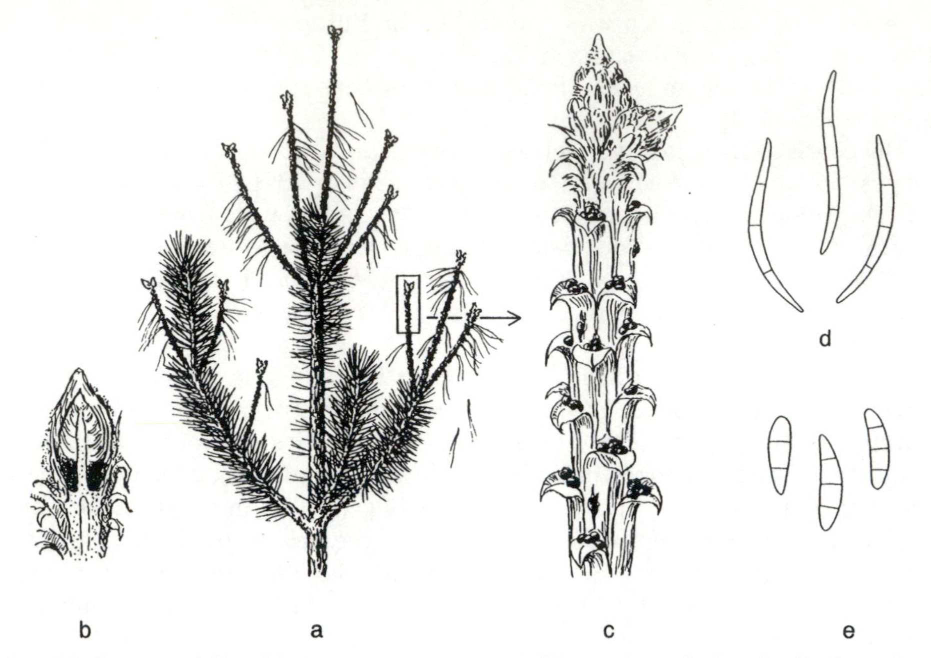

Fig. 56 *Gremmeniella abietina*. **a** symptoms on *Pinus nigra*, **b** longitudinal section through a bud with early symptoms at the bud base, **c** killed shoot tip with pycnidia, **d** conidia, **e** ascospores

2–4 celled conidia, 15–23 × 3–4 μm in size (Table **I,16**). Apothecia are of the perfect state blackish-brown with 2-celled, hyaline ascospores, 10–17 × 3–4 μm.

Shoot Shedding of Pine

Cause: *Cenangium ferruginosum* Fr.
Syn. *Cenangium abietis* (Pers.) Duby

Cenangium ferruginosum is mainly saprotrophic in the bark of dead branches of *Pinus sylvestris, P. nigra, P. mugo,* and non-European pine species. In the past, its significance as an occasional weak parasite has been greatly overrated. The reason for this may be its frequent occurrence together with *Gremmeniella abietina* when the latter fungus can be considered to pave the way for it. Also, atypical epidemics can result from extreme weather conditions (high rainfall in the spring followed by a long period of drought).

In nature, the fungus occurs almost always as the teleomorph only. The apothecia are dark brown and, when moist, saucer-shaped and 1–2 mm across. They are produced on the bark of dead twigs in late autumn and ripen in the following spring through to the autumn. Their asci each contain eight hyaline, 1-celled ascospores, 11–13 × 5–7 μm in size (Fig. **57**). The rare anamorphic state

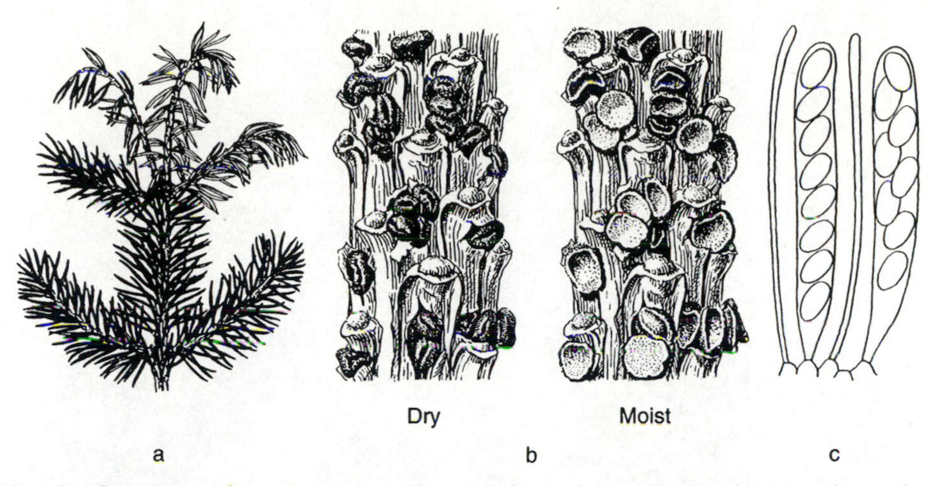

Fig. 57 *Cenangium ferruginosum*. **a** diseased shoot tips on Scots pine, **b** twig section with closed (left) and open (right) apothecia, **c** asci with ascospores and paraphyses

is represented by ellipsoidal conidia of the *Phomopsis* type; these measure 5–6 × 2–3 μm.

Juniper Rust

Cause: *Gymnosporangium sabinae* (Dickson) Winter
Syn. *Gymnosporangium fuscum* DC.

Juniper rust is important in gardens and parks in two respects: it occurs on various juniper species (particularly on *Juniperus sabina, J. chinensis, J. virginiana,* and *J. oxycedrus*) where it causes spindle-shaped swellings on the stem and twigs, and it is also found on pear (*Pyrus communis*), forming orange-red spots on the leaves and occasionally attacking the petioles, shoots, and fruits (Fig. **58**).

The haplophase of the fungus is spent on the pear where the first fruiting structures to appear are the pycnidia on the upper leaf surface, followed by the spherical, longitudinally split aecidia which form in groups on small swollen areas of tissue on the underside of the leaf. After the aecidiospores have been released, the mycelium dies out or, exceptionally, overwinters at the base of the buds.

The dikaryophase develops on the bark of one of the many juniper species, beginning with the teleutosori as no uredosori are produced. These are brown, wart-like outgrowths which, on taking up moisture, swell to form cone- or tongue-shaped, yellow-brown structures, 1–2 cm long, with a fleshy gelatinous consistency. The typical 2-celled teleutospores give rise to basidia and hence in turn to the basidiospores which can complete the cycle by infecting the leaves of the haplontic-state host. Unlike the haplophase, the dikaryophase on juniper is

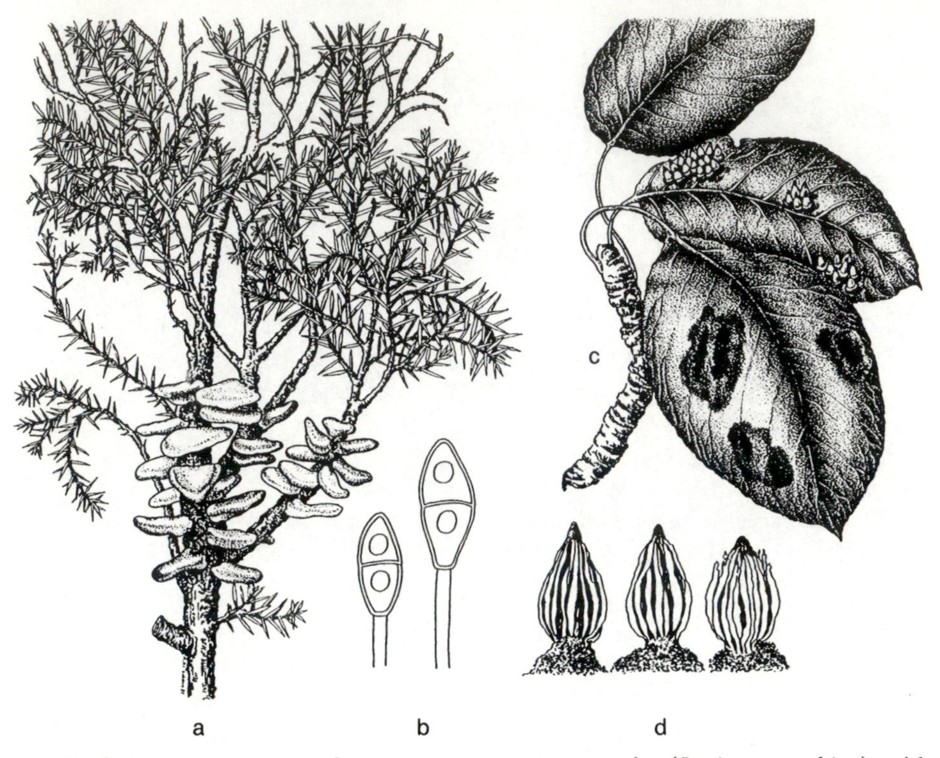

Fig. 58 *Gymnosporangium sabinae*. **a** symptoms on savin (*Juniperus sabina*) with teleutosori, **b** teleutospores, **c** symptoms on pear leaves, **d** mature aecidia

able to perennate in the form of mycelium, so that a fresh crop of teleutospores can be produced annually [17].

On juniper, *G. sabinae* can kill slender branches and also cause damage on pear in cases of heavy attack, weakening the tree considerably and predisposing it to attack by secondary pathogens. The value of the fruit may also be reduced. The most effective way of preventing this kind of damage has proved to be the removal of one or other of the host plants, at the very least maintaining a distance of more than 500 metres between them. If it is not possible to interrupt the disease cycle in this way (e.g. where junipers are to be retained in the interests of nature conservation), pear trees can be protected to some extent with prophylactic fungicidal sprays.

Other ***Gymnosporangium*** *Species* [69]:

- *Gymnosporangium clavariaeforme* (Jacquin) DC., leaf rust of hawthorn, in its dikaryophase, causes spindle-shaped twig swellings on *Juniperus communis* and *J. nana*, with orange-red, cone-shaped teleutosori. In the haplophase, the aecidiospores occur mainly on *Crataegus* and *Amelanchier* species, developing on the leaves and fruits.
- *Gymnosporangium confusum* Plowr. causes spindle-shaped twig swellings in

its dikaryophase on *Juniperus sabina, J. oxycedrus, J. virginiana,* and other juniper species. The teleutosori are brownish-red, ragged outgrowths. In the haplophase it occurs on *Cotoneaster integerrima, Crataegus laevigata,* and on *Cydonia* and *Sorbus* species, forming yellow spots, some turning red with aecidia.

- *Gymnosporangium juniperinum* (L.) Fr. comprises a number of biological species. It forms reddish yellow spots with cylindrical, 3–4 mm long pseudoperidia and yellowish aecidiospores on the undersides of the leaves of *Sorbus aucuparia*. The dikaryophase occurs on needles of *Juniperus* species; in the haplophase, the fungus is known as 'cluster-cups.'
- *Gymnosporangium tremelloides* (A. Braun) R. Hartig forms various host-specific forms. The dikaryophase on *Juniperus communis* forms flattened, shell-shaped teleutosori; in the haplophase, reddish yellow, thickly swollen spots with aecidia which develop on leaves of *Malus sylvestris, Sorbus aucuparia,* and other *Sorbus* species.

Kabatina Shoot Killing of Cupressaceae

Cause: *Kabatina thujae* Schneider & Arx

An attack by *K. thujae* manifests itself by the death of shoots and twigs; several species in the genera *Thuja, Chamaecyparis*, and *Cupressus* are affected. The most severe damage seems to be suffered by *T. occidentalis*, small plants of which can become completely bare. However, entire plants seldom die, and damage is more usually confined to a yellowish brown discoloration of the shoot tips. In nurseries, the fungus causes more damage on soils which are shown to be poorly supplied with nutrients; in particular, calcium and magnesium.

Similar damage results from the effects of frost or from the lack of light on suppressed, crowded twigs. It is therefore necessary to look for the presence of fruit bodies. Confusion is also possible with the symptoms of attack by the larva of a moth which mines the shoots of *Thuja*.

The pustule-like acervuli of the fungus, which are laid down beneath the epidermis and then burst through it, develop at the base of killed twigs. They are 50–170 μm across and brownish-black, containing hyaline to pale brown phialide-like conidiophores which bear a succession of acrogenously formed hyaline conidia, ovoid to ellipsoid in shape and measuring 5–8 × 2.5–3.5 μm (Fig. **59**).

In pure culture, this fungus, which belongs in the Melanconiales, forms a very slow-spreading mycelium which is at first pale coloured, later dark brown and forming small humps on which numerous single-celled, hyaline blastospores arise. Older cultures mostly bear sporodochial pustules like those which form on the plant.

Kabatina thujae var. *juniperi*, which occurs on *Juniperus chinensis* and other juniper species, is recognized as a separate variety of the fungus described above. Morphologically this is very close to *K. thujae,* but its cultural characteristics distinguish it from the basic species on *Thuja* [181].

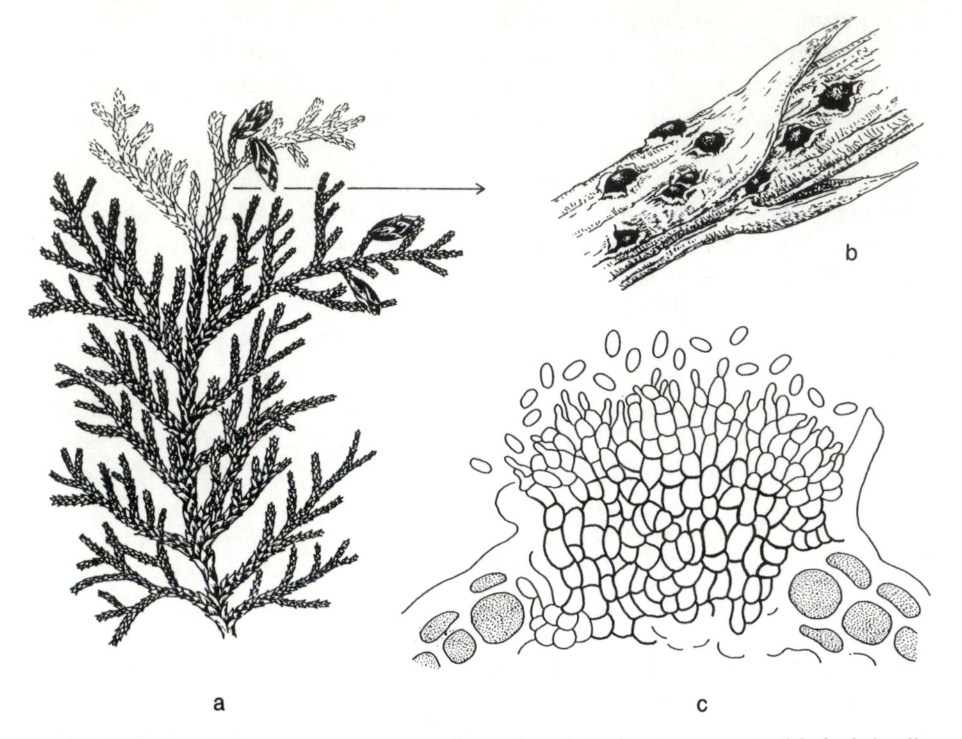

Fig. 59 *Kabatina thujae*. **a** symptoms on thuya frond, **b** shoot segment with fruit bodies, **c** cross-section through a fruit body with conidia (c after Schneider and v. Arx 1966)

Other Parasitic Fungi on the Cupressaceae:

- *Botryodiplodia (Lasiodiplodia) theobromae* Pat. [208] is the causal agent of crown wilt, stem canker, and seedling blight in *Cupressus sempervirens* in the Mediterranean area [30].
- *Botrytis cinerea* Pers., familiar as the cause of grey mould, attacks shoot tips in conditions of high air humidity (Fig. **53**).
- *Pestalotia funerea* Desm. can girdle single branches, but is otherwise more usually a saprotroph following other damage. Conidia are 21–29 × 9.5 μm, 5-celled with the 3 median cells darker (Table **II,1**).
- *Phomopsis juniperovora* Hahn kills branches on various cultivated forms in the genera *Chamaecyparis, Cupressus, Juniperus,* and *Thuja*; in nurseries it is particularly damaging to *J. virginiana.* Pycnidia are lenticular, black, paler in the centre, and on needles and on parts of stems not yet lignified. 'α-spores' are spindle-shaped and 6.5–12 × 2.2–3.5 μm in size. 'β-spores' are long and thin and 25–45 × 1–2 μm in size (Fig. **32d,e**). The fungus can be confused with the saprotrophic, smaller spored *P. occulta* (Table **I,17**), and is best differentiated by means of cultural characteristics [44,92].
- *Pithya cupressina* (Fr.) Fuckel kills the tips of shoots of various juniper species, mainly *J. virginiana.* Parts of the shoot become pale green, later

grey. Apothecia are disc-shaped, 2–3 mm across, and bright orange-red. Ascospores are roundish, and 9–10 μm in size (Table **I,18**).

- *Seiridium cardinale* (Wagener) Sutton & Gibson (Syn. *Coryneum cardinale*) causes the death of twigs and bark on various Cupressaceae [159]. It can be very damaging on cypresses, especially *Cupressus macrocarpa* and *C. sempervirens*, in warmer or milder countries, e.g. Italy, Yugoslavia, Great Britain [206], and Switzerland, where it causes a slow but progressive dieback. It is best distinguished by the 4-celled, brown conidia, 21–30 × 8–9 μm in size (Table **I,19**). Vectors are various species of beetle.

Pollaccia Shoot Blight of Poplar

Cause: *Venturia macularis* (Fr.) E. Müller & Arx
Anamorph: *Pollaccia radiosa* (Lib.) Bald. & Cif.

This worldwide disease, which affects only aspen, and White and Grey poplars, involves leaf necrosis and the death of young shoots. The fungus produces irregularly shaped, dark-bordered blotches on the leaves. Under favourable conditions, it spreads from the leaf via the petiole into the bark of young, unlignified shoots. This results in the typical symptoms of wilting and withering; the shoot with its attached leaves turns brownish-black and curves downwards in the shape of a hook as if damaged by frost. Repeated annual attacks result in considerable distortion and bushy growth. Young plants of susceptible clones are particularly at risk, and can die from recurrent attack over a number of years (Fig. **60**).

A few days after the appearance of the first leaf necrosis, the olive-green, velvety sporulating mats of the fungus can be seen in the centres of the leaf spots and later on the killed shoots. The long-ellipsoidal, pale brown conidia are formed on short conidiophores; they are 18–26 × 5–8 μm in size, irregularly 2 septate or occasionally 3 septate. They can be confused with the conidia of *Cladosporium* species although these are borne on considerably longer conidiophores. During winter, the asci develop inside dark brown, spherical perithecia on leaves lying on the ground. The 2-celled ascospores are not released until the spring, but seem to be less important in initiating fresh infection than conidia which are formed by overwintering mycelium on the killed shoot tips.

The species and hybrids of the poplar Section Leuce vary considerably in their susceptibility. In general, the forms close to *P. tremula* suffer worse damage than those close to *P. alba*. If the less susceptible clones are chosen, chemical control is not necessary.

Myxosporium Twig Blight of Birch

Cause: *Myxosporium devastans* E. Rostrup

Various fungi can kill twigs and shoot tips on birch, along with abiotic factors, and one of most important of these is *M. devastans*, especially on *Betula*

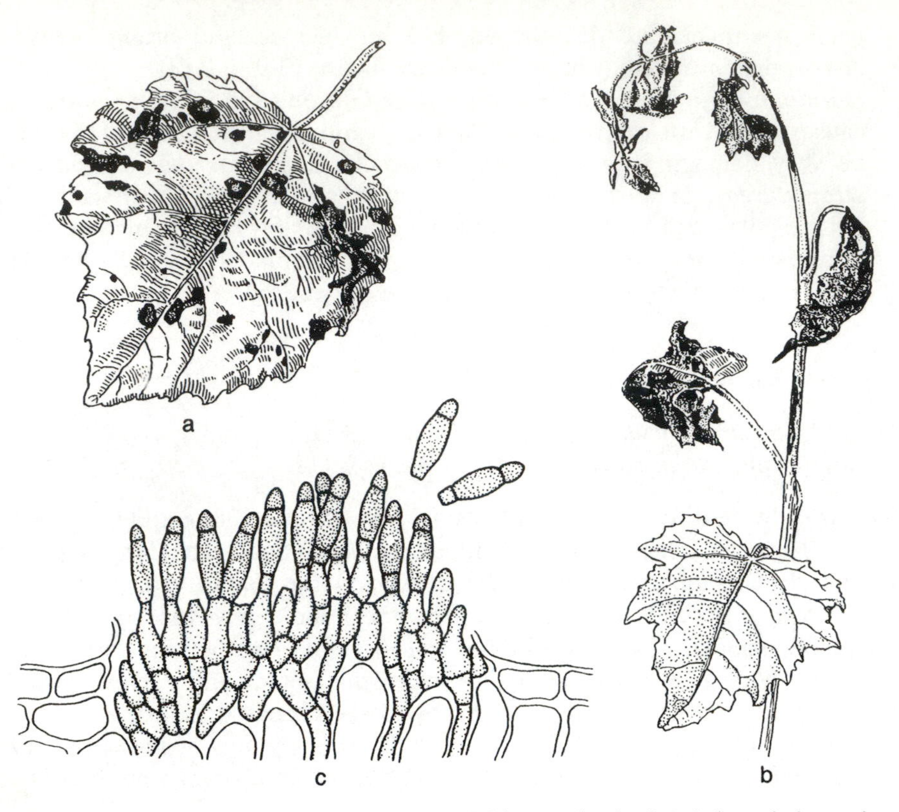

Fig. 60 *Pollaccia radiosa*. **a** leaf necroses on a White poplar leaf, **b** infected shoot tip, **c** cross-section through a fruit body with conidia

pendula. It occurs mainly in nurseries and typically causes the death of the tip of the leading shoot or of the uppermost laterals of 2 to 5-year-old plants. Compensatory development of short shoots on the lower surviving twigs produces a bushy appearance.

Observations in nurseries and of artificial inoculations show that the disease occurs only in particular circumstances which are conducive to an attack. These include: poor soil aeration; inadequate nutrition; excessively long delays in planting after heeling in; and problems with planting. Finally, attacks of the rust fungus *Melampsoridium betulinum* also encourage the disease.

Since twig death can also result from abiotic damage (frost, drought etc.), diagnosis depends on the presence of the fungal fruit bodies. These are brownish black and wart-like and, when ripe, contain hyaline, ellipsoidal to ovoid conidia, 6–10 × 2–3 μm in size (Table **I,20**).

Other Fungi on Twigs of Birch [69]:

- *Cryptosporium betulinum* Jaap is a saprotroph with conidia 30–50 × 3.5–4 μm in size (Table **II,13**).
- *Melanconium betulinum* Kunze & Schm. is a common saprotroph on dead

twigs of various birch species. Conidia are brown, 1-celled, narrowly ovoid and 13–18 × 5–7 μm in size (Table **I,21**).
- *Trimmatostroma betulinum* (Corda) Hughes is the most common colonizer and 'pruner' of dead, thin twigs. Sporodochia are cushion-shaped, brown or black. Conidia are in chains, with variable shape and septation, warty, and 10–20 × 5–13 μm in size.

Fireblight of Pomoideae (Rosaceae)

Cause: *Erwinia amylovora* (Burrill) Winslow

Fireblight is one of the most dangerous diseases of apples, pears, and other members of the rosaceous subfamily Pomoideae. Although this bacterial disease is of little significance in forestry, it is sometimes a major problem in gardens and parks where it attacks various ornamentals. Chief among these are cotoneaster, firethorn (*Pyracantha*), and the rowans and whitebeams (*Sorbus* species). The frequency of hawthorn (*Crataegus*) in hedgerows gives it a primary role as an infection source and thus as a means of spreading the pathogen further [67].

The symptoms of the disease are a rapid wilting of the leaves and inflorescences which then turn brown or black. Single shoots may die with their tips bent over like hooks. In very susceptible plants, the disease spreads rapidly into older branches and then into the stem. In humid conditions, diseased parts exude droplets of a sticky bacterial slime, at first white, later discolouring to a dark brown.

The bacterium spreads during the growing season, mainly by means of pollinating insect vectors but also by birds, wind, and rain. The risk of infection is high in sultry weather, particularly after stormy showers or hail (because of injury to the bark).

To protect orchards and nurseries, hawthorns and other susceptible host plants in the vicinity should be removed or, if used as windbreaks, trimmed back to prevent flowering. In regions where fireblight is a notifiable disease, the plant protection service should be informed without delay at the first sign of any occurrence. New plantings of susceptible hawthorn species should not be made. There is considerable variation in susceptibility within the host genera, and there is therefore scope for replacing diseased ornamentals with resistant species (e.g. *Sorbus intermedia*) or varieties.

Marssonina Leaf and Shoot Blight of Willow

Cause: *Drepanopeziza sphaerioides* (Pers.) Höhn.
Anamorph: *Marssonina salicicola* (Bresad.) Magnus

This disease, named after the imperfect state of the fungus, is responsible for leaf spots and shoot necroses as well as for canker-like changes to the unripened, green twigs of various willow species. On the leaves, the necroses are irregularly rounded, pale to dark brown and, depending on the host species, distributed singly (*Salix rigida*) or in clusters (*S. fragilis*) on the blade. Infection, which is

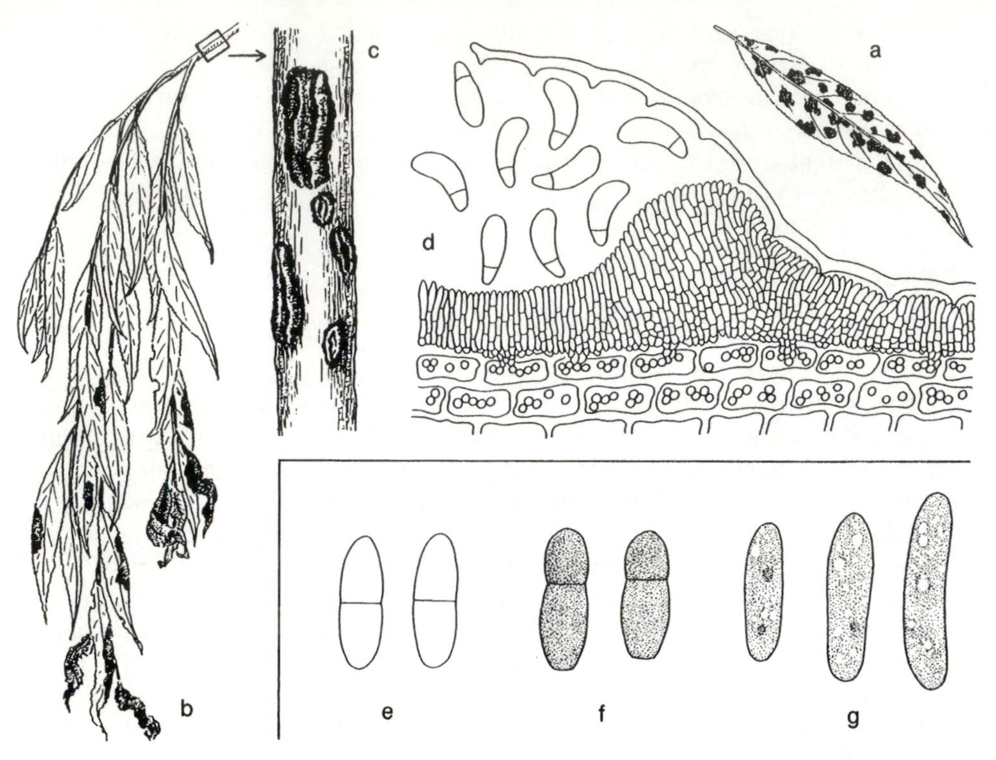

Fig. 61 Leaf and shoot tip diseases of willow. **a-d** *Marssonina salicicola*: **a** leaf spots on *Salix fragilis*, **b** infected shoot tips on *Salix alba*, **c** bark cankers, **d** cross-section through an acervulus; **e** conidia of *Diplodina microsperma*; **f** conidia of *Pollaccia saliciperda*; **g** conidia of *Colletotrichum gloeosporioides* (d,g after Butin 1960)

initiated in spring from the overwintering mycelium on the bark, can become so heavy that the leaves dry up and fall prematurely. Infections on the 1- and 2-year-old twigs are characterized by brownish black lesions, 1–3 cm long, which cause the bark to rupture and become scabby (Fig. **61a–d**). If the shoot tips are attacked, they shrivel up, and bushy growth results. Most commonly attacked are the pendulous forms of *Salix alba* and its hybrids [34]. Spraying with copper oxychloride has proved an effective control measure [173].

An unmistakable feature of a *Marssonina* attack is the appearance in summer of the black stromatic acervuli bearing the hyaline conidia which measure 15–17 × 5–8 μm and are septate in the lower third. It is not uncommon to find *Pollaccia saliciperda* (Fig. **61f**) occurring together with *M. salicicola,* and this can increase the extent of shoot tip killing. The edges of the scabby patches are often colonized by a saprotrophic and therefore unimportant *Phoma* species.

Other Fungi on Shoots and Branches of Willow [34,69,174]:

- *Colletotrichum gloeosporioides* Penz., the perfect state of *Glomerella miyabeana* (Fuckel) Arx, causes leaf spots, bark necroses and death of shoots.

It occurs on ornamental willows, and is the 'black canker' of basket willow cultivation. Conidia are cylindrical to ellipsoidal, hyaline, 13–22 × 4.5–7 μm in size (Fig. **61g**), and can be controlled in osier beds with fungicides.

- *Cytospora salicis* (Corda) Rabenh. is the imperfect state of *Valsa salicina* (Pers.) Fr., which appears very rarely in nature. It is a weak parasite on dying twigs, with a grey stromatic disc and centrally placed white pore. Conidia are sausage-shaped, 4–6 × 1.5 μm in size (Table **I,22**), and emerge in reddish spore tendrils.
- *Diplodina microsperma* (Johnston) Sutton (Syn. *Discella carbonacea*), the imperfect state of *Cryptodiaporthe salicella* (Fr.) Petrak, is a weak parasite on shoot tips and thinner branches. Pycnidia are black and cushion-shaped; conidia are 2-celled, hyaline, spindle-shaped, and 13–17 × 4–7 μm in size (Fig. **61e**).
- *Myxofusicoccum salicis* Died. f. *microspora* Died. is common on dead branches and slender branches of various willow species. Stromata are several chambered; conidia are ellipsoidal and 4–5 × 2–3 μm in size (Table **I, 23**).
- *Pollaccia saliciperda* (All. & Tubeuf) Arx, the imperfect state of *Venturia saliciperda* Nüesch, as well as leaf spots, causes death of shoot tips and bark necrosis on ornamental willows. Conidia are barrel-shaped, septate in the upper third, brownish, and 17–23 × 6–8 μm in size (Fig. **61f**). Damage can be prevented by choice of variety. Control with fungicides is possible.

Non-parasitic Dieback of Branches

The symptoms of branch death are often striking, particularly if the dead branches stand out in contrast to an otherwise green crown. There is considerable uniformity in the appearance of branches killed by different diseases and disorders, but there are a great many such causes. Syndromes which give rise to the fewest diagnostic problems are those where known pathogens (e.g. *Ophiostoma ulmi, Endocronartium pini,* or *Cryphonectria parasitica*) can be found. For abiotically induced damage, reliance must be placed on the history of the disease, especially regarding records of particular events (e.g. extremes of weather).

Quite often, however, the death of branches depends on the combined effects of abiotic and biotic factors; and it is then not always possible to distinguish clearly between those which predispose the plant to disease, those which cause the disease, and those which merely accompany it.

The following belong to the more common non-parasitic branch and crown problems whose causes are sufficiently well understood:

Stagheadedness caused by crown alterations. This is a result of opening up previously closed stands; the increased penetration of sunlight stimulates the flushing of numerous new shoots on the lower stem, and these eventually form a new, conical crown below the original one. The process leads to a deficiency in the supply of nutrients and thus to the death of scattered branches. This pattern occurs only in species, unlike beech and spruce, that have the ability to refoliate the main stem and branch bases in this way; pedunculate oak is a classic example (Fig. **62a**).

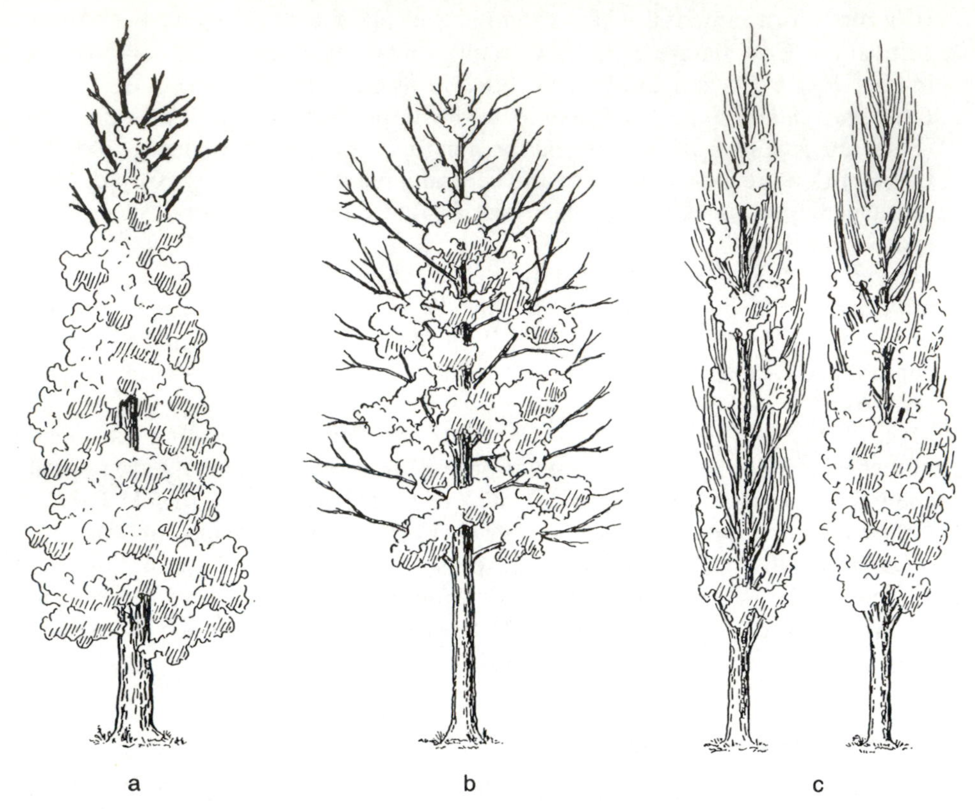

Fig. 62 Dieback of broadleaves. **a** stag-headed oak after opening up, **b** branch dieback over the whole crown of a poplar after water-table lowering, **c** dieback of Lombardy poplar after winter cold and secondary fungal infection (*Cryptodiaporthe populea*)

Stagheadedness after periods of drought, e.g. in birch. This involves the death of fine twigs, later of thicker branches as well; it occurs often with secondary infections with *Melanconium betulinum* (conidia brownish, 15–18 × 6–8 μm (Table **I,21**)) or *Myxosporium devastans* (conidia hyaline, 7–9 × 3–4 μm (Table **I, 20**)).

Crown degeneration after lowering of the water table. This involves dieback of larger branches over the whole crown with foliage retained only in scattered clumps at branch bases. It occurs especially in older poplar stands (Fig. **62b**). Similar symptoms can occur from the persistent effects of road deicing salt.

Crown degeneration due to sealing off gaseous exchange to the roots. (e.g. asphalting, dumping soil, compaction by heavy machinery). As a result of oxygen deficiency, root and bark rots develop, with consequent death of branches. It is especially common with beech and occurs in newly constructed car parks, building sites, etc.

Death of leaders and twigs from winter cold. This combined pattern of killing is typical in Lombardy poplar (Fig. **62c**), often with subsequent attack by

Cryptodiaporthe populea. A similar leader dieback also occurs in conifers (e.g. spruce) with frost as the predisposing factor followed by attack by *Cytospora kunzei*.

Death of twigs caused by road deicing salt. The symptoms are similar to the dieback due to a drop in the water table level, but regeneration is more vigorous; chloride determination is the means of differentiating the two. It occurs only along streets and paths on numerous broadleaves.

Ice damage. This is the result of branches breaking or bending under the weight of frozen rain. It occurs especially in larch and pine and in extensive, even-aged stands.

Pruning and Natural Pruning

It is well known that in the course of their lives many tree species lose their lower branches and develop a more or less branch-free bole. The same process takes place—if less obviously—in the crown. In forestry, the process is referred to as 'natural pruning.' This phenomenon is particularly striking in oak, beech, or ash; in contrast, spruce and Silver fir retain their dead branches for a long while.

In broadleaves, two essentially different types of branch shedding can be recognized. The first type occurs only in certain species (e.g. oaks, poplars, and willows) and is initiated by the tree itself through the formation of an abscission layer of parenchyma at the base of the twig. This results in the shedding of the twig which leaves behind a dished, usually roundish scar. In the case of pedunculate oak, the casting of large numbers of abscised twigs is a sign of increased stress, e.g. from a long period of dry weather.

The branch shedding of abscised twigs should not be confused with wind damage, by which 1-year-old shoots can be wrenched off (e.g. in spruce). Squirrels similarly remove twigs by biting, but in this case toothmarks can be seen.

The second, more common type of natural branch pruning involves certain wood decay fungi which become active when the branch ceases to function actively. The branch becomes so decayed at its base that the wind, or a jolt, breaks it off. In slender branches, the fungi involved are ascomycetes or Fungi Imperfecti, most of which are already present as endophytes in the bark before the branches die; the disintegration of thicker branches is brought about mainly by basidiomycetes [43]. In conifers, this branch pruning proceeds relatively slowly (pine, larch) or does not occur at all (spruce, Douglas fir)—the result of strong resin impregnation of the branch base.

Oak can serve as an example of fungus-aided natural pruning: its lower branches, dying from a lack of light or nutrients, are regularly colonized and decomposed by *Colpoma quercinum* (Pers.) Wallr. This weak parasite, present in all oak thickets, is distinguished by its lip-shaped, white-bordered apothecia, which are up to 15 mm long, and open more or less widely according to their moisture content (Fig. **63a,b**). The imperfect state, known by the name of *Conostroma didymum* (Fautrey & Roum.) Moesz, quite often develops before the perfect state is formed and can be recognized by its cylindrical to ellipsoidal

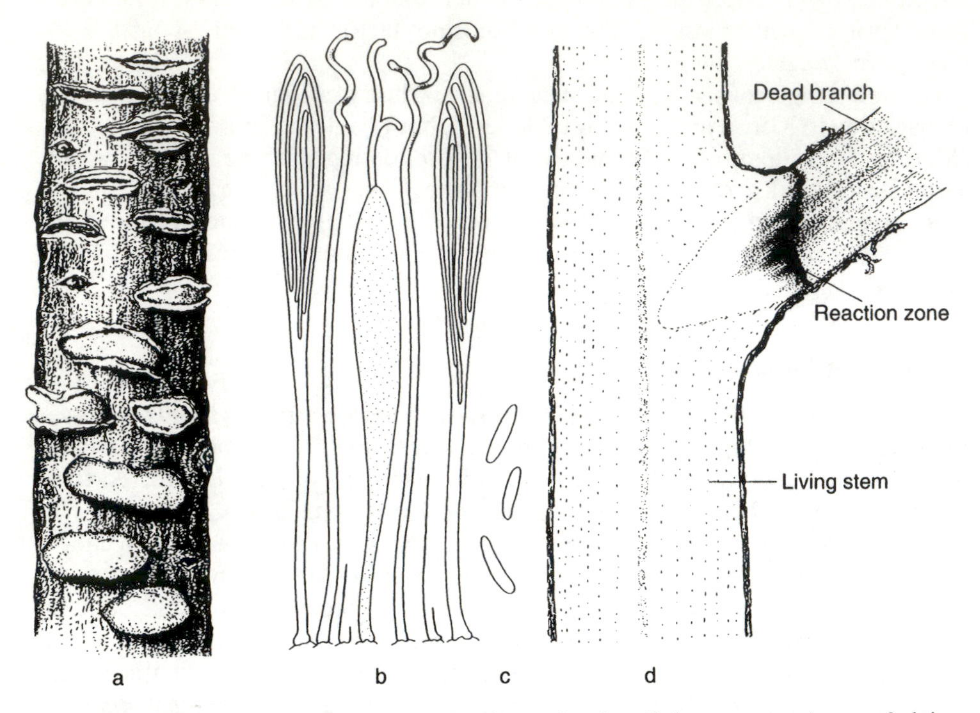

Fig. 63 Natural pruning in oak. **a** natural pruning by *Colpoma quercinum* of dying lateral twigs, **b** asci with paraphyses, **c** conidia of the imperfect state; **d** longitudinal section through a young oak stem with dead, walled-off side branch

conidia which measure 5–7 × 1.5 μm (Fig. **63c**). Other fungal species are also involved with *Colpoma quercinum* in the decay, occurring to some extent in a regular succession and combination [43]. Beech, ash, maple, and other species similarly have host-specific fungal communities that bring about this type of branch pruning. Thus, the process can be regarded as an normal part of tree development rather than as a pathological event.

Usually, the shedding of a branch is preceded by the formation of a physical and chemical barrier (reaction zone) across the sapwood at the base of the branch to protect the xylem of the main stem from attack by fungi and bacteria. In broadleaves this protective zone is characterized by the deposition of phenolic materials in the parenchymatous tissue, and the plugging of the vessels by tyloses or deposited substances, indicated by the brownish discoloration in the region of the branch base (Fig **63d**). In conifers, a similar protective effect is achieved by resin impregnation of the wood and the formation of a resinous cone within the wood.

The recognition of this process of 'compartmentalization' is important in practical tree care if living or dead branches are to be removed. In broadleaves, care should be taken not to cut too close to the main stem, as otherwise the region in which the protective zone is formed will be removed or adversely affected [188].

The exact position where pruning cuts should be made is still a subject for research and appears to depend partly on the species of tree involved and the size and formation of the branch. In all cases, excessively flush pruning, which damages the 'branch bark ridge' or the branch collar if one is present, is likely to encourage invasion by wood-rotting fungi, or at least the development of an undesirable amount of staining in the stem wood. At the other extreme, there is a risk in leaving a stub because it can serve as a food base for canker organisms such as *Nectria cinnabarina*, sapwood pathogens such as *Chondrostereum purpureum,* or secondary decay fungi, which can thereby enter the main stem. This risk may be low for some species, such as oaks, but it is significant in the case of many other trees, including cherry and maple species.

Other aspects of tree pruning can also influence the risk of attack by pathogens and decay fungi, including the time of year, the number of pruning cuts made on an individual tree, and the decision whether to use wound paints [189]. It seems that, for many species, multiple pruning wounds may be harmful, and that the autumn and winter may be an unfavourable period, but much research still remains to be done in this area and the practitioner must consult the up-to-date specialist literature.

5 *Bark Damage*

The Tree's Defence Reactions

Bark differs from the relatively short-lived parts of a tree, such as leaves or needles, in that it is a permanent organ which is not shed or replaced, except by a gradual process of surface sloughing and cell replacement. As a whole, therefore, the bark is as old as the stems and branches that it covers. As such, it is exposed throughout the life of the tree to potential damage from environmental extremes, parasitic attack, or wounding. If such damage occurs, the underlying wood can be exposed to attack by decay organisms, with potentially lethal consequences if the main stem is involved. This threat is countered by a range of defence and repair mechanisms which operate in both bark and sapwood and have been an essential element in the evolutionary success of trees.

Bark damage can take three main forms: (1) mechanical wounding, (2) injury from extremes of weather, and (3) biotically induced necrosis. The defensive reactions of the tree to all these forms of damage are similar and involve biochemical and physiological alterations and the formation of new cells. If the damage does not penetrate to the cambium, the dead zone becomes 'walled off' by the proliferation of cells in the surrounding bark parenchyma to form a new layer of tissue; an induced periderm or 'wound periderm'. The underlying vascular cambium continues to form new wood and phloem as usual. Such superficial wounds heal rapidly and remain externally visible only until the dead portions of bark are sloughed off as new cortical tissue expands beneath them.

If the damage penetrates to the vascular cambium, any repair processes can occur only around the rim of the area so affected, since the only living tissue present in this area is the underlying sapwood, the cells of which are incapable of proliferation. At this rim, cambial development is restored, producing new bark and wood—also called wound callus or wound tissue—that closes roll-like over the unprotected wood surface. Within a period ranging from one to many years, depending on the extent of the damage and the vigour of the tree, total occlusion may take place, but a scar generally persists as evidence of the earlier trauma.

Bark damage caused by certain pathogenic micro-organisms may recur seasonally in an alternating cycle of callus formation and killing of the new tissue, and thus prevent occlusion of the lesion, which takes on a swollen and roughened appearance. Such lesions are termed 'cankers,' but the same term is sometimes also applied to other forms of bark necrosis. Thus, an 'annual canker' is an area of necrosis delimited by a single rim of living callus growth, whereas a 'diffuse canker' is an undelimited area of necrosis that develops when the defensive responses of the host are entirely suppressed.

If the tree fails to close a wound or lesion promptly, there is a danger that certain micro-organisms may infect the unprotected tissue; either bark tissue—which can become necrotic to a greater or lesser extent—or wood, which can be destroyed by wood-rotting basidiomycetes. Mechanical bark wounds are particularly dangerous because they can completely remove patches of bark, exposing the underlying wood to such attack, with consequent risk to the life of the tree and to its economic or aesthetic value.

Non-parasitic Bark Damage

There are many non-parasitic agents that can damage bark, some of which produce characteristic types of damage or scars. In other cases, however, the external symptoms have no diagnostic value, particularly in the later stages of development when the initial damage may have been overcome by the tree's largely non-specific responses. If the cause needs to be ascertained in such cases, the underlying wood must be examined for any characteristic defects (cf. p. 141).

Mechanical bark damage. This is one of the most commonly observed types of damage, especially on older trees. In most cases, man is directly or indirectly to blame: trees in plantations are damaged by extraction, felling, or pruning; street trees suffer from impacts with vehicles, and particularly prominent trees often become targets for the carving of names and symbols.

Hail damage can be classed together with other weather-induced injuries. The impact of hailstones can crush and thus kill cells. As the cambium is usually unharmed, the wounds can close over in the same year and do not become infected. The damage occurs most often on younger branches which are growing out from the tree more or less at right angles. A diagnostic feature is the mainly one-sided distribution of the injuries.

Lightning damage is also an effect of the weather, although it can involve a far larger area of bark than hail damage. Where the lightning discharge travels superficially over the bark surface to the roots, the resulting damage is usually minor and involves only the destruction of a strip of outer bark tissue or 'lightning furrow', which soon heals. If, instead, the lightning follows the moist cambial zone, the result can be that strips of wood are torn out, or that parts of the crown are blown apart, creating infection courts for pathogens or decay organisms. The risk of a tree being struck by lightning depends on its species, on the relative size of neighbouring members of the stand, and on its location. Oak, elm, ash, and poplar, together with tall conifers, are particularly at risk while beech, hornbeam, and birch are less often struck.

In many cases lightning damage is non-catastrophic, causing gradual dieback with no immediate or distinctive symptoms. Such damage may affect individual trees or groups of trees in plantations, or hedgerows. It may resemble the effects of root disease, but the roots generally remain alive until the final stages of decline. The presence of a 'lightning ring' in the stem cross-section may be another diagnostic feature, and if a number of trees are involved, their annual ring patterns will show a simultaneous decline in growth.

Sunscorch is caused by excessive solar heating of the cambial zone, and becomes noticeable later when some of the bark dies and flakes away. Only

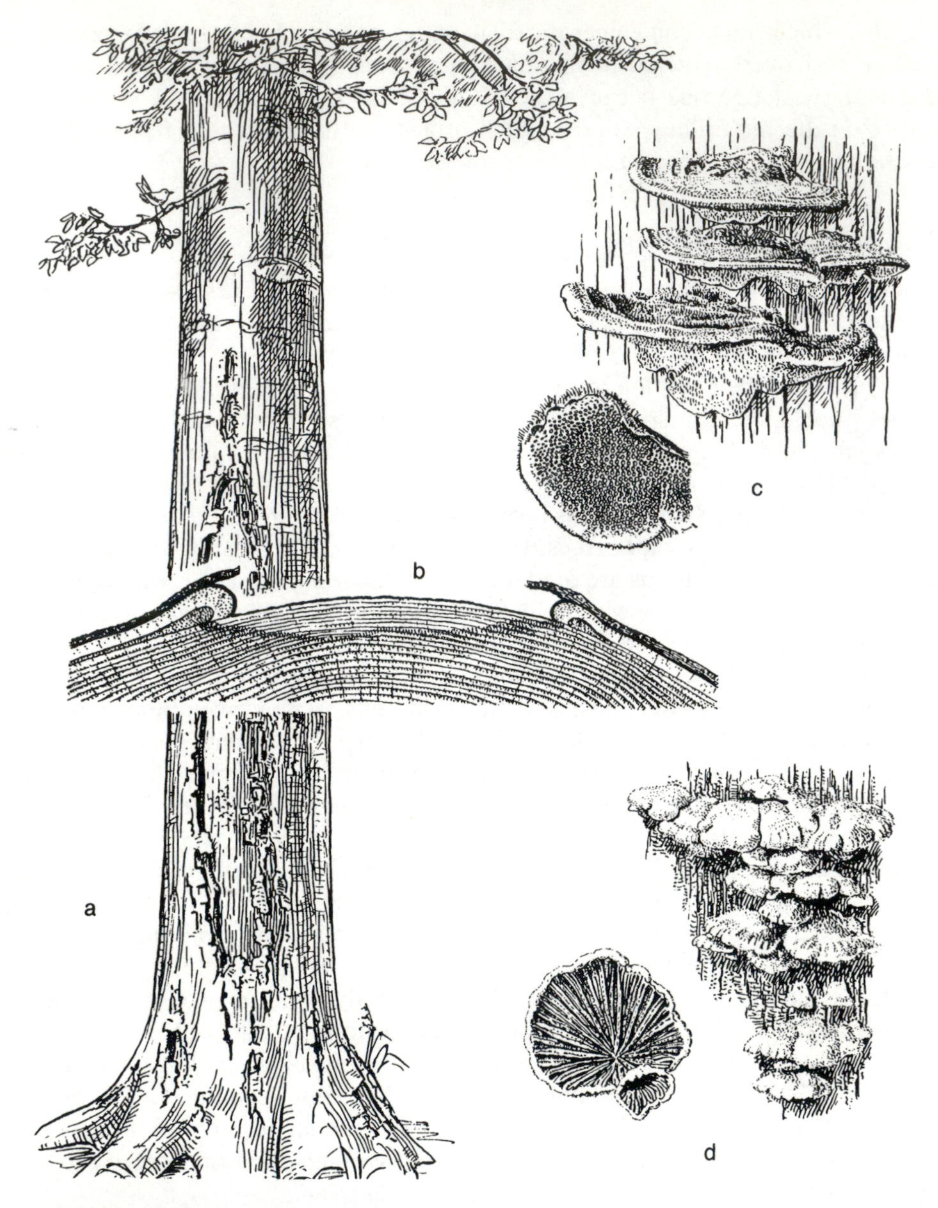

Fig. 64 Sunscorch on beech. **a** symptoms on stem, **b** cross-section (detail) through a sunscorch-damaged stem with early callusing, **c** fruit bodies of *Trametes hirsuta*, **d** fruit bodies of *Schizophyllum commune*

thin-barked species such as beech, Horse chestnut, alders and spruces are damaged. Mature trees exposed by adjacent clearing of the stand on their southwestern aspect are particularly at risk. If the necroses are small, the wounds are soon callused over by the tree. Although sunscorch itself is an abiotic form of damage, it can pave the way for infection by wood decay fungi in cases where extensive areas of bark are involved. Old sunscorch injuries almost invariably exhibit a characteristic fungal flora which in beech, for example, includes *Schizophyllum commune* and *Trametes hirsuta* (Fig. **64**). The stem decay may become severe, and trees affected in this way soon lose their economic value; nevertheless, they should be retained in the stand as the shade they cast protects neighbouring trees from a similar fate. Where it is not possible to avoid opening up a stand on the southwest side, the danger of sunscorch can be reduced by applying an opaque paint to the newly exposed, lower parts of stems. When heavy standard specimen trees are planted, the stems are quite often wrapped in jute or painted with mud or china clay to protect them against damage from sunscorch and drying winds.

Sunscorch damage can also occur without the death of cambial cells. In such sublethal injury, only the outer bark cells are affected. This results in the formation of a rough, superficially splitting 'pathological bark' which earlier led to the supposition that such trees were particularly rough-barked 'sports', especially in the case of beech.

True frost cracks arise when bark splits open longitudinally after intense radiation frosts and wide temperature fluctuations. Such cracks, which are usually located at the base of the stem and mostly on the south- or southwest-facing side, can reach up to a metre in length. Usually they start to callus over in the following growing season and, in young trees, the margin of new tissue around the wound is often a conspicuous feature. As a rule, the wood which was present at the time of the wounding is not injured. The most commonly damaged trees are young individuals of *Chamaecyparis, Thuja*, poplar (Fig. **65d**), and Red oak.

Frost-damaged bark sometimes forms so-called **frost plates** rather than true frost cracks. Typically, these are broad areas of dead bark extending up to 2 m above the ground. They arise mainly on the south side where the cambium is strongly warmed during the daytime, becomes active, and then is injured by unusually low night-time temperatures. Such damage is often followed by fungal infection, leading to extensive bark necrosis or wood decay. Susceptible species include beech, maples, and several of the more sensitive poplar clones.

False frost cracks originate in the wood rather than the bark, and so develop from the inside outwards. A crack can form incrementally across a number of annual rings in the wood before it reaches the bark. The initiation points for false frost cracks are old, occluded cambial wounds or decayed wood. The only role of frost here is to re-open bark cracks ('seams') which have callused over (cf. Fig. 85).

Tension or expansion cracks are confined to the bark only. They arise during the growing season if the tangential expansion of the bark (dilation growth) fails to keep pace with the radial growth of the wood. As a result, the bark may split open over lengths of up to 1 m. Trees affected in this way include fast growing species and individuals with an eccentric radial growth form. Cracks in broadleaves (maple, beech, whitebeam, poplar, Horse chestnut) usually close again with no further damage. In poplar (e.g. the clone 'Muhle-Larsen') tension cracks result in unsightly staining and structural failure in the adjacent wood [177].

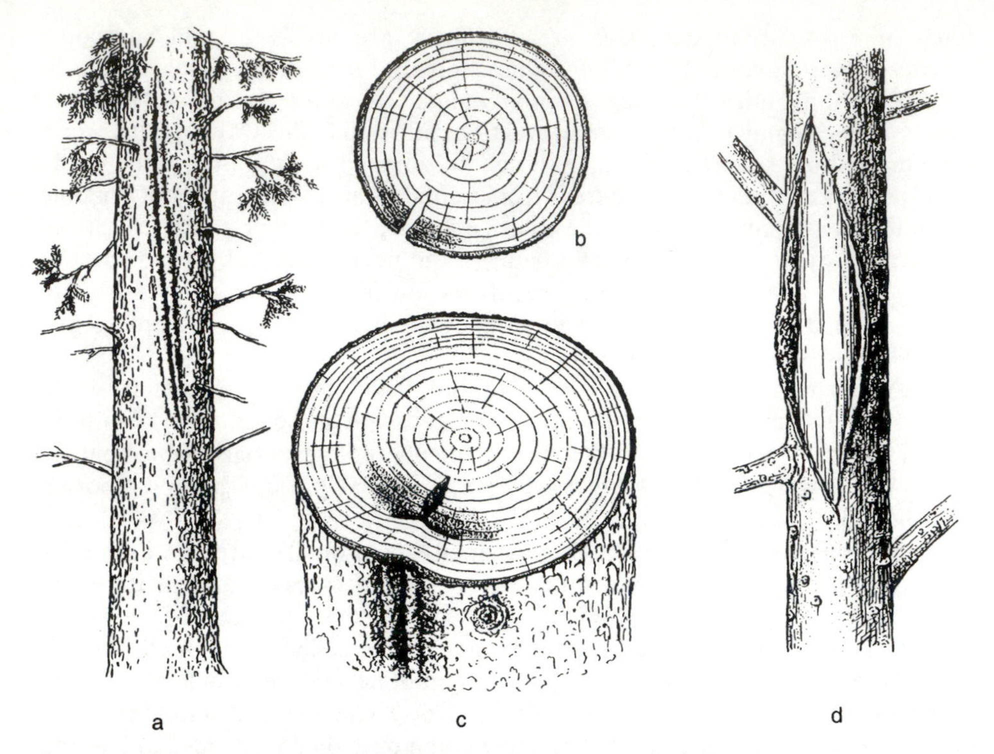

Fig. 65 Bark cracks. **a-c** drought cracks on spruce: **a** occluded drought crack on a spruce stem, **b** young radial crack before occlusion (stem cross section), **c** radial crack and seam' 5 years after occlusion; **d** a 1-year-old frost crack on a 3-year-old hybrid Black poplar (a after Caspari and Sachsse 1990)

Drought cracks are radially aligned splits in the stem. They occur in 15–16-year-old stands of fast-growing Norway spruce and appear as rapidly callusing, axial cracks in the bark and outermost annual rings, reaching 0.5–2.5 m in length. They arise in late summer during long periods of persistently dry weather and have been attributed to extreme suction tensions in the xylem [46]. Thus the splits start in the wood, rather than the bark. They are initially small, but they can lengthen until they finally tear open the overlying bark. Other conifers are also affected. Trees damaged by such cracks should always be removed (Fig. **65a–c**).

Phacidium Disease of Conifers

Cause: *Phacidium coniferarum* (Hahn) DiCosmo
  Syn. *Potebniamyces coniferarum* (Wilson) Hahn
  Anamorph: *Phacidiopycnis pseudotsugae* (Wilson) Hahn
  Syn. *Phomopsis pseudotsugae* Wilson

This canker disease can occur on trees of various ages. On 1–7-year-old saplings, it can either kill the leading shoot or girdle a localized part of the young stem, thus killing the distal part of the stem and its side shoots. Attacks on older trees are generally restricted to diamond-shaped bark necroses about 20 × 10 cm in extent. These soon dry out and the plate-like patches of dead bark are lifted away by callus rolls which develop at the margins of the lesion. In the thin-barked Douglas fir, such damage can become clearly apparent within one year after infection; in the thicker-barked Japanese larch, the disease is less noticeable so that it can easily be overlooked in the early years (Fig. **66**). The fungus almost always appears on the killed bark in its conidial state. It forms black pycnidia 0.3 mm across, from which large numbers of ovoid conidia, 5–9 × 2–3 μm in size, emerge in damp weather. The fruit bodies of the associated perfect state, which are disc-shaped apothecia, are produced only occasionally.

Phacidium coniferarum is found in an unusually wide range of ecological situations. Most frequently it occurs as a saprotroph, colonizing and breaking down twigs which have died from lack of light, so in this form it can be regarded as beneficial. On drought-stressed young trees, it acts as a weak parasite and, as a seasonal parasite during the dormant season, it can even invade healthy bark tissues. Finally, it has also exploited an ecological niche in woody tissues where—known by the name of *Discula pinicola*—it occurs as the cause of a blue stain in the stem wood of pine [165].

Host plants for the fungus include Douglas fir, larch, pine, and Silver fir. However, in central Europe, damage of economic significance has been seen

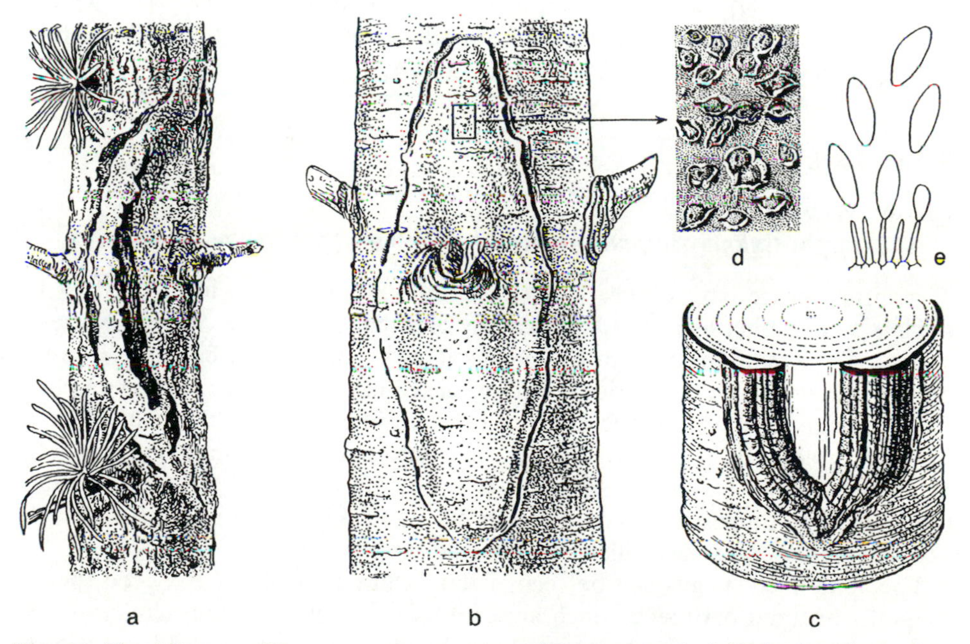

Fig. 66 *Phacidium coniferarum*. **a** symptoms on Japanese larch, **b** bark plate formation on Douglas fir, **c** well-advanced callusing, **d** pycnidia in dead bark tissue, **e** conidiophores with spores of the imperfect state

only on Douglas fir and Japanese larch. In northern Europe, Scots pine can also be attacked and outside this region, deaths of *Larix russica* have been noted.

Douglas fir becomes particularly susceptible to this disease at certain stages of its silviculture, given predisposing conditions. Thus, the first spate of attacks—often fatal—is seen in the first year after planting if the resistance of the young trees has been impaired by prior water loss. To help maintain resistance to attack, the period between lifting and transplanting should be kept as short as possible, and plants should not be heeled in for too long. Where possible, Douglas fir plantations should be established under a reasonably dense cover of older trees; by underplanting in large areas or maintaining side shelter in smaller ones. Another critical stage is the first brashing which creates wounds through which this ubiquitous fungus can enter, although the development of infections can take place only in dormancy when the tree is unable to seal off the wounds quickly enough. For this reason, winter brashing, which in some countries is favoured in the interests of obtaining decorative foliage, brings with it a certain risk. The risk can, however, be avoided if the bottom 10 cm of every branch is left attached to the tree. If high quality timber is to be produced, it is recommended that the stubs that are left are taken off in the following growing season. Another prophylactic method which has proved successful is the use of wound sealants. These are applied to fresh winter brashing/pruning wounds.

Similar bark damage can be brought about by *Leucostoma kunzei* (Fr.) Munk ex Kern. As well as Douglas fir, larch, Silver fir, and various species of pine are attacked. This weak parasite, which usually appears in its imperfect state (*Cytospora kunzei*) has spherical fruit bodies buried deep in the bark tissues. These produce conidia, usually in spore tendrils, which are cylindrical, slightly curved, and measure 4–5 × 1.5 μm (cf. Table **I,22**).

Spruce Bark Disease

Cause: *Nectria fuckeliana* Booth
Anamorph: *Cylindrocarpon cylindroides* Wollenw. var. *tenue* Wollenw.

The term 'spruce bark disease' ('Fichtenrindenkrankheit') is usually taken to mean a stem cankering disease, principally of Sitka spruce, less often of Norway spruce. The symptoms are those of a classic perennial canker characterized by numerous concentric callus ridges around a wound which does not close over. Even though an infected tree can reach a considerable age, the value of the timber is reduced, and occasionally the stem snaps. Infection and ensuing necrosis of bark on Norway spruce may occur after careless removal of dead branches. The fungus is also parasitic on Silver fir, killing young shoots as well as forming bark necroses which can become large.

This long-lived fungus can be recognized externally from its foxy red, pear-shaped, clustered perithecia, each about 0.5 mm across, formed mainly at the edges of the wounds. The ascospores are 2-celled, fusiform, and 13–16 × 5–6 μm in size [21]. In the absence of fruit bodies, the fungus can also be identified by isolation and culturing on artificial nutrient media. In culture, diagnostic features are microconidia of the *Cephalosporium* type and multiseptate macroconidia (Table **II,14**), 33–40 × 4–5 μm in size, of the *Cylindrocarpon* type [22].

Crumenulopsis Stem Canker of Pine

Cause: *Crumenulopsis sororia* (Karsten) Groves
Anamorph: *Digitosporium piniphilum* Gremmen

The first signs of an attack are the lifting of plates of bark, increasing resin exudations, and small, dome-shaped irregularities on the lower stems. Necrosis and cracking develops at these points and cankerous lesions develop from them. The cankers have callus ridges at their edges which can be killed back repeatedly each year. The canker, which occurs on *Pinus contorta, P. nigra,* and other pines, leads to a reduction in timber value and a slow dieback of the trees. In the main, it is weakly growing individuals that are attacked.

The cause of the disease is the fungus *Crumenulopsis sororia*, recognized by its greyish black, disc-shaped fruit bodies, 1–1.8 mm in diameter. Its asci are 80–100 × 11 μm in size and are surrounded by filiform paraphyses. The smoky ascospores, 8 in each ascus, are at first 1-celled, later 4-celled and 17–23 × 5–6 μm in size. The associated imperfect state often occurs along with the apothecia and, with its typical finger-shaped, multicelled conidia, is equally identifiable as the causal fungus (Fig. **67**).

Direct control of this canker fungus on the host plant does not appear to be possible. It is therefore all the more important to employ preventive measures such as the choice of vigorous provenances suited to site. Infected trees should be removed promptly from the site to guard against further infections [202].

Pine Stem Rust (Resin-top Disease)

Cause: *Cronartium flaccidum* (Alb. & Schwein.) Winter
or: *Endocronartium pini* (Pers.) Hiratsuka

These fungi, members of the Uredinales, are among the most damaging pathogens of Scots pine. Since susceptibility increases with age, symptoms are most commonly seen and are most striking on older pines when, after appearing sickly for some years, the top of the tree dies so that its bare branches stand out above the still green part of the crown. Before the leader dies, attack on the cambium leads to the malformed development of the infected bark and wood at its base, and these tissues finally become impregnated by a copious flow of resin ('resin-top'). Less strikingly and less often, young trees are attacked. The bark swellings are accompanied by orange-yellow aecidial pustules, especially at the bases of the whorls on the main stem, by which the causal agent can be identified with certainty. The recorded hosts are *Pinus sylvestris, P. halepensis, P. mugo, P. nigra, P. pinaster, and P. pinea* (Fig. **68**).

If the infected parts are examined microscopically, intercellular fungal mycelium is found in both the outer bark and the bast (phloem). From here the fungus grows along the rays, penetrating the wood to a depth of about

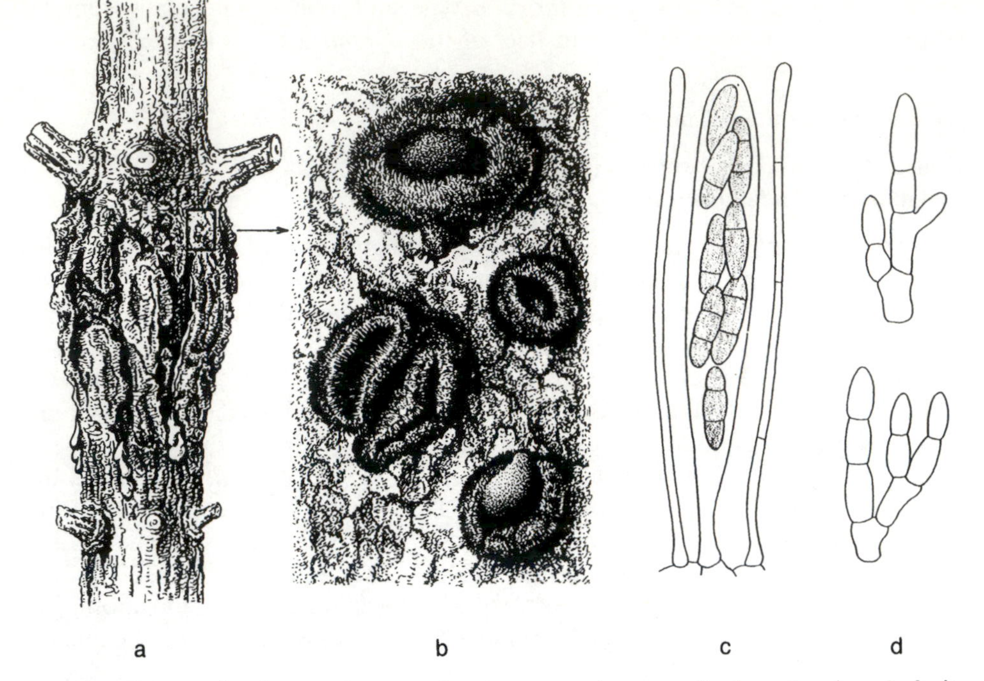

Fig. 67 *Crumenulopsis sororia*. **a** canker-type symptoms on Lodgepole pine, **b** fruit bodies (apothecia) on dead bark, **c** ascus with ascospores and adjacent paraphyses, **d** conidia of the imperfect state (b,c,d after Stephan and Butin 1980)

10 cm, killing the ray cells and causing the tissues to become resin-soaked. Each year, the mycelium spreads out further from the diseased parts of the bark, killing more bark tissue. As this induces only a weak callusing reaction by the tree, the affected part of the stem becomes unilaterally flattened and, in time, strongly deformed.

Two closely related rust fungi are involved in this canker disease whose symptoms on the pine cannot be differentiated. One of these, *Cronartium flaccidum*, alternates between hosts and is found as several biological types [80] in the southern part of Europe. The alternate host there is represented by the white swallow-wort (*Vincetoxicum hirundinaria*) or other herbaceous plants on which the uredo-, teleuto-, and basidiospores are formed. The other rust, *Endocronartium pini,*, previously known as *Peridermium pini,* occurs predominantly in the northern part of Europe. It often causes epidemics there because it can pass from pine to pine without a change of hosts, being unlike most haploid monokaryotic rusts in that it produces only aecidiospores; these can infect the same host. In Britain, there are morphological differences between the populations of the fungus in Scotland and eastern England.

One possible way of preventing attacks by *C. flaccidum* is to destroy the local populations of its dikaryotic-state hosts. With *E. pini*, the species more common in northern Europe, such a method cannot be employed because it requires no alternate host. In the absence of any known means of direct control, the only treatment available for diseased stands is to remove infected stems as well as trees that are particularly at risk by having multiple infections. Pine stem rust is therefore primarily a thinning and utilization problem to be resolved stand by stand. The indications are that it would be possible to raise less susceptible types by the selection of individuals but no practical guidance on this is yet available.

White Pine Blister Rust

Cause: *Cronartium ribicola* J.C. Fischer

The symptoms of this bark disease, also known in German as 'Strobenrost,' after the trivial name of the principal host, are very similar to those of pine stem rust; swellings, resin exudations and orange-yellow aecidia develop on the twigs and stems (Fig. **69**). However, the most common site of infection is at the base of

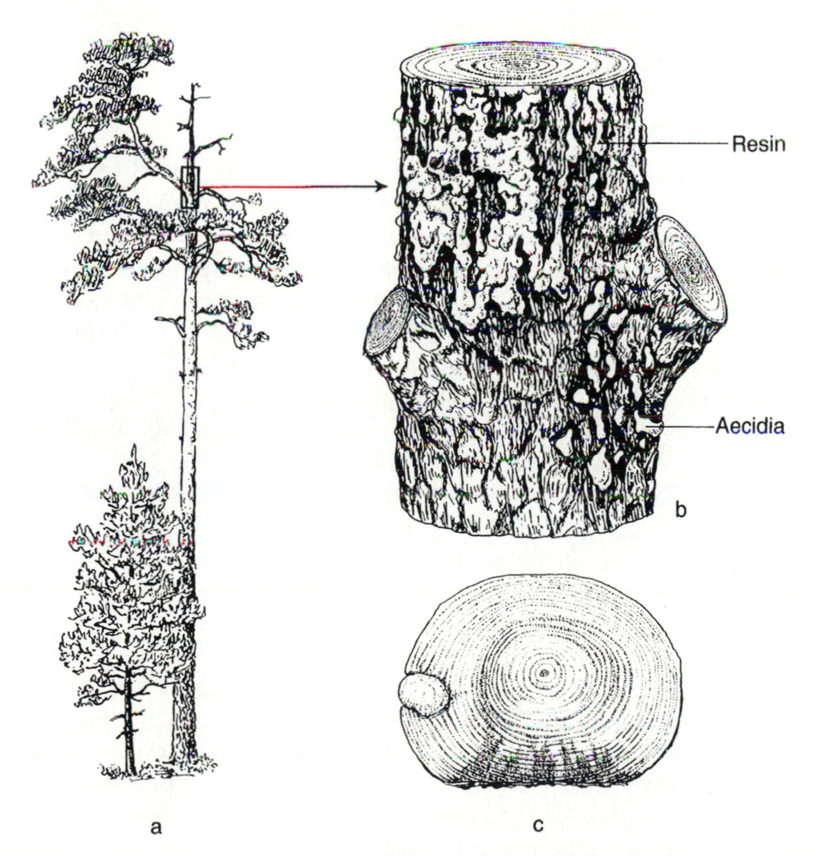

Fig. 68 *Endocronartium pini*. **a** diseased pine with dead top, **b** portion of dead top with resin flow and aecidia, **c** cross-section through a resin-soaked stem

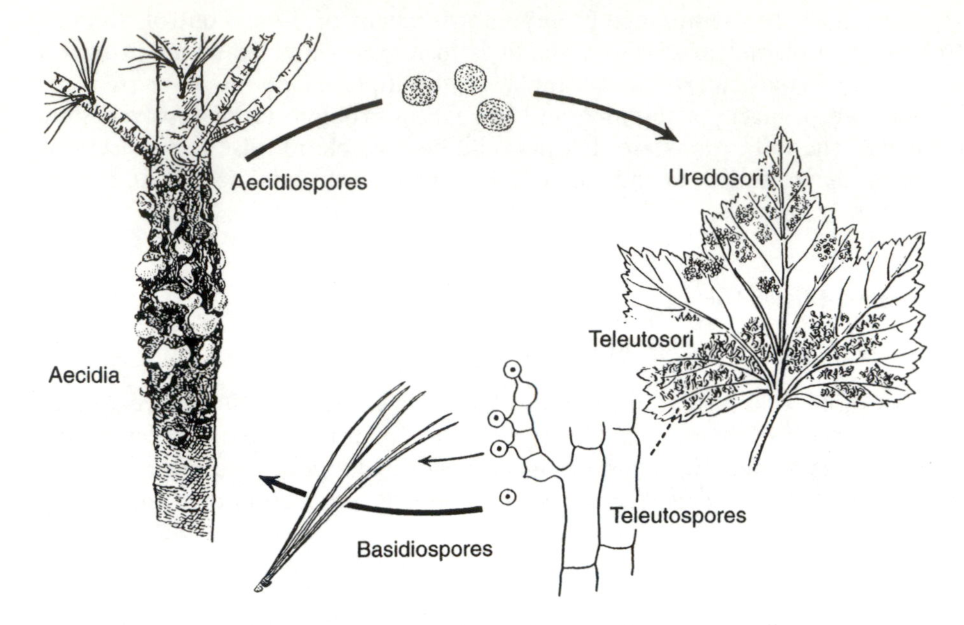

Fig. 69 The 1-year life cycle of *Cronartium ribicola* with corresponding symptoms on Weymouth pine (left) and currant (right)

the stem, rather than the top of the crown, so that it does not often lead to the stag-headedness that is so typical of the resin-top disease. When young plants are attacked, the result is stunted and bushy growth (so-called 'Bubikopf'—bobbed hair—formation) and a yellowish green needle discoloration. In old trees, a stem infection can persist for 20 years before the stem dies. The most definite indication of a White pine blister rust infection in such cases is the presence of resin-soaked and often sunken, necrotic patches of bark.

Cronartium ribicola occurs only on five-needled pines. As well as *P. strobus*, nearly all the other North American five-needled pines like *P. flexilis, P. lambertiana,* and *P. monticola* are severely attacked. On the other hand, the Eurasian species, like *P. cembra* and *P. peuce*, from whose natural range the fungus originates, are less at risk. Particular attention has been paid to the Bhutan pine, *P. wallichiana,* as it is almost the silvicultural equal of Weymouth pine though less rust-susceptible. Of the dikaryotic-state host plants, the cultivated forms of currants and gooseberries are particularly susceptible to the rust.

Cronartium ribicola is an obligate, heteroecious rust fungus. The disease only occurs where a haplontic-state host (*Pinus* species) and a dikaryotic-state host (*Ribes* species) are present in the same area. The blister-like aecidia develop on the haplontic-state host, and the spores from these infect the leaves of gooseberries or currants. An infection on the dikaryotic-state host is marked by brownish uredosori on the upper surfaces of the leaves and—in autumn—by small, peg-like, yellowish brown teleutosori also on the upper leaf surfaces (known in German as 'Säulchenrost;' pillar rust). These later give rise to basidiospores which continue the development cycle when they once again reach the pine. After infection has taken place, the fungus spreads symptomlessly across the needles and into the bark where it completes the cycle by producing pycnio-and aecidiospores in the following year or some years later.

The spread of White pine blister rust across large areas of the world has a remarkable history and is one of the classic examples of plant disease epidemics. *Cronartium ribicola* was originally endemic only in the region of the Siberian and alpine Arolla pine, without posing any threat to their survival. The situation changed when the very susceptible *P. strobus* from North America began to be cultivated on a larger scale. Once contact was made between the Siberian Arolla region and plantings of Weymouth pine in Europe, the blister rust disease broke out so severely in the latter that further cultivation of Weymouth pine at first looked questionable. By the importation of diseased plants into the northern states of America, the fungus was also established in pine stands there. In these native stands *P. monticola* and *P. strobus* were the main species affected, with many susceptible species of *Ribes* acting as alternate hosts. Today White pine blister rust is one of the most important diseases of five-needled pines on both continents.

Considerable economic damage can be caused by the rust in European plantations of Weymouth pine. Young plants often die, and older ones become sickly and remain so, often for years, so that the expected yield of timber is not realized. And at the same time the damage to cultivated *Ribes* species, which can lead to significant crop reductions, should not be overlooked.

The recommended method of preventing the disease is the same as for all heteroecious rust fungi. This is to keep the two host plants at a distance from one another, to remove the dikaryotic-state host from Weymouth pine growing areas (where wildlife conservation considerations allow), or to plant only those *Ribes* species that are rust-resistant. In new plantings, a safety margin of at least 500 m between the pines and currants or gooseberries should be maintained. This applies especially to nurseries, where the use of fungicides should be also be considered if the risk of infection is particularly high. In the forest, attention should be paid to achieving a closed canopy as soon as possible so that the lower branches, which are the first to become diseased, are shed early. Alternatively, these branches can be pruned. Infected stems should be completely removed in order to inhibit further spread of the fungus. In dense stands it seems possible to dispense with all control measures, and to leave the task of thinning to the rust itself. In such a case it would only be necessary to salvage the dying trees.

Larch Canker

Cause: *Lachnellula willkommii* (R. Hartig) Dennis
Syn. *Trichoscyphella willkommii* (R. Hartig) Nannf.

The most noticeable feature of this disease are localized deformations or open lesions on the branches and stems which, in cross-section, can be recognized as cankers (Fig. **70a,b**). The fungus can kill younger branches, while on older branches or stems it induces the development of more or less evenly shaped sunken lesions by periodically killing the wound tissues that repeatedly form at the lesion margins, and thus usually preventing occlusion. If a branch becomes infected very close to its base, the canker can extend to the bark of the stem and then becomes of economic significance. In such cases, the branch dies quickly but remains recognizable in the centre of the stem lesion for a long time afterwards [130].

A further distinctive feature of larch canker are the orange, 1–4 mm diameter, saucer-shaped fruit bodies, fringed with white hairs. The hymenium consists of numerous asci, each with 8 hyaline, 1-celled ascospores, 16–25 × 7–8 μm in size [28].

In Europe, larch canker is a widespread and typical disease of European larch. *Larix gmelini* and *L. laricina* are also susceptible. Japanese larch (*L. kaempferi*),

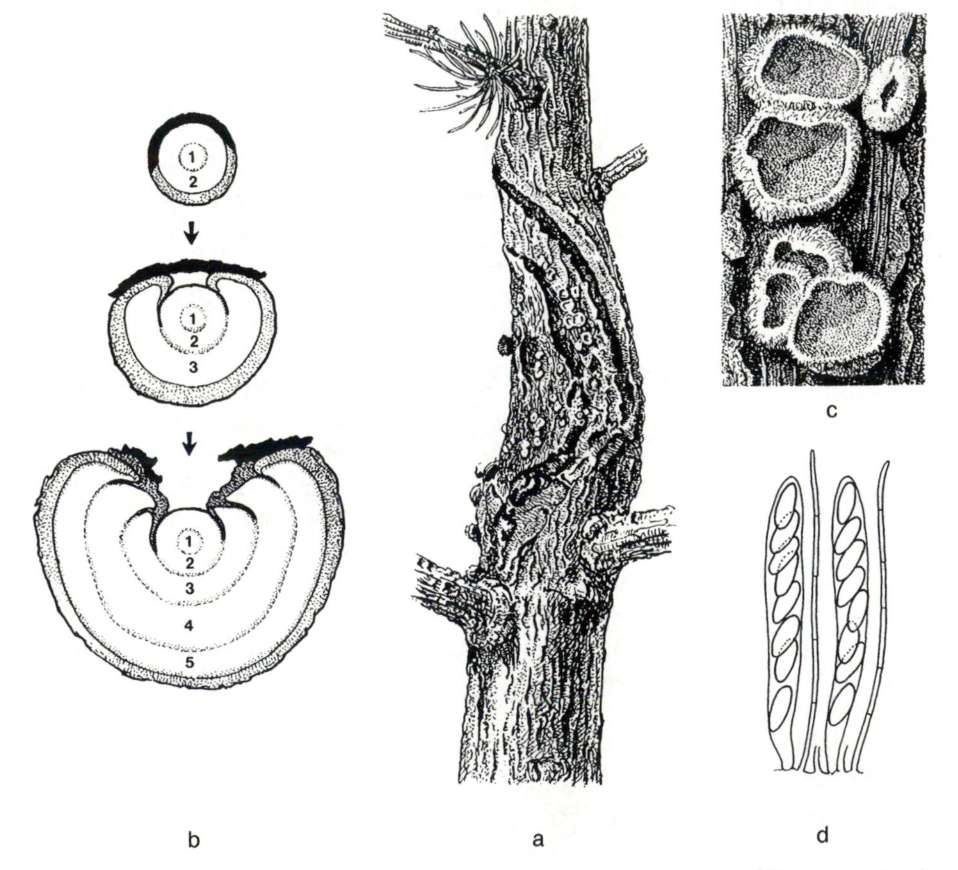

Fig. 70 *Lachnellula willkommii*. **a** canker formation on a small stem of European larch, **b** cross-sections through various stages in the development of larch canker, **c** fruit bodies (apothecia) on dead bark, **d** asci with ascospores and paraphyses

on the other hand, is thoroughly resistant in Europe, although it is attacked in its native land.

Certain larch provenances are reported to show relatively infrequent attack by the canker fungus, and this may indicate resistance or tolerance which could be important for the control of larch canker. The occurrence of the disease is also influenced by site microclimate. For example, one of the more important prerequisites for the disease is stagnant moist air, particularly in autumn.

*Other **Lachnellula** Species* [28,57]:

- *Lachnellula flavovirens* (Bresad.) Dennis occurs on dead, bark-covered branches of juniper and pine; and on Cembran pine, usually on branches killed by snow mould. Fruit bodies are orange-yellow, saucer-shaped, and 1.5–3 mm across with brown marginal hairs.
- *Lachnellula occidentalis* (Hahn & Ayers) Dharne (Syn. *L. hahniana*) occurs on dead bark-covered branchlets of larch, which are often confused with *L. willkommii* but with somewhat smaller ascospores, 14–22 × 4–6 μm in size. The fungus is exclusively saprotrophic.
- *Lachnellula subtilissima* (Cooke) Dennis occurs on dead, bark-covered branches of Silver fir, and less often on spruce or pine. Fruit bodies are 1–2.5 mm, cup or saucer-shaped; disc is orange-yellow with white marginal hairs.

Beech Canker

Cause: *Nectria ditissima* Tul.
Anamorph: *Cylindrocarpon willkommii* (Lindau) Wollenw.

The development of beech canker begins with the formation of flattened, necrotic depressions in the bark which are succeeded by irregular spindle-shaped branch thickenings (Fig. **71**). In summer, canker-infected branches are identifiable by the yellowish green colour of their leaves, which results from the impairment of water and nutrient transport. Diagnosis is possible even in winter, when the swollen branches are clearly visible.

The cause of beech canker, the ascomycete *Nectria ditissima*, has strikingly red-coloured perithecia on a similarly coloured basal stroma, which form in the spring at canker margins. They are spherical to ovoid, about 0.5 mm across and, when ripe, become filled with numerous asci, each containing eight 2-celled, hyaline ascospores, 14–18 × 6–7 μm in size. Quite often the associated imperfect state is also found, characterized by its white sporodochia and *Cylindrocarpon* type conidia. The spores are long-cylindric, slightly curved, 50–80 μm long, and 2–4 septate Table **(II, 15)**.

Infection occurs mainly via leaf scars or twig stubs, producing the initial small necroses. Repeated annual cycles then occur, in which wound tissue develops but is killed by the fungus, resulting in the production of a canker which usually remains open and can be many years old. Control is not possible, so that the removal of severely diseased trees is the only practicable option for reducing the impact of the disease.

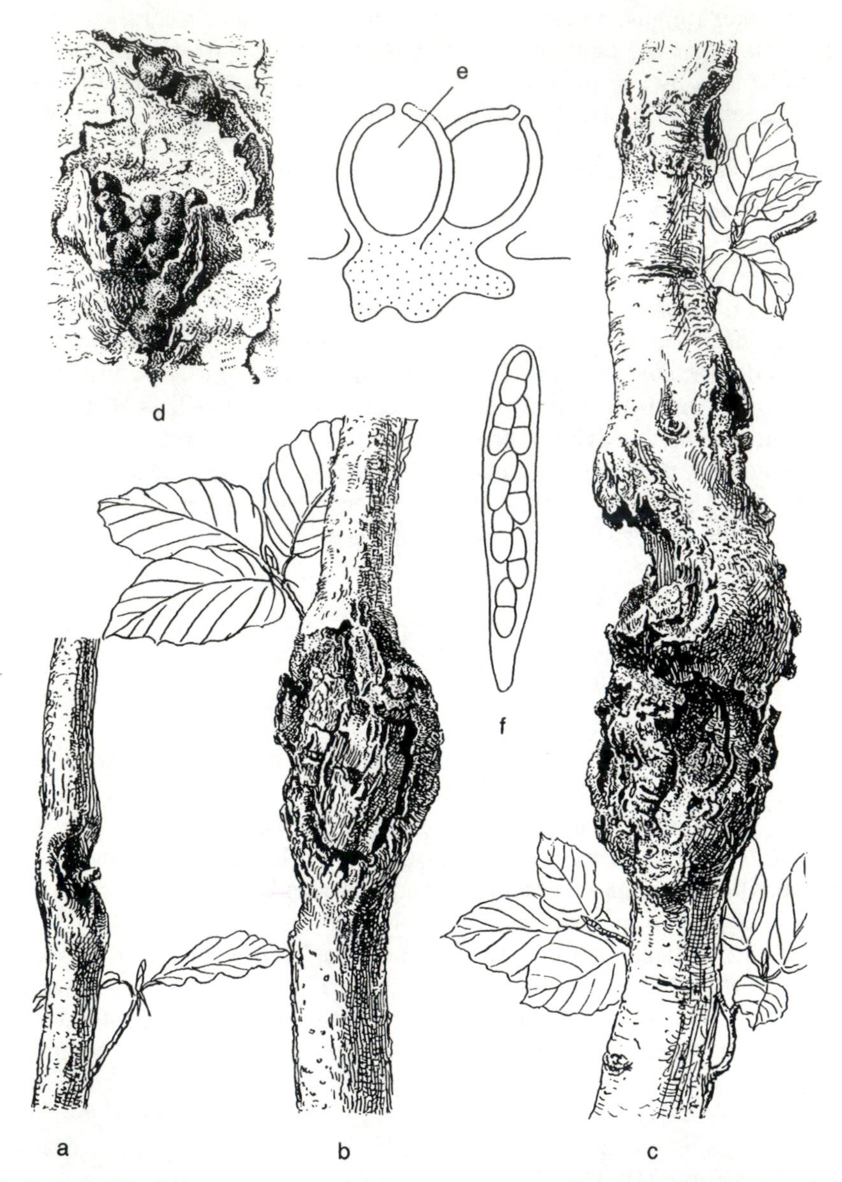

Fig. 71 *Nectria ditissima*. **a-c** various stages in the development of cankers on beech, **d** fruit bodies (perithecia) erumpent through dead bark, **e** cross-section through two perithecia with a basal stroma (diagrammatic), **f** ascus with ascospores

Other hosts of *N. ditissima* exhibiting more or less typical cankers occur in the following genera: *Alnus, Betula, Corylus, Fraxinus, Sorbus,* and *Tilia.*

Beech Bark Disease and Related Disorders

Causes: *a complex of abiotic and biotic factors*

In all the disorders under this heading, necrotic patches develop in the bark and can be recognized externally from the slimy, dark exudations ('tarry spots') that appear at various heights on the stem, mainly in spring or autumn. These eventually dry up or become washed away by rain, but they leave behind persistent pale patches on the bark surface, devoid of algae or lichens. The necrotic patches, which can be revealed as brown or orange discolorations under the bark surface, can range from a few millimetres to a metre or more in length. In some cases, the development of very extensive or coalescent necroses can lead to girdling of the stem, at which advanced stage the leaves suddenly wilt or turn prematurely yellow.

Small bark necroses, no larger than the palm of the hand, are mostly occluded by the tree without further damage, but thereafter remain visible externally for some years as bark scars. Internally, they can be seen as T-shaped defects in a cross-section of the wood, and their year of origin can be dated precisely by an annual ring count. Larger bark necroses lead to the shedding of patches of bark so that the wood is laid partly bare. In such cases, wood destroying fungi often appear, leading to a rapid decay of the wood. The most common species in central Europe are *Fomes fomentarius* (Fig. **99**) and *Fomitopsis pinicola* (Fig. **98**), while in Britain *Bjerkandera adusta* and *Stereum* species are more typical invaders. Stems attacked by such decay fungi are liable to snap after some years (Fig. **72**).

A number of micro-organisms can be isolated from the necrotic tissues, none of which is strongly pathogenic, and it is therefore considered either that their presence is purely secondary, or that they are weak parasites which invade tissues weakened by one or more other agents. The most frequently reported of these micro-organisms is the fungus *Nectria coccinea* (Table **I,26**). Among the abiotic agents thought to play either a predisposing or a primary role, particular importance is attached to water stress which may result from certain site conditions (wet sites with shallow spreading roots) or by extremes of weather (hot, dry summers). The most common biotic factor is the presence of heavy infestations of the felted beech coccus (*Cryptococcus fagisuga*), and it has been suggested that abiotic stress and this insect may have interchangeable roles in predisposing the bark to *Nectria* infection and hence to the development of necroses [125].

A clear-cut involvement of *C. fagisuga* as a predisposing agent for fungal infection is seen in the beech bark disease of *Fagus grandifolia* in North America [68]. There, the disease appears in forests on an epidemic front, which follows the spread of the insect from the sites of its introduction around 1890. This contrasts with the situation in Europe where cause and effect are obscured by

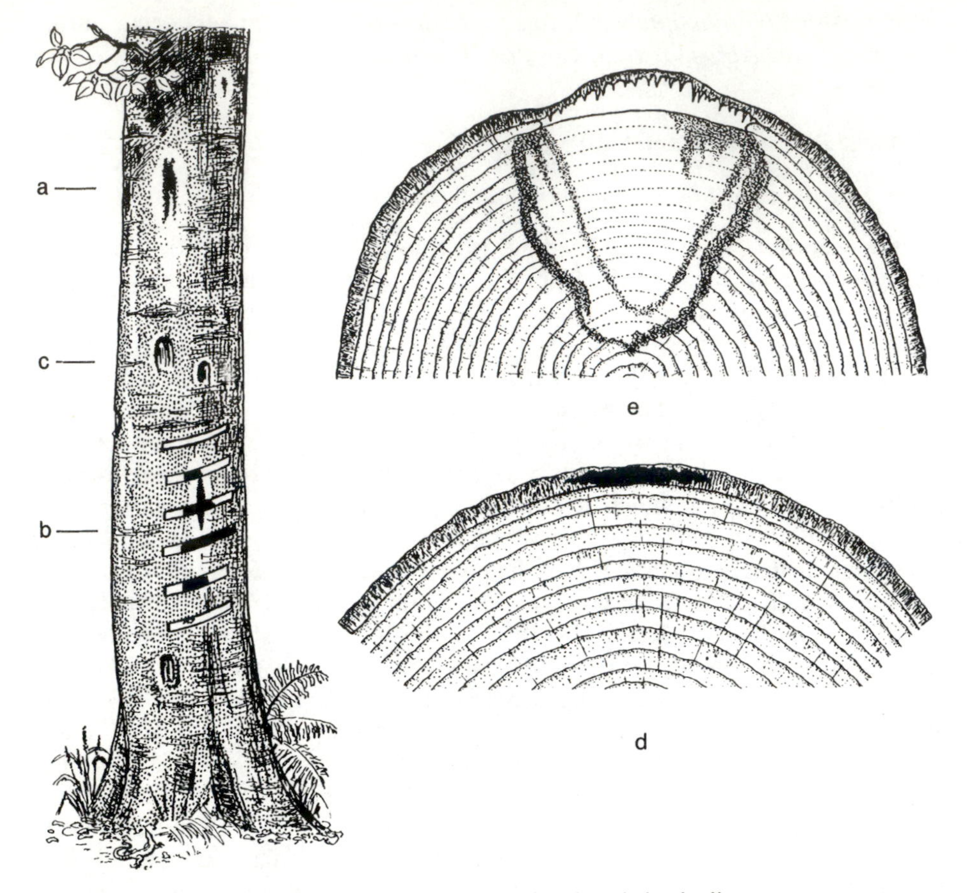

Fig. 72 Slime flux syndrome of beech, including beech bark disease. **a** tarry spots on the bark, **b** method of determining the extent of the bark necrosis by scoring the bark with a timber scribe, **c** occluded bark lesions, **d** early stage of bark necrosis, **e** late stage with incipient white rot in the wood

the endemic presence of the insect. Nevertheless, the young beech plantations of southern England, established in the mid-twentieth century, have provided new niches for the insect, and the resulting local flare-ups of infestation have been invariably followed by typical slime-flux development, associated with *N. coccinea* infection. [107,126]. In North America the fungus is most often *N. coccinea* var. *faginata*, though other *Nectria* species may take its place.

Bark necrosis resulting from the *Cryptococcus-Nectria* association can occur on beech of all ages, but tends to be most frequent on young trees which become infested by the insect for the first time. Necrosis associated with abiotic stress is more frequent on trees over 60 years old.

For trees on which the predisposing agent is *C. fagisuga*, and which have high amenity value, direct control by the use of insecticides or mechanical removal of the insect is feasible. In commercial plantations, the main consideration must be to avoid loss of timber from secondary decay. This makes it important to recognize the necroses at an early stage. If a single stem bears several necrotic patches larger than palm size, decay can be expected to develop soon. Such trees should be salvaged as soon as possible. The same applies to stems with a very dense covering of coccus.

The death of patches of bark, and slime fluxing can also be seen on other types of tree, e.g. maple, birch, alder, and Red oak. These symptoms are general indications of a disturbance to the tree's water relations, in which damage to the roots or to the stem base is often the primary cause.

Black Bark Scab of Beech

Cause: *Ascodichaena rugosa* Butin
Anamorph: *Polymorphum quercinum* (Pers.) Chev.

This bark disease, only elucidated in recent years [36], was previously supposed to be caused by a lichen or saprotrophic imperfect fungus. It is characterized by patches or strips of a black coating on the bark, found mainly on the lower part of the stem of beech of various ages. The fungus occurs on various species of beech, as well as *Fagus sylvatica*, and also various *Quercus* species on which it can maintain itself only on bark up to 10 years old which has not yet developed a rhytidome.

The cause of black bark scab is the fungus *Ascodichaena rugosa*, a member of the Order Rhytismatales. On beech bark it occurs almost exclusively in its imperfect state, forming clusters of pycnidia. In their young stage, these are coffee-bean-shaped, black, and 300–450 × 300 μm in size; they are bedded on a thin, meristematic, persistent stroma on which new fruit bodies are produced each year. With this process, older fruit bodies remain attached to younger conidiomata so that in time a rough, 'goose-pimple' -like fungal crust is formed. Inside the pycnidia, egg-shaped hyaline conidia are formed, 18–24 × 13–14 μm in size, which in damp weather in summer are released or picked up and distributed by slugs and snails (Fig. **73**).

Ascodichaena rugosa is parasitic, but it invades only the cells of the bark cork layer. It takes up nutrients by means of haustoria which are formed within those cells in the phellem that have been most recently formed from the phellogen and which still contain nuclei and cytoplasm. The tree reacts by increasing the rate of cell formation, which results in the roughening of the bark surface. An attack is of no consequence either to the tree or to the forester; control measures are therefore inappropriate. Slugs and snails may effect a certain degree of biological control by grazing the pycnidia but at the same time, of course, also spread the spores about.

Other Bark Fungi of Beech [69]:

- *Asterosporium asterospermum* (Pers.) Hughes is a frequent primary colonizer of dead bark on branches and stems. Acervuli are subepidermal, 1–2 mm

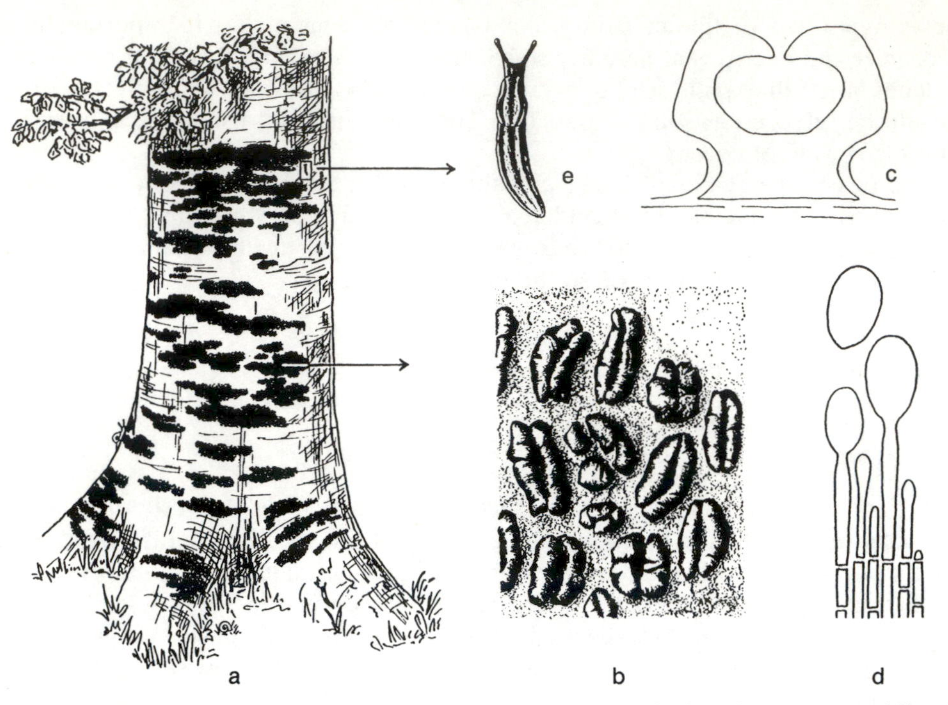

Fig. 73 *Ascodichaena rugosa*. **a** attack on the lower stem of a beech tree, **b** pycnidia of the imperfect state, **c** cross-section (diagrammatic) through a pycnidium, **d** conidiophores with conidia, **e** a slug - a vector for spore dispersal

across, rupturing the epidermis with a slit filled with a mass of blackish spores. Conidia are four-armed, several celled, dark brown, and 40–50 μm in size (Table **II,16**).

- *Fusarium avenaceum* (Fr.) Sacc. causes bark of young beech to break up into longitudinally cracked plates, possibly after prior injury from pollution. Conidiomata are cushion-shaped, and 1–2 mm across. Macroconidia are sickle-shaped, several celled, and 35–55 × 3.5 μm (Fig. **9b**) in size.
- *Fusicoccum galericulatum* Sacc. is a primary colonizer of dying branches. Stromata are cushion-shaped, 1–2 mm across, black with immersed loculi. Conidia are hyaline, 1-celled, ellipsoid, 12–14 × 4–5 μm in size (Table **I, 24**). *F. macrosporum* Sacc. is similar with spores 35–45 × 8–10 μm in size (Table **II, 17**).
- *Biscogniauxia nummularia* (Bull.: Fr.) O. Kuntze (Syn. *Hypoxylon nummularium* Bull.:Fr.) is one of the fungal species which are associated with the phenomenon of strip cankering of beech [104], and is initially a xylem colonizer; stromata are erumpent from inner bark or occasionally on decorticated wood, orbicular to elliptic, brown to dark brown, carbonaceous, and 2–3 cm in diameter. Ascospores are dark at maturity, 11–14 × 7–10 μm in size [132].
- *Libertella faginea* Desm. is a xylem colonizer which commonly fruits on dead bark of beech branches lying on the ground. Conidia are curved like

sickles, 14–17 × 1.5 μm, and often emerge from the bark in the form of golden red spore tendrils (Table **I, 25**). It belongs to the ascomycete genus *Diatrypella*.

- *Nectria coccinea* (Pers.) Fr. colonizes dying bark of stems or thicker branches (see under 'beech bark disease', above). Perithecia are ovoid, brick red to dark red, 0.3 mm across, and in clusters of from 5–30 on a similarly coloured stroma. Ascospores are 2-celled, finely warty, 12–17 × 5–7 μm in size (Table **I, 26**).

Fusicoccum Bark Canker of Oak

Cause: *Fusicoccum quercus* Oudem.

This fungus typically produces an annual canker whose development is usually completed within a single year. The first signs of attack become noticeable in the early spring with the appearance of elliptical, reddish yellow discoloured patches on the bark, reaching 5–15 cm in length and often centred around a twig base. With the start of cambial growth, the lesion becomes delimited and later occluded by wound tissue. Occasionally, the fungus spreads out again from the edges of the wound in the following year, inducing a perennating bark canker with irregular wound healing.

Two- and 3-year-old seedlings, saplings, and pole-stage trees of *Quercus robur* and other oak species are attacked. Also, shoot tips on older oaks can be attacked, and these are particularly noticeable when the dead shoot tips protrude from the green crown. When seedlings are attacked they are quite often killed, but saplings and young shoots are killed only if the necrosis girdles them. In some cases, e.g. when shoot tips are attacked, colonization by the fungus seems to follow sublethal frost damage.

As well as occurring parasitically, the fungus occurs as a saprotroph in the bark of dying, thicker branches and so may be more common than might be suggested by the occurrence of the disease.

The causal fungus has so far been seen only in its conidial state; this manifests itself in two different types of fruit body and spore forms. From spring until summer the 'summer form' develops, forming pycnidia of the *Phomopsis* type which are 0.5–1.5 mm across; from September, cushion-shaped, multilocular overwintering stromata of the *Fusicoccum* type, up to 3 mm in size, are formed. In both kinds of fruit body there are either 'α-spores,' 12–17 × 3.5–4.5 μm in size, or smaller, similarly ellipsoidal 'β-spores,' measuring 5.5–6.5 × 2.5–3.5 μm. Both kinds of spore are germinable (Fig. **74a–f**).

In monocultures, especially young stands, and in nurseries, the fungus can occur epidemically. The main risk exists on sandy soils with little water-storage capacity, and so it is advisable not to plant oaks on very porous soils. The predisposition of younger plants by dry weather in the spring could be reduced by irrigation. The use of copper fungicides has been considered but has not been adequately tested.

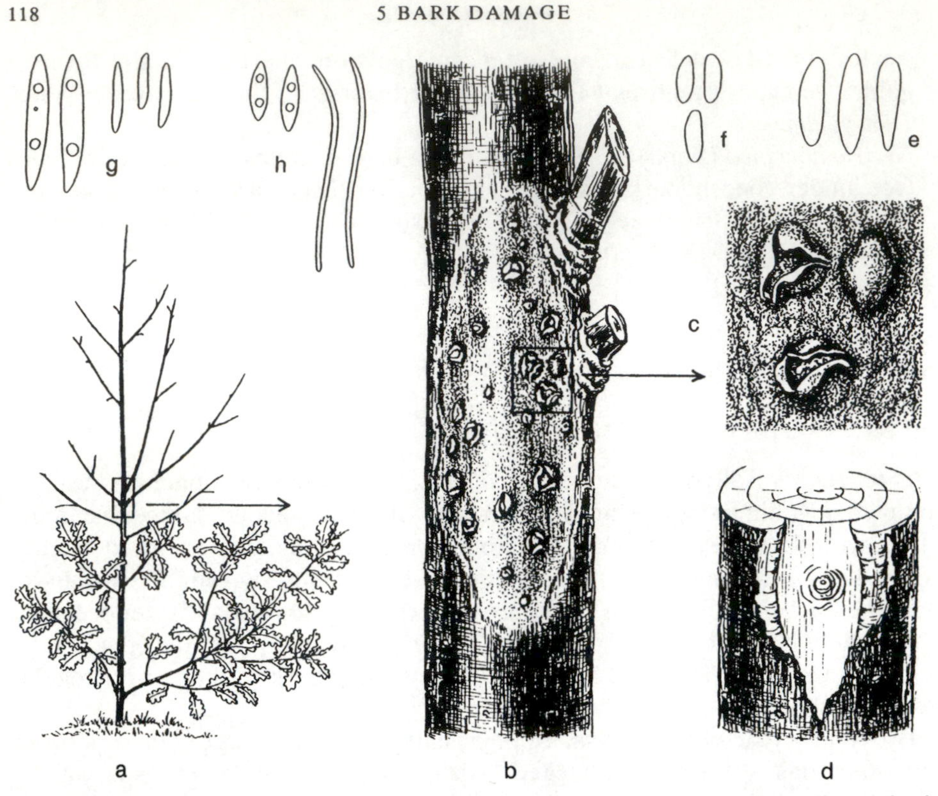

Fig. 74 Oak bark fungi. **a-f** *Fusicoccum quercus*: **a** symptoms on an oak sapling, **b** bark necrosis with pycnidia on a 3-year-old stem, **c** ruptured pycnidia in dead bark, **d** the beginning of wound callusing, **e** α-conidia, **f** β-conidia; **g** α- and β-conidia of *Phomopsis quercina*, **h** α- and β-conidia of *Phomopsis quercella*

Other Bark Fungi on Oak [37,69]:

- *Colpoma quercinum* (Fr.) Wallr., ascomycete has fruit bodies which are lip-like with white edges, and up to 15 mm long. Ascospores are filiform, up to 90 μm long. The conidial form is *Conostroma didymum* (Faultry & Roum.) Moesz. The fungus is an endophyte in living bark or a weak parasite on suppressed, dying branches (Fig. **58a–c**).
- *Cryptosporiopsis grisea* (Pers.) Petrak is the imperfect state of *Pezicula cinnamomea.* Macroconidia are cylindrical, hyaline, and 28–38 × 9–12 μm in size. It is a weak parasite on branches and saplings and economically important mainly on Red oak (Fig. **80e,f**).
- *Diplodia mutila* Fr. apud Mont. (teleomorph: *Botryosphaeria stevensii* Shoemaker) causes a branch canker and dieback on various species of oak; in *Quercus suber*, it is also a cause of stagheadedness [127]. Conidiomata are uni- or multilocular; conidia are initially 1-celled and hyaline, later 2-celled and brown, measuring 27–31 × 12–13.5 μm [208].
- *Biscogniauxia mediterranea* (de Not.) O. Kuntze (Syn. *Hypoxylon mediterraneum* [de Not.] Miller) (anamorph: *Botrytis sylvatica* Malençon), causes

the 'charcoal disease' of cork oak in the Mediterranean region [129]; trees are most susceptible when in poor conditions, e.g. on poor soil, or when damaged by drought and fire. The fungus colonizes the inner bark and sapwood, mostly at the base of the stems. The stromata, which are formed in the cambial zone, are black, thin, circular or elliptical, 2–5 cm in diameter, and have numerous embedded perithecia; ascospores are ellipsoidal, dark brown, and 16–23 × 6–10 μm in size [132].

- *Phomopsis quercella* (Sacc. & Roum.) Died. is saprotrophic on dying plants; α-spores are ellipsoidal and 8–10 × 2–3 μm; β-spores are vermiform and 25–30 × 1–1.5 μm (Fig. **74h**).
- *Phomopsis quercina* (Sacc.) Höhn. is a saprotrophic fungal associate of *Fusicoccum quercus*. α-spores are spindle-shaped and 13–21 × 3.5–4.5 μm; β-spores are 10–15 × 1.8–2.2 μm in size (Fig. **74g**).

Ash Canker

Cause: *Pseudomonas syringae* subsp. *savastanoi* pv. *fraxini* Janse
or: *Nectria galligena* Bres.

The name 'ash canker' embraces two different diseases of ash bark which both produce perennating lesions. In the one case bacteria are the cause of the disease; in the other, certain fungal species are responsible. The two diseases can be distinguished by differences in symptoms.

Bacterial canker of ash outwardly bears some resemblance to the bacterial canker of poplar, similarly starting with the appearance of swollen areas on twigs and stems which later crack longitudinally. As a result of the death of cambial cells, subsequent hypertrophic growth, and repeated killing of callus tissue, blackish, convoluted areas of bark are formed (Fig. **75c**). The causal agent is the bacterium *Pseudomonas syringae* subsp. *savastanoi* pv. *fraxini*, which reaches the bark parenchyma through wounds, lenticels, or leaf scars. Further development of the disease can be encouraged by certain species of fungi such as *Fusarium lateritium* and *Phoma riggenbachii*. To limit spread of this bacterial canker it is recommended that affected trees are removed [112,113].

Young cankers can easily be confused with the overwintering gallery system of the ash bark beetle (*Leperisinus varius*). The bark proliferations that this beetle causes do not destroy the cambium. They are known in German as 'Eschenrosen' (ash roses) or 'Käfergrind' (beetle scab).

In contrast to the bacterial canker, the lesions formed by Nectria canker of ash are evenly—almost symmetrically—shaped. Thus, they resemble all the other Nectria cankers as well as those caused by *Lachnellula willkommii* (Fig. **70a,b**). On most older ash trees this results in the formation of oval bark cracks which start from broken twigs or similar entry points and eventually develop into crater-like, open, cankerous lesions up to 30 cm wide. These are typical 'target' cankers, with concentric ridges of callus tissue which form annually before being killed by the fungus during the dormant season. The number of these ridges allows the age of the canker to be determined. This is a long-term conflict

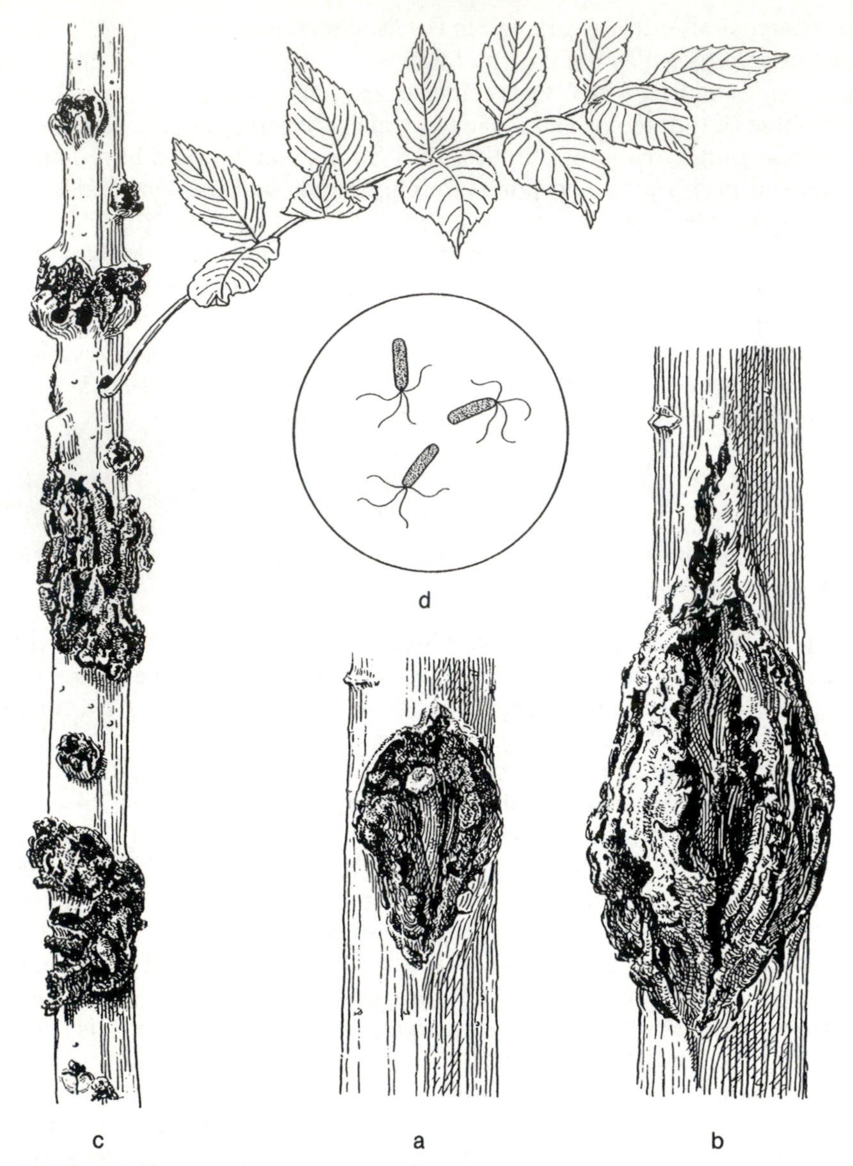

Fig. 75 Bacterial cankers. **a,b** various stages in canker development on a hybrid Black poplar caused by *Xanthomonas populi*; **c** *Pseudomonas syringae* damage on ash, **d** the bacteria which cause the damage (d after Janse 1981b)

between host and parasite, in which both usually survive for many years. It is the forester who suffers the loss, as he has to reckon with a lower grade of timber.

The *Nectria* species that causes this canker is *N. galligena,* a widely distributed ascomycete which attacks other types of tree in a similar fashion; these include maple, birch, Horse chestnut, willow, and especially apple. It has been suggested that the fungus which occurs on ash is a host-specific form: *N. galligena* f.sp. *fraxini* [76].

The fruit bodies, recognizable with a hand lens, are red, pinhead-sized perithecia, located mainly on the edges of the killed annual callus ridges. The ascospores are 2-celled, hyaline, and 14–22 × 6–9 μm in size [151]. Later, the dead wood in the open canker is colonized by various saprotrophic fungi among which *Trematosphaeria* species and *Phoma* species are particularly frequent.

Chestnut Blight

Cause: *Cryphonectria parasitica* (Murrill) Barr
Syn. *Endothia parasitica* (Murrill) And. & And.

This pathogen, which was introduced into Europe from North America in 1938, causes a serious bark disease which has already led to major damage in Sweet chestnut stands in Italy, southern France, Spain, and Switzerland. The first symptoms of disease to appear are reddish brown spots which later become sunken and split open longitudinally. Older, rough-barked stems show a longer term symptom in which the dead bark characteristically splits open, and a fan-shaped, cream-coloured mycelium can be seen between the bark and the wood. If the death of a branch occurs during the growing season, part of the crown is seen to wilt, caused to some extent by a wilt toxin. The infection ends with the death of the affected branch or stem. The result of this is that the rootstock throws up shoots which can remain healthy for many years.

This fungus, which belongs to the Ascomycotina, invades the bark via wounds, and kills it either superficially or down to the cambium. Later, ochre yellow pycnidia develop on the bark. These have rod-shaped conidia, 3.5 × 1.2 μm in size. The grouped perithecia are the same colour as the pycnidia; they contain numerous asci each with eight 2-celled, 7–11 × 3.5–5 μm ascospores (Fig. **76**).

In Europe, *C. parasitica* is of economic significance only on Sweet chestnut, although other members of the Fagaceae, such as *Quercus pubescens*, *Q. ilex*, and *Q. petraea*, may be attacked weakly. These other tree species must not be overlooked as they can act as infective secondary hosts. To control chestnut blight, the chemical destruction of rootstock sprouts in infected stands is recommended in some countries. A more effective long-term strategy is the planting of resistant chestnuts which have been produced by selection from the European stock and from crossings with the resistant east Asiatic *Castanea crenata*.

Direct biological control of the pathogen is a more recently developed approach, in which hypovirulent strains of the same fungus are used therapeutically. The hypovirulence is usually caused by the presence of a virus-like agent in the fungus, and this can be transmitted to the pathogenic form already present in a canker. The present widespread occurrence of such strains of reduced pathogenicity is very probably one reason why the death rate of cankered trees in Europe has slowed down. In this instance, Nature herself has succeeded in developing a regulatory mechanism which ensures the co-existence of both parasite and host.

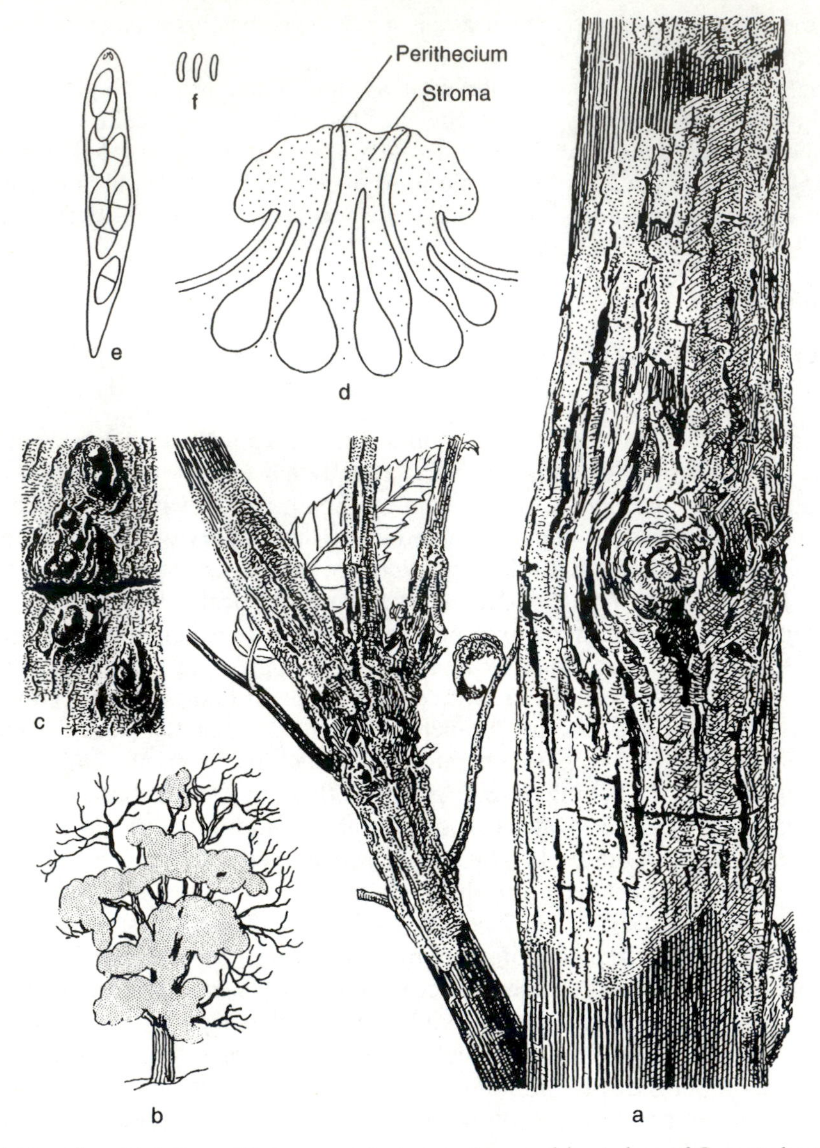

Fig. 76 *Cryphonectria parasitica*. **a** symptoms on stem and branches of Sweet chestnut, **b** general view of a diseased tree, **c** perithecia and pycnidia bursting through the dead bark, **d** section (diagrammatic) through a part of a stroma with perithecia, **e** ascus with ascospores, **f** conidia

Other Bark Fungi of Sweet Chestnut:

- *Cryptodiaporthe castanea* (Tul.) Wehm. (anamorph: *Fusicoccum castaneum* Sacc.) causes a dieback and canker of shoots and thicker branches; perithecia are in groups, with spindle-shaped to clavate asci, and 2-celled, hyaline, ascospores measuring 10–16 × 2–3 μm; conidia are 1-celled, spindle-shaped, and 6–11 × 2–4 μm [54].

- *Diplodina castaneae* Prill. & Delacr. causes a dieback with brown necroses on the bark of thinner branches. Pycnidia are embedded. Conidia are 2-celled, spindle-shaped, and 6–7 × 1–1.5 μm in size.
- *Phytophthora cambivora* (Petri) Buism. causes 'ink disease' in older chestnut stands in, for example, Italy, southern France, Spain, and Switzerland. Externally visible symptoms are the yellowing of leaves and the failure of fruit to set. A knife cut into the stem base will reveal a tongue-shaped area of blackish brown discoloured cambium, spreading upwards. Practicable control measures are not known [162].

Bacterial Canker of Poplar

Cause: *Xanthomonas populi* subsp. *populi* (Ridé) Ridé & Ridé
Syn. *Aplanobacter populi* Ridé

The lesions produced by this bacterial disease are typically very irregular, convoluted, erumpent cankers on the stem and branches of a tree, sometimes persisting for many years (Fig. **75a,b**).

After infection, which as a rule takes place via leaf, bud scale, and stipule scars, a slimy exudate appears on young shoots as the primary symptom in the spring, emerging mainly from small bark cracks. Further development depends on the susceptibility of the poplar in question. The most susceptible clones often develop large necroses in which callus tissue repeatedly forms but is killed by the bacterium. In other clones the tree reacts with an increased and unregulated growth of cells so that the edges of the wounds become thickened and swollen. Often, extensive cankers arise where the bacteria have been spread by the bark-inhabiting larvae of certain flies.

Xanthomonas populi subsp. *populi* is a strictly host-specific pathogen, occurring only on members of the genus *Populus* [163]. Moreover, as indicated above, there are considerable differences in susceptibility amongst the individual species and clones of poplar. Thus, for example, *P. nigra* is completely resistant; also, most of the euramerican hybrids grown today are relatively canker-resistant. Examples of extremely susceptible clones include 'Brabantica' and 'Grandis' which nowadays are retained only in special collections. Similarly large differences are seen in the balsam poplar group. However, the occurrence of the canker is not exclusively related to clonal susceptibility, since it also depends on climatic conditions. In particular, the bacterium is heat-sensitive, so that it causes relatively little disease in parts of Europe where temperatures are high throughout the growing season.

To limit the spread of the disease, all affected trees must be destroyed since, when infection is high, even moderately resistant poplars can become diseased. Chemical control of poplar canker appears to be neither very promising nor economically acceptable. Therefore, the only recommendation is to plant resistant varieties. Information on the canker susceptibility of specific clones can nowadays be given as early as 4 weeks after the inoculation of 1-year-old plants [117].

Related Species:

- *Xanthomonas populi* subsp. *salicis* de Kam causes a canker on willow; symptoms are similar to those of bacterial canker on poplar [53].

Dothichiza Bark Necrosis and Dieback of Poplar

Cause: *Cryptodiaporthe populea* (Sacc.) Butin
  Anamorph: *Discosporium populeum* (Sacc.) Sutton
  Syn. *Dothichiza populea* Sacc. & Briard

This disease is characterized on young poplars by elliptical, brownish bark necroses on the stem and by the death of twigs or of the terminal shoot. Necroses on the main stem are usually found at the base of side branches. Typically, the lesions are small and become walled off and occluded in one growing season so that the disease in this form is an annual canker (Fig. **77**). In certain circumstances, the fungus can break out of its encirclement and spread into the neighbouring healthy tissue, giving rise to canker-like bark necroses. Damage of this type is found, for example, on the shoots at the tops of old poplars and is a feature of Lombardy poplar, on which it occurs in combination with frost damage (Fig. **62c**).

Another aspect of the symptomatology of *Dothichiza* on poplar is the development of bark necroses on older stems, mainly around the scars of abscised twigs. On stems that have not developed a rhytidome, these necroses can be seen on the bark surface as approximately palm-sized, elliptical, sunken patches. On bark with a thick cork layer they are visible only if the bark is removed, exposing the cambium. Such lesions are known is German as 'Pappelgrind' (poplar scab), or 'Braunfleckengrind' (brown-spot scab). Small lesions on vigorous trees are soon callused over, but if the damage occurs as extensive patches it can lead to the death of the whole tree. This type of damage is encouraged by water shortage in the bark, and waterlogging of the site also seems to play some part. Disease development also depends on clonal susceptibility: for example, *P.* × *euramericana* 'Robusta' is particularly susceptible.

The imperfect fruiting stage of the fungus appears in spring and summer, when papillate pycnidia, approximately 1 mm in diameter, erupt from the dead bark. In humid weather, these release hyaline conidia, 10–13 × 7–9 μm in size, in dark-coloured spore tendrils. After spore release the pycnidia persist as black cavities in the bark. The associated perfect state, which is characterized by spherical long-necked perithecia, is formed only on the dead shoot tips of old poplars and then only when the attack has continued for a second year, causing the bark scars to re-open.

The causal agent of this bark disease is highly host specific, occurring only on members of the genus of *Populus*. Within the genus, the various species and clones differ greatly in their susceptibility. Thus, Black poplar and most of its hybrids are generally very prone to *Dothichiza* attack; in contrast, aspen, and White and Grey poplars are only slightly susceptible and the members of the balsam poplar group are hardly ever attacked, owing to the presence of specific fungistatic substances in the bark.

In view of the specific varietal nature of susceptibility to the disease, choice of clone is the most important and most environmentally friendly way of protecting against an outbreak. In addition, care should be taken not to allow recently planted poplars to suffer from a shortage of water, as loss of moisture decreases the defensive capability of the bark tissue. Finally, poplar plantations should not be established in the vicinity of old poplars as these often harbour residual centres of disease. If these various prophylactic options are followed, chemical protection can be dispensed with.

Other Bark Fungi of Poplar [33,69]:

- *Chalaropsis populi* Veldeman causes elongated bark necroses on the stems of various poplar species; in culture both endogenous phialospores and roundish ovoid aleuriospores are formed. No perfect-state fruiting form is known [215].
- *Cytospora chrysosperma* (Pers.) Fr. and *C. nivea* Sacc. are regular colonizers of dying or already dead thin branches of various poplar species or clones; conidia are small, sausage-shaped, hyaline, and mostly expelled from the bark in reddish tendrils (Fig. **77e**).

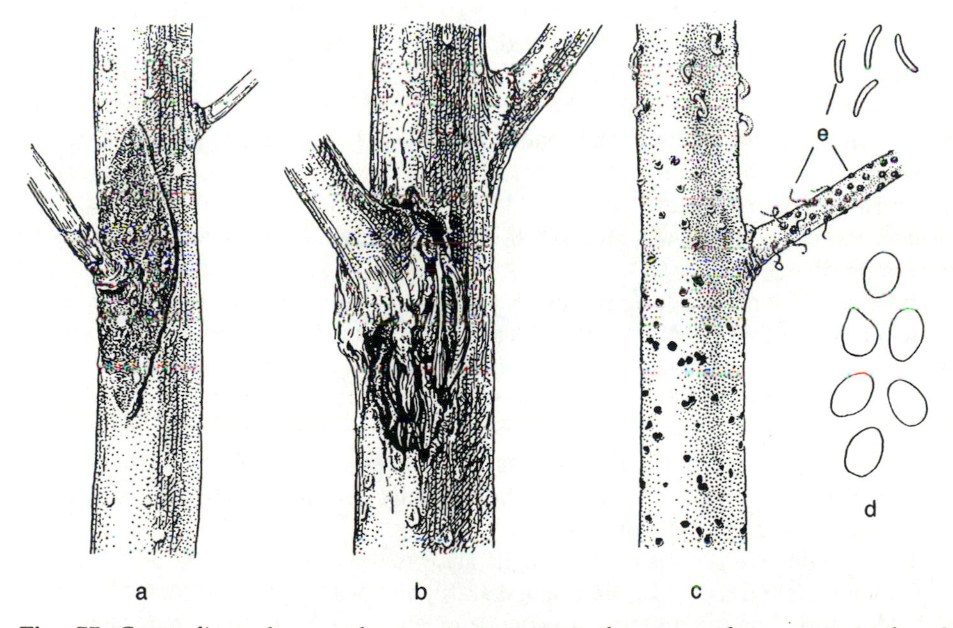

Fig. 77 *Cryptodiaporthe populea*. **a** symptoms on the stem of a young poplar, **b** advanced callusing, **c** dead poplar stem with spore tendrils (above) and empty pycnidia (below) of the imperfect state, **d** conidia; **e** pycnidia, spore tendrils and conidia of *Cytospora nivea*

- *Hypoxylon mammatum* (Wahlenb.) Miller is the feared cause of a bark canker on *Populus tremula, P. tremuloides,* and their hybrids. It is widespread in North America. In Europe, it occurs only sporadically in mountainous regions [157]. Ascospores are blackish brown, and 20–33 × 9–12 μm in size (Table **I, 27**).
- *Neofabraea populi* Thompson occurs on the bark of aspen and balsam poplars, and causes circular necroses which later become more extensive; infections are through lenticels. Conidia are sickle-shaped, and 20–30 × 4–6 μm in size [172].

Canker Stain of Plane

Cause: *Ceratocystis fimbriata* (Ell. & Halsted) Davidson f. *platani* Walter

This disease of bark and xylem vessels—known in German as 'Platanenkrebs' (plane canker) or 'Platanenwelke' (plane wilt)—occurs only on plane trees. It has been known in Europe only since 1972, the causal fungus having very probably been introduced from North America. At present it is still confined to scattered areas in France, Italy, Spain, and Switzerland; however, the possibility that the fungus is spreading farther north cannot be ruled out.

The first obvious symptom of an attack is the appearance of small, violet-brown depressions in the bark. At an advanced stage, axially elongated, anastomosing bark necroses are found, mostly extending from the base of the stem with the eventual platy cracking of patches of bark. By this stage, the colonizers of the bark and wood are mostly secondary, e.g. *Fusarium* sp., *Pestalotia* sp., *Chondrostereum purpureum,* or *Schizophyllum commune.* Internally, a typical symptom is the occurrence of bluish brown stains arranged radially in the wood. If the bark is badly damaged, the discolorations can extend as wedge-shaped areas into the centre of the stem. This staining has given rise to the English name for the disease. Also, at certain stages of development, the disease can cause a leaf wilt, though this affects only odd branches and cannot always be observed.

The causal agent, the fungus *Ceratocystis fimbriata* f. *platani*, enters the tree via wounds and first kills the bark cells and the cambium. Subsequently, by way of the xylem rays, the mycelium also spreads through the sapwood, in which the vessels become blocked by tyloses. Thus, the fungus is both a bark parasite and a vascular wilt agent. After spreading in an axial direction, the fungus can again grow out via the rays at various heights on the stem, reaching the bark, where it again kills strips of tissue. The infection usually continues until the tree dies, a process which can take from 3–6 years.

The reproductive organs of the fungus arise mainly on the cut wood surfaces of pruned or felled trees. In the asexual state, various 'imperfect' forms of fruit body develop. For diagnostic purposes, the most useful is the *Chalara* form, which forms chains of endoconidia, measures 10–20 × 4–6 μm, and can grow out from increment cores, which are taken from diseased trees and incubated in the laboratory [217]. The perfect state is characterized by dark perithecia about

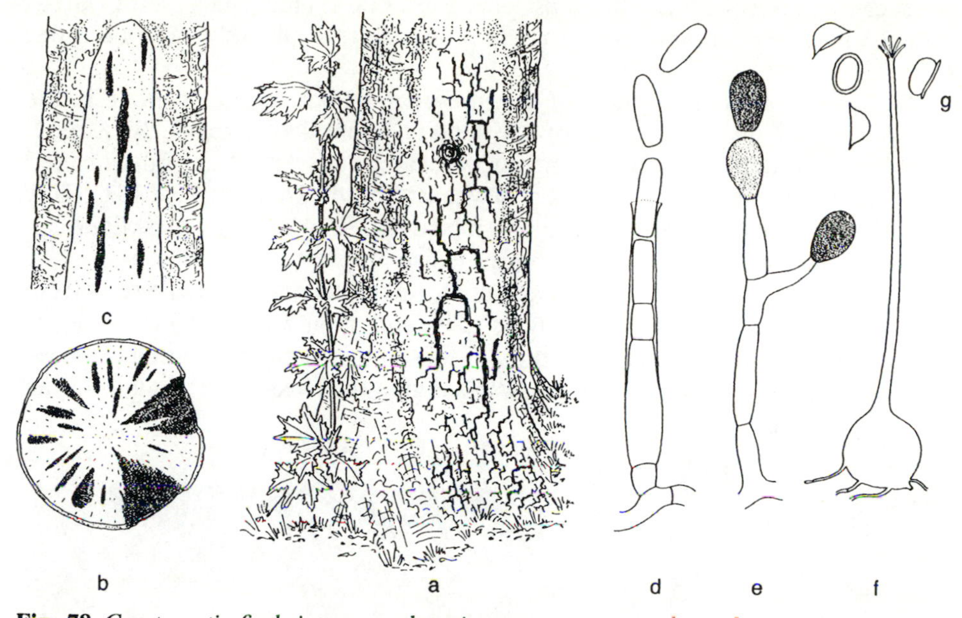

Fig. 78 *Ceratocystis fimbriata* var. *platani*. **a** symptoms on plane, **b** stem cross-section showing stained wood, **c** tangential stem section showing the stain as streaks, **d** phialide with conidia of the *Chalara* imperfect state, **e** conidiophore with chlamydospores, **f** perithecium, **g** ascospores

1 mm high with long necks and kidney-shaped ascospores, 4–7 × 2–2.5 μm in size (Fig. **78**).

The pathogen is usually carried from tree to tree during pruning operations and its spread can, therefore, be reduced by hygiene. Disease control can also be attempted, both preventively and therapeutically, by the injection of the outer sapwood with systemic fungicides, although these have not yet been tested sufficiently for general use. New centres of infection should be eliminated as soon as possible by the destruction of the infected trees. Finally, to prevent widespread dispersal of the fungus, the European Community guidelines, and any national plant health inspection regulations must be observed.

Stereum Canker Rot of Red Oak

Cause: *Stereum rugosum* (Pers.) Fr.

The symptoms of this disease are flask- or club-shaped swellings and mostly one-sided depressions in the lower part of the stem. Later the bark splits and dies. On closer examination, a dead branch stub, which has played a part in the infection process, can usually be made out in the centre of the lesion. This disease occurs mainly on Red oak on warmer sites with ample rainfall; occasionally the fungus is also seen on European species of oak [5] and also on beech and hornbeam. Other host plants are birch, alder, hazel, and willow, though on these the fungus occurs as a harmless saprotroph and as a wood rotter (Fig. **79**).

The most certain indication of the presence of a Stereum canker is the appearance on the killed bark of the closely appressed, irregularly shaped, crust-like fruit bodies. They are grey-brown, later ochre and their multilayered hymenium becomes stained blood red when injured. Their surface is initially smooth, becoming cracked with age, and it bears 1-celled basidia with cylindrical, hyaline basidiospores, 10–14 × 4–6 μm in size [70].

Stereum canker of Red oak is a perennial canker, as can clearly be seen from a cross-section of the stem; the margins of wound tissue formed by the tree are repeatedly killed back during the dormant season by the fungus so that a broad, open lesion arises, which even in the long run does not become occluded. In the

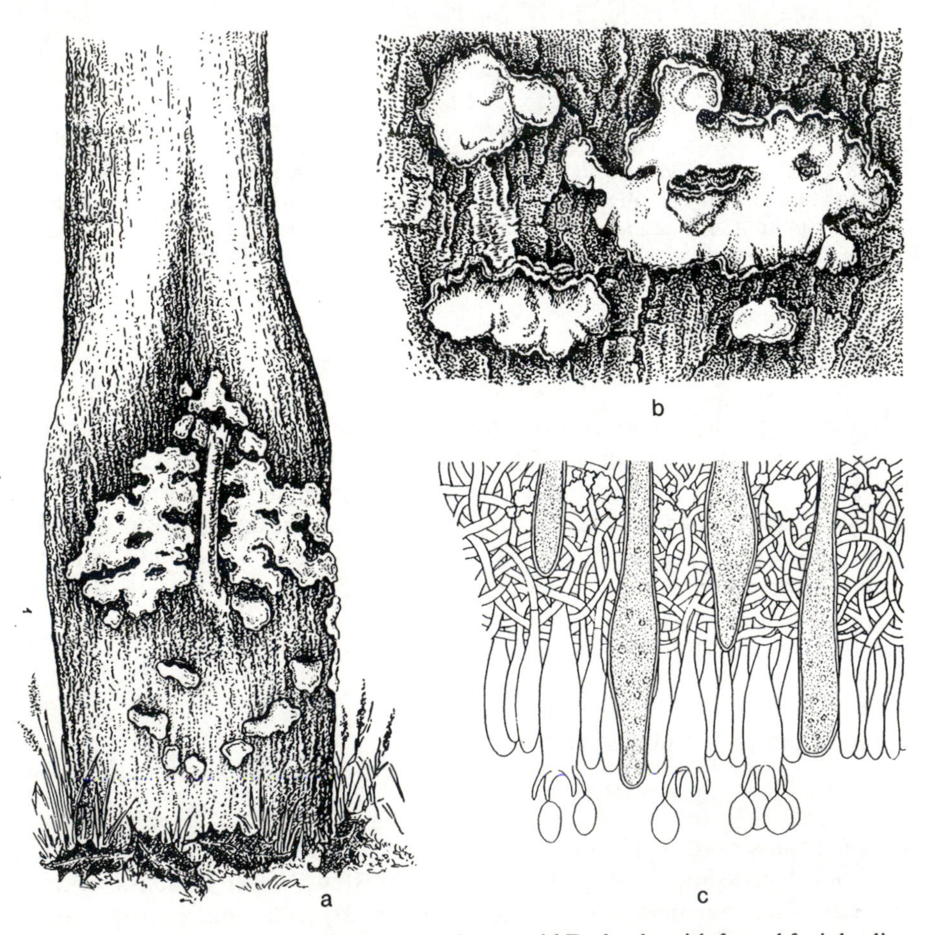

Fig. 79 *Stereum rugosum*. **a** symptoms on a 30-year-old Red oak, with fungal fruit bodies, **b** fruit bodies on dead bark, **c** cross-section (detail) through the hymenial layer with basidia, sap-conducting pseudocystidia and crystals (c after Jahn 1971)

vicinity of the lesion, a white rot also develops but is delimited by an effective reaction zone and so penetrates only a short distance into the stem wood. Since *S. rugosum* causes both a canker and a wood rot in Red oak, the disease is classified as a canker rot.

As the fungus enters the stem as a wound parasite through dead branches, the only possibility for preventing the canker seems at the moment to be the prompt removal of any twigs which are growing near the ground. However, larger pruning wounds can, themselves, be points of infection.

Pezicula Canker of Red Oak

Cause: *Pezicula cinnamomea* (DC.) Sacc.
Anamorph: *Cryptosporiopsis grisea* (Pers.) Petrak

The main external symptoms of this bark canker disease, which was elucidated as recently as 1981 [41], are the development of flask-shaped distortions of the stems and the death of large patches of bark. In places, the bark collapses inwards, cracks open, and finally crumbles away. Such necroses can extend tongue-like on one side of the stem from ground level up to a height of 2 m when drops of slimy liquid exude from above the affected parts. Attacks occur on Red oaks which lie in the age range 20–45 years, and whose susceptibility to disease is apparently increased by disturbances to the water economy. Such disturbances occur on certain types of soil or site as a result of dry periods.

Pezicula canker of Red oak is a perennating disease which can persist for many years without causing the premature death of the tree. This long term survival results from the inability of the fungus in many cases to overcome host resistance sufficiently to penetrate to the cambium. The strong resistance reaction of the tree becomes evident in the irregular formation of wound tissue at the edges of the lesions. Lesions where the fungus grows only in the outer layers of bark result in an increased rate of periderm development, forming a thickened 'pathological' bark.

The cause of the disease is the fungus *Pezicula cinnamomea*, a member of the Helotiales, which is distinguished by top- to disc-shaped, 0.5–1 mm, cinnamon-coloured apothecia. These occur in groups clustered on the surface of mostly 2-year-old bark necroses. Later these are superseded by saprotrophic species so that it is difficult to demonstrate the actual cause of the canker. At certain stages in its development, the associated imperfect fruiting state is formed by the fungus; this is distinguished by cushion-shaped fruit bodies, 0.5–1.0 mm in size, which tear open the epidermis raggedly when they burst through at maturity. In the initial phase, rod-shaped microconidia, measuring 8–9 × 1.5 μm, are found; subsequently, macroconidia are formed, 28–38 × 9–12 μm in size, almost cylindrical, and yellowish in colour (Fig. **80**).

There is no way of curing a diseased tree, since the fungus is no longer accessible once it has become established in the bark. And there may be little sense in a direct external attack on the fungus, as it is one of the agents involved in the natural pruning of Red oak. It is better to employ silvicultural methods

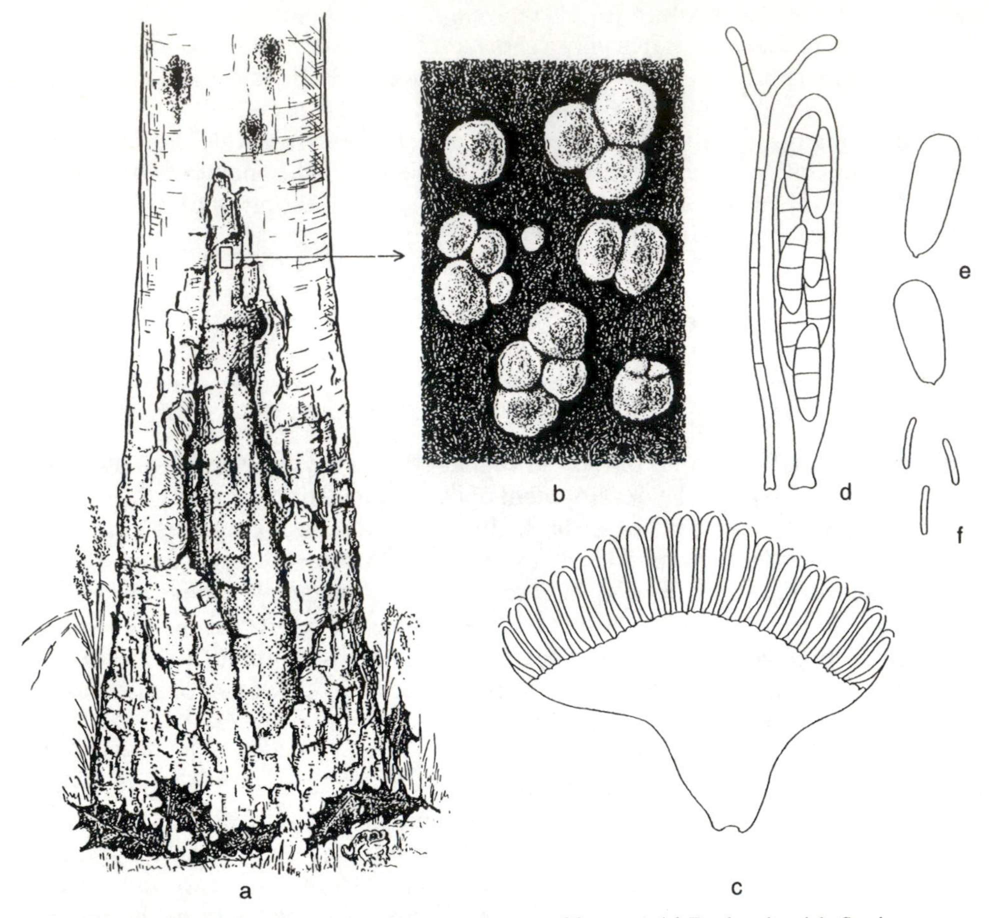

Fig. 80 *Pezicula cinnamomea*. **a** symptoms on a 30-year-old Red oak with fluxing, tarry spots (above), and bark necrosis (below), **b** fruit bodies (apothecia) on dead bark, **c** cross-section through an apothecium (diagrammatic), **d** ascus with ascospores adjacent to a paraphysis, **e** macroconidia, **f** microconidia

which guarantee vigorous growth. Thus, planting should be avoided on sandy soils at risk from drought. As for the utilization of the timber, badly affected trees should be removed promptly from the stand. If the care and maintenance of the countryside is the prime concern, these measures can be dispensed with, as even diseased Red oak trees remain alive for many years.

Similar symptoms of ill health can be caused by *Armillaria* spp. (when there is no canker-type resistance reaction by the tree) or by *Stereum rugosum* (when the typical fruit bodies will be present on the bark). If the cambium is discoloured and dies, the condition may be ink disease, caused by a *Phytophthora* species. In this genus of root-infecting fungi, *P. cinnamomi* is a major pathogen of oaks and other tree species, particularly in warmer regions such as southern France [56].

Related Species:

- *Pezicula livida* (Berk. & Broome) Rehm; fruit bodies are like those of *P. cinnamomea.* It causes bark necroses and girdling lesions on weakened young larch and pine plants and is saprotrophic on Silver fir and Norway spruce. The imperfect state is *Cryptosporiopsis abietina* Petrak, with macroconidia measuring 24–32 × 11–13 μm (Table **II, 18**).

Coral Spot

Cause: *Nectria cinnabarina* (Tode) Fr.
  Anamorph: *Tubercularia vulgaris* Tode

Coral spot occurs commonly in forests, on woody broadleaved plants in parks and gardens, nurseries, and recently planted avenue and street trees. The symptoms of the disease, which often appear in early summer, include the sickly appearance of current shoots on which the leaves wilt and soon wither. The bark of an affected branch is necrotic over part of its length, with the necrosis often spreading from the stub of a dead side branch or a pruning wound. The disease is most readily identified from the fruit bodies which break out from the bark during the winter months. At first the sporodochia appear in the form of waxy, soft, pinhead-sized warts, cinnabar red in damp weather, pale reddish in dry weather. When examined under the microscope large numbers of elliptical to cylindrical conidia, 5–7 × 2–3 μm in size, are found, abstricted both apically and laterally from long conidiophores. The associated perithecia appear in spring from the same positions; they are also red and contain asci in each of which 8 ascospores, 2-celled and measuring 14–18 × 6.5 μm, ripen (Fig. **81**).

The fungus is a wound parasite which penetrates the twigs or stem through bark injuries or branch stubs and there spreads out further both in the bark and in the wood. As the result of toxin production and vessel plugging, foliage wilts suddenly and the wood becomes stained a greenish to brownish colour.

Nectria cinnabarina occurs predominantly as a saprotroph on dying and cut branches. On plants damaged by frost or weakened by shortage of water it can, however, act as a parasite and by its rapid spread be very damaging. The risk of attack is high for young trees which have suffered major water loss before or during planting. Attacks occur most often on lime, Horse chestnut, elm, and hornbeam, and especially maples which include some particularly susceptible cultivated forms. In the case of limes, lethal attacks have been observed, particularly after the pruning of tall stems.

Among the control measures against coral spot that have been suggested, the greatest emphasis is put on various preventive practices. These include the provision of a balanced supply of nutrients and water for the plant. Even slight drying of the roots at transplanting can result in an increased liability to attack. As far as possible, tree pruning should be carried out in dry weather in summer, at which time further protection against infection can be achieved by the use of a wound protectant material. Infected plants should be cut back

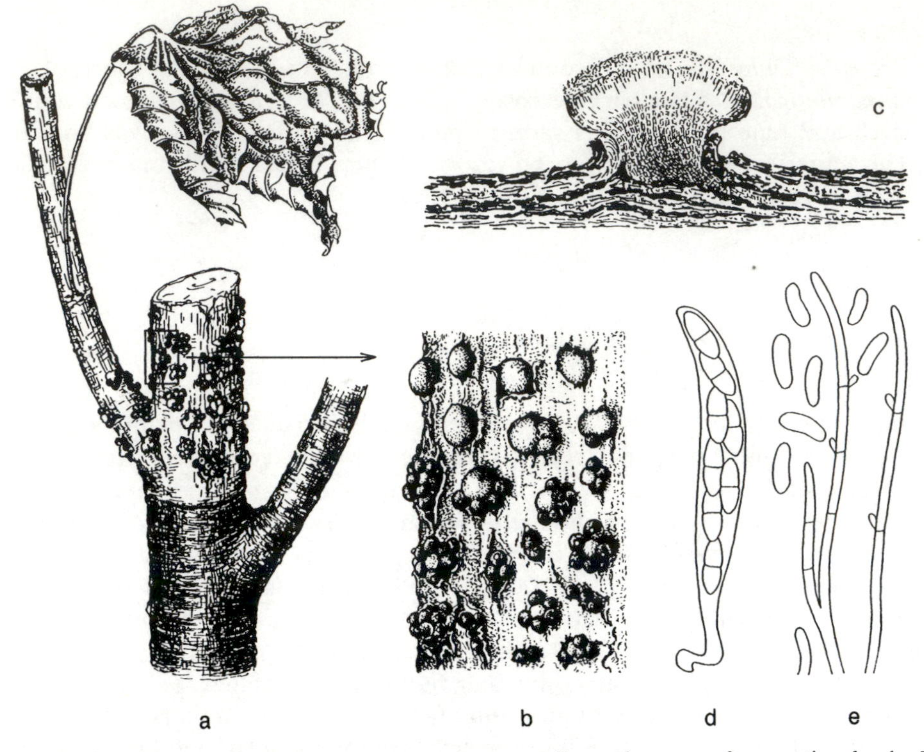

Fig. 81 *Nectria cinnabarina*. **a** symptoms on a small maple stem after cutting back, **b** fruit bodies of the imperfect state (above) and perfect state (below), **c** cross-section through a fruit body of the imperfect state, **d** ascus with ascospores, **e** conidiophores with conidia (c after Ferdinandsen and Jørgensen 1938/39, e after Booth 1959)

into sound wood. In nurseries, care must be taken to remove infected material from the site immediately. Regarding the use of fungicidal sprays, there is still too little known.

Coral spot is not infrequently the subject of refunds and requests for replacement of nursery stock. In such cases one should begin by understanding that fungal spores are always present and are ubiquitous, and in practice cannot be excluded. For this reason, the question is not: 'When was the plant colonized or infected?' but: 'At what time and as the result of what conditions or improper plant handling was the predisposition to disease brought about?'

Sooty Bark Disease of Sycamore

Cause: *Cryptostroma corticale* (Ell. & Ev.) Gregory & Waller

The characteristic feature of this disease, which occurs occasionally in England, France and a few other countries, is the spectacular flaking away of the periderm to reveal a brownish black, powdery spore mass within the bark tissues. This sporulation phase often occurs over much of the bark on the main stem, or is confined to axial strips. Although this is the most typical symptom, it occurs late in the disease process. An earlier symptom is the wilting of leaves followed by the dieback of individual branches or the entire crown. The woody tissue inside a wilted branch or main stem is stained dark yellow or green. Trees attacked in this way usually die within a year or two, but the infection is sometimes restricted to a discrete 'strip canker.'

Acer pseudoplatanus is by far the most commonly attacked species, though other species of maple can also be attacked. Members of the genera *Carya, Aesculus,* and *Tilia* are also recorded as hosts.

The cause of the disease is the hyphomycete fungus *Cryptostroma corticale* [83]. The fungus first invades the woody tissues and moves out to the cambium and bark. A black, extensive stroma is then formed within the bark and consists of a 'floor' layer and 'roof' layer which are linked by 1 mm high pillar-like strands. The spaces between are filled with the powdery spore masses. The conidia, which are formed on cylindrical sporophores, are ovoid, at first hyaline, later pale brown, and measure $4–6 \times 3.5–4$ μm [69]. The sooty brown spore masses are held together by a network of sticky, unbranched, capillitial threads. In human medicine the spores are known as a cause of asthma.

It is thought that *Cryptostroma corticale* is primarily saprotrophic and that it causes sooty bark disease only in association with certain environmental factors. Thus, to date, severe outbreaks have been seen only in years following particularly hot summers. Experimental work has confirmed that the growth of the fungus is encouraged by high temperatures and by water stress in the host [60].

Localized bark necrosis on *A. pseudoplatanus* can arise following dry summers in the absence of infection by *C. corticale*. Such lesions may contain weak fungal parasites which can cause further extension of the necrosis [84].

6 *Wilt Diseases*

Dutch Elm Disease

Cause: *Ophiostoma ulmi* (Buism.) Nannf.
or: *Ophiostoma novo-ulmi* Brasier

Dutch elm disease ranks as one of the most damaging of tree diseases, with many millions of elms having fallen victim to it during this century [32]. The elm populations of Europe, North America, and parts of Asia have been affected by two successive pandemics, the second and more devastating of which has two centres of origin; in the North American and Eurasian continents.

A characteristic symptom of the disease is the wilting and discoloration of the leaves during the growing season. The extent of the disease within the crown and its rate of development vary according to the aggressiveness of the form of the fungus that is involved, as well as the genetically and environmentally determined susceptibility of the host. Initial attack may be confined to a single branch, but in cases where one of the aggressive forms of the fungus (*O. novo-ulmi*) is involved, or where host susceptibility is very high, the entire tree frequently dies. Death may occur within as little as 2 months in trees which receive multiple infections of *O. novo-ulmi*. Trees which survive the first season may die within 2–5 years.

The internal symptom of Dutch elm disease in an affected branch is a dark discoloration of the spring-wood laid down in the year when infection took place. In cross-section, this discoloration appears as a brown stain, often in the form of discrete spots, within one or more annual rings. In tangential section, the brown discoloration of the springwood vessels appears in axially oriented streaks. If the branch has been killed, the cambium will also show discoloration (Fig. **82**).

The causal agents of this parasitic wilt disease, members of the ascomycete genus *Ophiostoma*, grow as mycelia and in a yeast-like phase within the sapwood, especially inside the vessel lumina, and secrete a wilt toxin (cerato-ulmin) which interferes with the water economy of the tree. This leads to tylosis formation in the vessels, normally a defensive response that mechanically blocks rapid microbial invasion, but which occurs to such an extent in this disease that it further hinders water transport.

Until recently, all forms of *Ophiostoma* causing Dutch elm disease were regarded as a single species, *Ophiostoma (Ceratocystis) ulmi* (Buism.) Nannf., a non-aggressive and an aggressive strain of which were later distinguished [81]. More recently, Brasier [25] has raised the aggressive strain to the rank of a species in its own right as *O. novo-ulmi* Brasier, on the basis of morphological, physiological, and molecular differences.

The non-aggressive, more weakly pathogenic *O. ulmi* is now believed to have been responsible for the first pandemic of the disease in Europe and

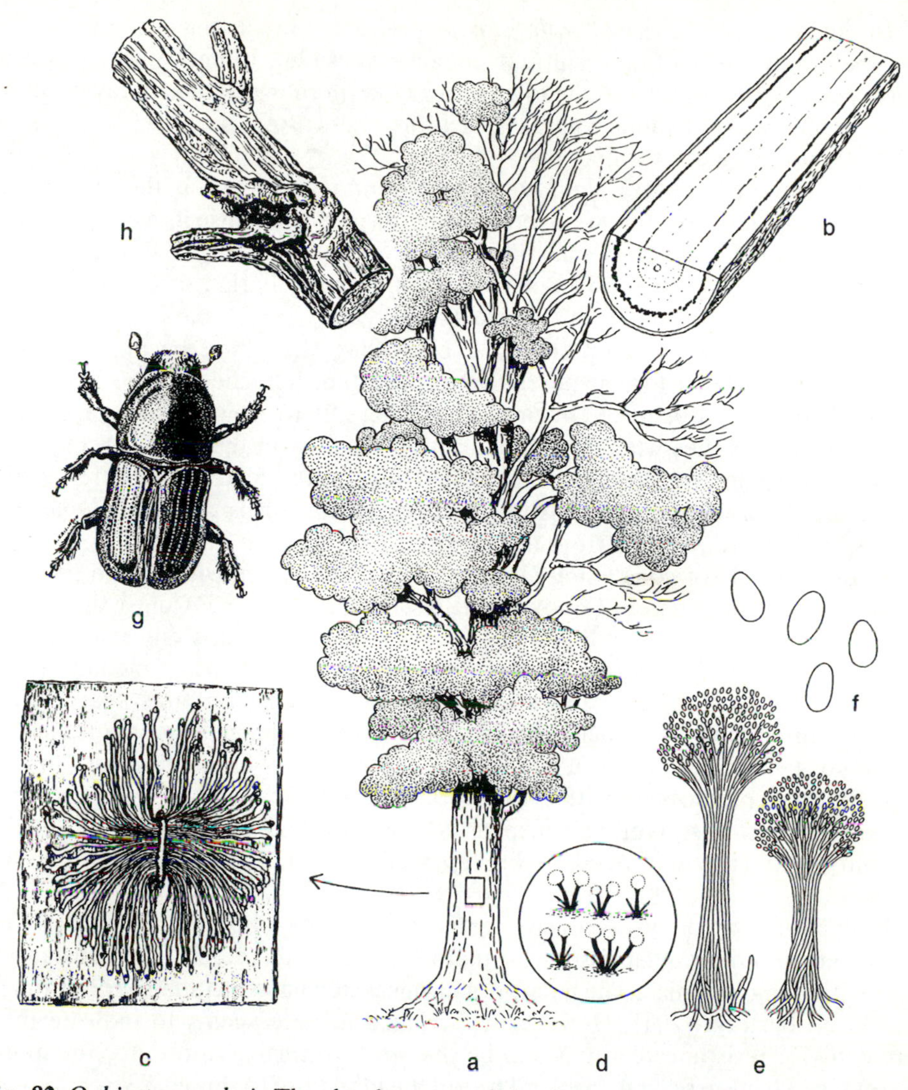

Fig. 82 *Ophiostoma ulmi*. The development of the fungus and the disease on elm; **a** general view of a diseased tree, **b** longitudinal and transverse section through an infected branch with typical staining, **c** larval galleries of the elm bark beetle with the *Graphium* state of the fungus (d and e) and conidia (f), **g** the small elm bark beetle, **h** old feeding damage at the base of a twig

North America in the 1920s–1940s. The highly pathogenic aggressive strain, or subgroup (now *O. novo-ulmi*), is responsible for the current second pandemic of the disease to which most types of elm that had some resistance or tolerance to *O. ulmi* have also fallen victim. It occurs in two forms; one that has invaded Europe from North America via England, and another spreading westward from a centre of origin somewhere in southeastern Europe or Asia [24].

In the laboratory, the two *Ophiostoma* species can be distinguished from one another on the basis of their cultural characteristics [26]. In nature, both species form peg-shaped coremia of the form genus *Graphium* which bear slimy droplets of large numbers of hyaline conidia at their tips. Less often, perithecia with necks 250–500 μm long are formed.

Like a number of other *Ophiostoma* and *Ceratocystis* species, the Dutch elm disease pathogens are remarkable for being dispersed by insect vectors; in this case by various species of elm bark beetle. After the fungus has killed the cambium, it begins to grows out into the still-living bark tissues. The dying branches and stems meanwhile attract egg-laying female bark beetles. The larval galleries that develop from the egg-laying sites then become invaded by the fungus which forms spores that can become attached to the bodies of the immature adult beetles. On emergence, these fly to twig crotches of, as yet, undamaged branches where their maturation feeding on the bark can result in spore transfer and hence xylem infection. These crotch feeding sites are mostly to be found on relatively thin twigs. The aggressive forms of *Ophiostoma* can also spread via root grafts from the diseased into a healthy tree.

Various means of controlling Dutch elm disease are known, though most of these are of limited value. For example, insecticidal spraying against the beetle vector has been attempted but is not always successful and is environmentally undesirable. A safer variant of this approach is the use of attractants (e.g. pheromones) to lure the beetles into insecticidal traps, but this shows little promise of effectively reducing populations. In trees that are not too severely attacked, the spread of the fungus can be controlled by fungicides, introduced into the stem or roots by various injection devices. However, this can become impracticable or prohibitively expensive if control is not effected early in an outbreak, and may need to be repeated, with cumulative injury to the sapwood.

Biological control, using micro-organisms that may be antagonistic to the pathogen within the host tissues, or that may induce increased host resistance, may offer some promise. The micro-organisms tested have included various fungi [178] and bacteria [204]. However, experience to date seems to indicate that breeding for resistance continues to be the most promising option for retaining the elm as a landscape and parkland tree in the long run. To this end, selections of individuals from native populations have been made, and hybrid elms have been bred, incorporating natural resistance derived from Asiatic elm species [203].

Oak Wilt

Cause: *Ceratocystis fagacearum* (Bretz) Hunt
  Anamorph: *Chalara quercina* Henry

This infectious disease, which so far is known only in North America, manifests itself during the growing season, causing the leaves to wilt and then to turn brown, mainly from the margins. Premature leaf fall also occurs and can involve leaves which have not discoloured. In Red oak, the symptoms can affect the entire crown, usually starting at the top; in white oaks, which include *Quercus robur* and *Q. petraea*, the damage is often confined to single branches.

The disease is induced by *Ceratocystis fagacearum* whose conidia are transported rapidly up through the stem in the transpiration stream of the outermost annual ring. As a defence reaction, tyloses are formed in the vessels which in turn disrupt the water-conducting system. After the death of the tree or branch, the fungus penetrates deeper into the sapwood and also out into the bark. Finally, spore-producing mycelial mats are formed on the cambium, by which fungus can be identified. The dispersal of the fungus is by spores which can be carried by certain species of beetle. Spread via root grafts can also occur.

To confirm a diagnosis it is usually necessary to isolate the fungus from infected wood and to identify it in culture. On malt agar it forms a brownish mycelium and conidiophores with characteristic cylindrical endoconidia, produced in chains and measuring 8–15 × 2–4 μm [211].

Red oaks are the principal species attacked, and these can be killed within a few weeks. White oaks are less often attacked, and they sometimes throw off the infection completely after several years. The potential danger to the cultivation of oaks in Europe has made it necessary to impose strict phytosanitary controls on the importation of oak timber from North America.

A quite distinct condition, which has been reported in central Europe through the 1980s is the so-called 'Eichensterben'—death of oak. This is characterized by the shedding of twigs, crown thinning, bark cracking, and longitudinal, brown discoloured strips in the bast and cambium. The precipitating factor is assumed to be severe winter frost in conjunction with predisposing drought years [98]. Similar damage has been observed earlier in eastern Europe and there too its occurrence is thought to be due to a complex of causes (extreme weather conditions, prolonged waterlogging, insect and fungal attacks).

Further Wilt Pathogens in the Genera Ceratocystis/Ophiostoma:

- *Ceratocystis coerulescens* (Münch) Bakshi causes both a blue stain in felled coniferous timber in Europe and a vascular disease in various maple species in North America, principally in *Acer saccharum* and *A. rubrum*; single branches or the whole tree may be killed.
- *Ceratocystis fimbriata* (Ell. & Halsted) Davidson f. *platani* Walter is the cause of canker stain of plane ('Platanenkrebs') (see p. 126). It causes wilting only at certain stages of the disease.
- *Ceratocystis laricicola* Redfern & Minter causes the dieback and death of *Larix decidua* by killing the bark and cambium; *Ips cembrae* is a vector. Attacks seem to be frequently associated with drought or other forms of stress [161].
- *Ophiostoma polonicum* Siem. causes dieback in several species of spruce, mainly *Picea abies*; it is spread by the spruce bark beetle, *Ips typographus*, among others [49].
- *Ophiostoma wageneri* (Goheen & Cobb) Harrington (anamorph: *Leptographium wageneri*) is the cause of a brown-black stain in the roots and

sapwood of Douglas fir and other conifers. It causes wilting, chlorosis, and needle-cast; distribution is at present only in North America [95].

Verticillium Wilt

Cause: *Verticillium albo-atrum* Reinke & Berth.
 or: *Verticillium dahliae* Kleb.

Verticillium wilt of woody plants is exclusively a problem of nurseries, gardens, and park and street trees. The most striking symptom is initially the wilting of leaves and shoot tips on scattered branches which can later die completely. In the sapwood, greenish brown stains are present which appear as patches or spots in cross-section, often following the outline of annual rings (Fig. **83a**). Microscopic examination reveals large numbers of hyphae, mainly in the springwood vessels, where they block water conduction as well as secreting wilt toxins. Susceptible tree species or seedlings can be killed in the first year of infection; in older or less susceptible plants a chronic form of the disease usually develops in which sparse foliage and the dieback of progressively more branches are typical.

For a positive diagnosis of Verticillium wilt, the causal fungus must be isolated from freshly diseased wood. On malt agar both species of *Verticillium* initially form a light-coloured mycelium which bears characteristic whorls of phialides from whose tips ovoid to ellipsoid conidia are abstricted. The commoner *V. dahliae* (Fig. **83b,c**) also forms dark-coloured, grape-like microsclerotia which eventually give the culture a greyish black appearance. Such microscelorotia are not formed

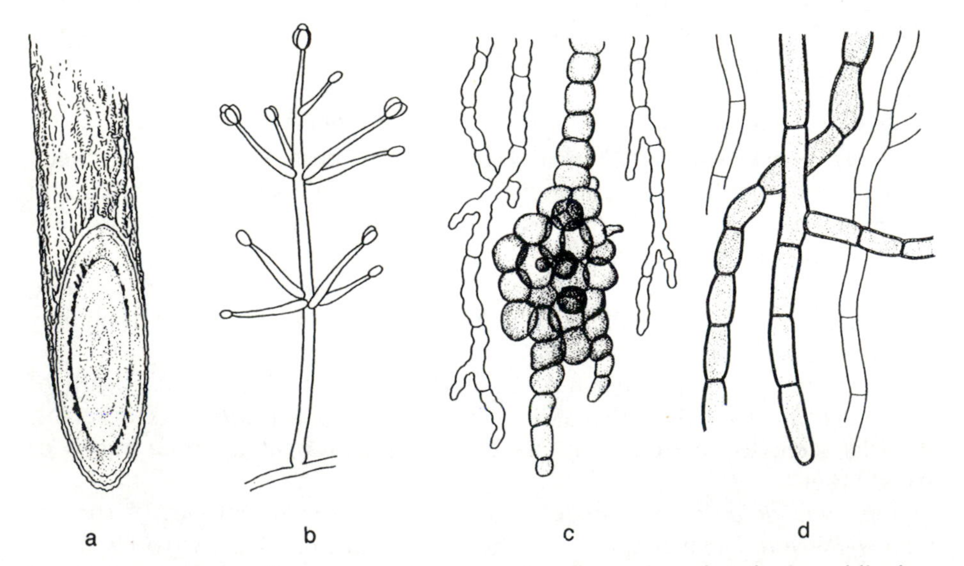

Fig. 83 Verticillium wilt. **a** cross-section through a diseased maple twig; **b** conidiophores of *Verticillium dahliae*, **c** sclerotium from a culture; **d** resting cells and normal hyphae of *V. albo-atrum* from a culture

by *V. albo-atrum* (Fig. **83d**) in culture, and the colonies remain white or appear grey only in places; instead, older cultures form thickened, dark brown hyphae which function as resting cells. Physiologically, the two species can be separated on the basis of their differing temperature requirements: *V. dahliae* still exhibits good growth at 30°C, while at this temperature *V. albo-atrum* ceases to grow.

Both *Verticillium* species comprise host-specific races and morphological varieties which occur throughout the world on a total of 270 different plant species. Of the more common woodland and park trees, the most frequent hosts are maples, ash, Sweet chestnut, limes, robinia, and rowan. Among the woody ornamentals, catalpa, Judas tree, Venetian sumach, Stag's-horn sumach, and almond are very susceptible. Infection takes place via wounds which arise when, for example, branches and roots are cut with pruning shears or by machines which have come into contact with infected tissue or with contaminated plant residues in the soil, on which the causal fungi can survive for years. Hygienic working practices and the avoidance of wound creation are therefore the most effective prerequisites for the avoidance of disease.

Curative treatment, though fundamentally difficult, has sometimes met with success in the case of solitary plants which have been treated with systemic fungicides based on benzimidazole. These can either be watered on or injected into the soil. Palliative treatment, such as copious watering in dry weather or adding organic materials (e.g. sawdust) which deplete free soil nitrogen has also been successfully attempted. New plantings should not be undertaken on ground which has previously carried *Verticillium*-susceptible plants [158].

Watermark Disease of Willows

Cause: *Erwinia salicis* (Day) Chester
Syn. *Pseudomonas saliciperda* Lindeijer

The cause of this wilt disease is the bacterium *Erwinia salicis*, which attacks a number of *Salix* species, the principal hosts being *S. alba* var. *coerulea*, *S. caprea*, *S. cinerea*, *S. fragilis*, *S. purpurea*, and *S. triandra*. Losses occur in amenity plantings and commercial crops as well as amongst wild trees. The most significant losses have been noted from the Netherlands and from England [148] where, because of its effect in reducing timber strength, the disease has serious consequences for the growing of *S. alba* var. *coerulea* for cricket bat manufacture.

The early symptoms on trees of any age are the wilting of leaves and young shoots from May to July. The wilted leaves may show a reddish discoloration. Young trees can be killed in the first year of attack. In older trees, the disease is at first restricted to single branches which soon die and dry up and for some time remain prominent in the green crown. After some years the main stem can become affected as well so that older trees also die in time. Another indication of the disease are dark patches and streaks which can be seen on the cross-section of affected branches. These can merge to form larger coloured rings or patches from which the disease gets its English name (Fig. **84**).

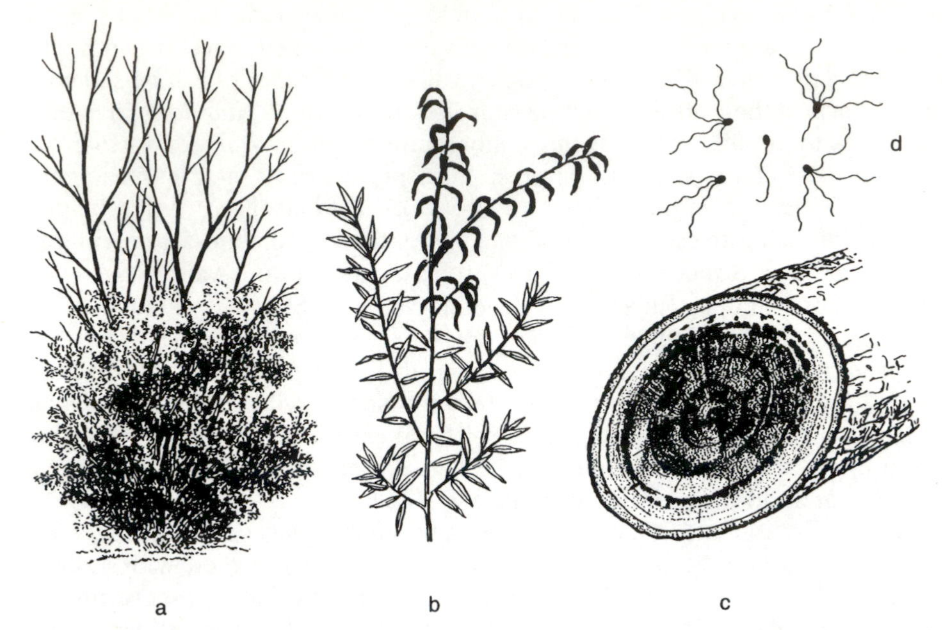

Fig. 84 Watermark disease of willow. **a** general view of a diseased *Salix alba*, **b** early stage showing the wilting symptom, **c** stem cross-section with spots arranged in rings, **d** the flagellate bacterium *Erwinia salicis* (c after Lindeijer 1932, d after Gremmen and de Kam 1970)

If the discoloured wood is examined under the microscope, large numbers of bacteria can be found in the partly occluded vessels. By means of special staining methods, their peritrichous flagella can be rendered visible. As *Erwinia salicis* also lives in the vessels of healthy willows and even as an epiphyte on the leaves, it has been suggested that the typical wilt syndrome is only expressed if the bacterial infection is accompanied by one or more predisposing factors.

Spread of the causal bacterium is effected by wind, rain, and probably also by insects. Spread over greater distances is by infected propagation stock. Therefore, the disease is controlled by the maintenance of disease-free propagating beds and by the felling and destruction of infected trees.

7 *Wood Damage in the Standing Tree*

ABIOTIC CAUSES OF DESTRUCTION AND STRUCTURAL DAMAGE

Forest fires can be one of the greatest causes of loss in forests since, at their most severe, they can wipe out entire stands by spreading through the forest canopy. Certain years come to be recognized as 'forest fire years' when persistently hot, dry weather brings an exceptionally high risk of damage. Information on fire risk factors and on prevention and control can be found in the specialist literature.

Storm damage can take the form of 'windthrow,' when the whole tree is uprooted or of 'wind snap,' when the root anchorage is relatively firm, but the stem breaks under wind pressure which exceeds its bending strength. Windthrow, which progresses through a stand due to restricted rooting, is said to be 'endemic,' while extensive windthrow due to severe storms, regardless of site conditions is described as 'catastrophic'. Storm damage destabilizes the forest ecosystem, sometimes favoring the build-up of pests, especially bark beetles that can harm surviving trees or logs awaiting extraction in coniferous stands [227].

Breakage due to ice or snow can occur when excessive accumulations build up in extreme winter weather. In the case of ice, accumulation occurs when supercooled rain freezes on contact with the tree. The weight of ice or snow can break leading shoots as well as branches. Pines, spruces, Silver fir, Cedar of Lebanon, and beech are particularly at risk.

In a **weak fork failure**, an acute-angled branch or a co-dominant stem with included bark at its base, splits away from the tree. In a normal more open crotch, the adjoining members are strongly united by wood, whereas in an acute-angled crotch there may be a bark-to-bark contact which lacks strength. This type of weak junction can be recognized by the incomplete formation of the 'branch bark ridge'. This ridge is a feature of normal branch junctions, being formed in a zone where the bark is forced to grow upward and outward. The tendency of weak forks to break can be a major safety hazard, particularly in streets and parks.

Lightning damage occurs mainly on single trees which protrude above the canopy, which stand on open ground or for some other reason are exposed to discharges of lightning. The damage may involve the breakage or splintering of stems or the vertical tearing apart of the sapwood. Another form of lightning damage can occur when whole groups of trees are killed by conduction of the electric charge to the roots.

False frost cracks are radial cracks which develop progressively from the interior of the stem, being initiated either at sites of old cambial injury or at pockets of fungal heart rot. It is possible that bacteria are also involved in their initiation. The internal cracks are recognized as 'false frost cracks'

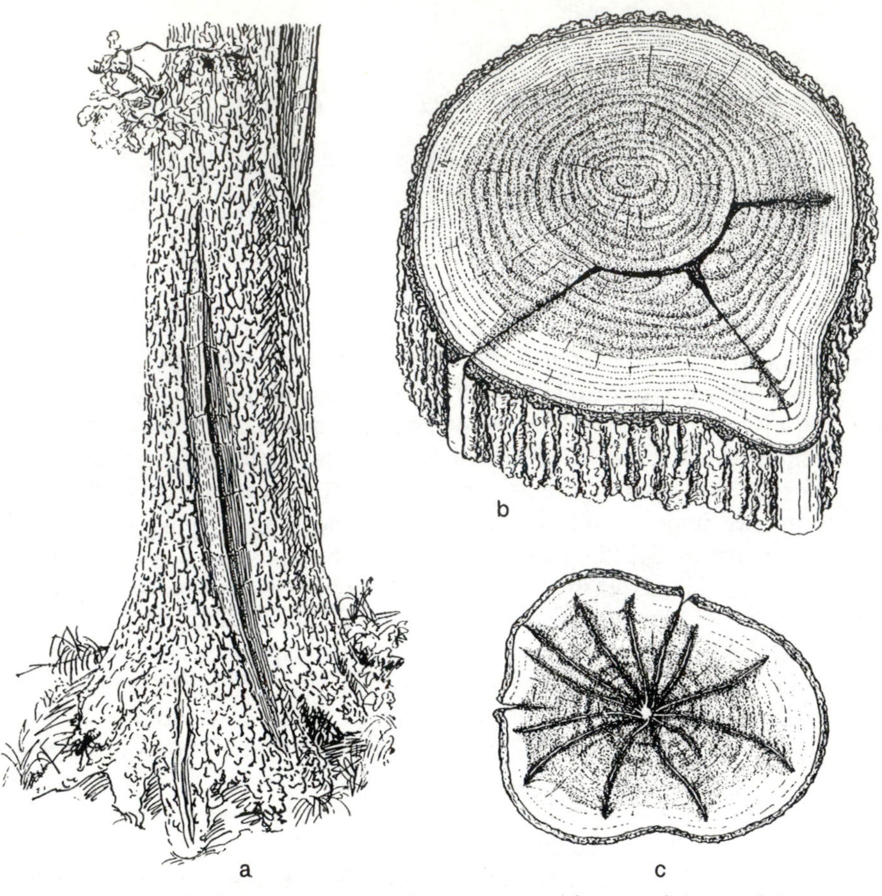

Fig. 85 Stem cracks in oak. **a** external symptoms with several 'seams,' **b** stem cross-section with an old cambial injury as the point of origin of stem cracks (wound-related cracks), **c** star-shake in Red oak extending from a central area of decay

when—perhaps after many years—they eventually extend out to the bark surface. After this happens, they are usually occluded by rolls of callus growth, but these can split open again in cold winters—often making a loud report. In regions where very cold conditions tend to recur, repeated splitting and callusing eventually result in the formation of prominent lip-like wound ridges or 'seams' (formerly called frost ridges), which can reach several metres in length (Fig. **85**). Radial cracks of this kind are common and of economic importance in oak, but they are found also in ash, elm, poplar, and Silver fir. The development of false frost cracks can be prevented by avoiding the breakage of branches and injury to the cambium [190]. True frost cracks involve only the bark, and the wood remains unharmed at the time that the damage takes place (cf. p. 101).

Ring shakes are extensive circumferential splits between adjacent annual rings or within a single ring. Complete separation of two wood surfaces often takes place only later when the stem is cut and dries out. This type of separation has been attributed to the presence of a structurally weak layer of mainly parenchymatous cells that has formed within a annual ring in the place of normal wood cells. Such cell layers are formed by the cambium in a defense reaction following injury and vary in extent according to the severity of the injury, sometimes occupying the entire circumference of the affected annual ring. These layers can be recognized in the timber of broadleaved trees at an early stage when they have not yet split into ring shakes, appearing as darkened rings or arcs in a cross-section of the stem.

Drought cracks arise at weak points within the wood during periods of extreme moisture tension during drought. The cracks extend secondarily to the bark, but then become occluded within one year, so that the underlying persistent cracks in the wood are hidden (cf. p. 102).

Scars in the wood are mostly the remains of old, occluded cambial injuries. For example, crosscut beech stems sometimes show T-shaped defects that contain included bark, indicating the possible sites of former beech bark disease lesions.

WOOD DISCOLORATIONS

Colour patterns in wood can result from normal processes such as heartwood formation, but there are also types of abnormal 'pathological' discoloration that occur only after particular events. In such cases, the discoloration may be at least partly due to the reaction of the tree itself, but it can also be directly caused by external factors. These may include invasion by micro-organisms such as fungi, but abiotic factors may be primarily responsible. The most frequent and important of these are mentioned under the following headings.

The general term '**protection wood**' has been applied to physically and chemically altered wood that is more durable than normal sapwood. In some species, the alteration can occur with the normal aging of sapwood, whereby it is converted into heartwood. If sapwood is injured before this transformation is due to occur, it can locally form a 'reaction zone' which is usually dark in colour. In broadleaves, e.g. beech or oak, the darkening is due to the formation of tyloses or gums in the vessels and to the deposition of heartwood substances. Protection wood may also develop within the new annual ring that is laid down after the injury has occurred, due to altered cambial activity around the affected site. This takes the form of a layer of traumatic xylem or 'barrier zone.' The reaction zone in the existing wood and barrier zone wall off ('compartmentalize') the damaged parts from the still-sound neighbouring tissues and protect them from the ingress of air, from water loss, and finally also from penetration by potentially harmful micro-organisms.

False heartwood—also called facultative heartwood—is the name given to a central column of dark wood in one of those woody species which do not normally produce a dark heartwood. As in reaction zones, tyloses or gums block the vessels, and brownish heartwood substances are deposited. The tendency to form false heartwood is species specific [192]. Two examples of species which produce facultative heartwood are beech, which does not normally form a distinct heartwood, and ash, in which the heartwood is normally pale. Factors which trigger the formation of false heartwood are thought to be the breakage of branches or injuries to the bark. Air then penetrates the adjacent wood or vessels, setting in train a genetically determined program of chemical changes peculiar to the species of tree.

The most familiar example of false heartwood occurs in beech, being known as 'red-heart.' It appears in cross-section either in the form of cloud-like concentric rings of different tints—'Wolkenkern' in German—(Fig. **86b**) or an irregular star-shaped pattern—'Spritzkern'— in which the points of the stained area are often associated with cracks in the wood (Fig. **86c**). When examined under the microscope, the dark zones are seen to be rich in vessel tyloses, while the paler areas between are only slightly altered. On drying, the discoloration becomes markedly paler. The triggers and starting points for false heartwood formation are broken branches, pruning cuts made too close to the stem, or other large areas of bark damage which impair the tree's ability to contain the zone of xylem dysfunction and thus allow the stain to extend over several years. The so-called 'Frostkern' (frost heart), which may be triggered off by particularly low winter temperatures, is said to develop more quickly.

Pathologically discoloured wood tends to be less resistant to microbial invasion than naturally formed heartwood. If invasion occurs, the micro-organisms

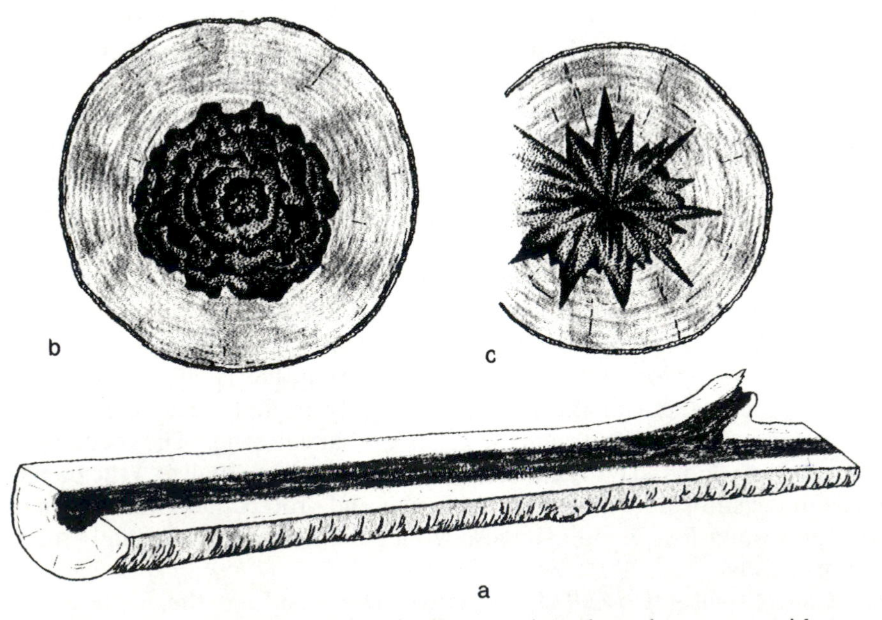

Fig. 86 Red heart of beech. **a** longitudinal section through a stem with central staining extending from a broken branch, **b** stem cross-section showing 'Wolkenkern' (cloudy-heart), **c** cross section with 'Spritzkern' (Splash-heart)

initially responsible are mainly deuteromycete fungi (e.g. *Phialophora* spp.) which utilize the cell contents but cause little or no breakdown of the cell walls. Their presence is often associated with an intensification of the wood discoloration (microbially induced wood stain). In the last phase of the succession, wood-decomposing basidiomycetes appear and enzymically break down the wood, a process which is also associated with a change in the wood colour [187].

Similarly, **wet heartwood** can be regarded as a form of wood stain that can develop without damaging the structure or quality of the wood. It occurs as a normal phenomenon in various tree species, initially developing in the central heartwood at the stem base as a regularly shaped wet zone, uniformly brown in colour. In contrast, 'pathological' wet heartwood appears uneven in cross-section, both in shape and in its brown coloration, and spills over into the functional sapwood, where it can impair water conducting capacity. When freshly exposed, it has an unpleasant sour smell due to the presence of bacteria. The trees most commonly affected are poplar, elm, and willow. Among the forest trees, Silver fir is particularly liable to develop wet heartwood, characterized by a rich bacterial flora, and is one of those species in which the sapwood may be affected. It is possible that this abnormality, whose cause is still largely unexplained, plays a part in the pathogenesis of Silver-fir decline [184].

Wood staining caused by fungi in the interior of the standing stem is mostly associated with the development of decay, although staining near wounds may be caused by non-decay organisms. The decay may either involve the breakdown of both lignin and cellulose, leading to a 'white rot,' or of cellulose alone, leading to a 'brown rot.' During the development of wood rotting fungi, further specific colour reactions in the wood can occur (e.g. red streaking from *Stereum sanguinolentum*, or a violet stain from *Heterobasidion annosum*) which will be discussed in more detail under 'Wood decay.'

WOOD DECAY

Wood decay is caused mainly by specialized fungi, and involves a number of phases, most of which continue over a number of years. Ecologically, it is important in the natural cycling of nutrients and in providing habitats for a very large number of animal species and micro-organisms. For the forester, wood decay represents an economic loss while, for the general public, the resulting risk of structural failure can be a considerable hazard to life and property as well as perhaps reducing the aesthetic value of trees. It is therefore vital that all trees in places frequented by the public, especially streets and parks, should be regularly inspected and maintained as necessary.

The **detection of decay** in trees requires the recognition of various external signs and the use of special devices. If undetected at a early stage, decay may progress until the tree has to be felled or is blown over. In some cases, the first external indication of the presence of decay is the appearance of fruit bodies of the causal fungus on whichever part of the tree is affected. Their precise identification may be important in the prognosis of decay development or the risk of breakage. In cases where fruit bodies are not present, other clues can sometimes be used. These may include the presence of externally visible decayed wood on old wounds, a deterioration in crown condition (density, colour or shoot extension), or the cracking and subsidence of soil around an unstable shifting root-plate.

If external signs indicate that decay may be present, it is important to determine whether it is extensive enough to justify remedial action. A long-established method is the examination of wood cores extracted with an increment (Pressler) borer. Unfortunately, the hole created by such an instrument, together with the surrounding zone of crushed wood cells, may provide an avenue for the spread of the decay fungi across the natural defensive barriers previously laid down by the tree. The risk may be slightly less if a non-compressive method is used to create a sample hole. Such a hole can be optically scanned by an endoscope.

It is usually preferable to use a method of locating decay that involves minimal injury. There are two types of instrument that can achieve this by using very narrow holes. The first type detects zones of altered electrical resistance (e.g. the 'Shigometer' and the 'Konditiometer,' the first of which uses a particularly narrow hole). The other type measures the penetrability of the wood, either by recording the resistance to the entry of a special drill bit, or by forcing a metal rod into a pre-drilled hole. Finally, gamma ray computer tomography involves no drilling at all and can produce accurate images of decay columns [91].

Most wood decay fungi begin their attacks at wounds which may be on the main stem, in the crown, or on roots, although a few of the root-infecting species can enter the unwounded surface directly. Decay is especially favoured beneath wounds that cannot be quickly occluded, such as broken branch stubs [188]. Pruning can also lead to the development of extensive decay, especially if it is carried out improperly. Basically, bark damage of any kind that involves the cambium is a prerequisite for infection by wood rotting fungi.

The **type of wood decay** is determined largely by the enzymic armoury and physiological tolerances of the fungal species concerned. The two main types that occur in living trees are white and brown rots. Brown rot fungi gain their carbon and their energy source by metabolizing sugars derived from the progressive breakdown of the cellulose chain molecule; a process that involves cellulase enzymes. These fungi leave the lignin largely unchanged. In contrast, white rot fungi break down the lignin, as well as the cellulose and hemicelluloses at a later stage. The enzymes involved in the sequence include phenoloxidases, cellulases, and hemicellulases, which are secreted mostly as ectoenzymes from the hyphae.

Wood attacked by brown rot fungi cracks and becomes crumbly as the cellulose breaks down. Eventually it breaks up into rectangular blocks when the cellulose is completely depolymerized. In the case of the white rots, the wood shrinks more evenly as the lignin is lost and does not begin to crack until a very late stage. A variant of the white rots is 'white pocket rot' (also called 'Wabenfäule' in German—honeycomb rot), which is characteristic of certain fungi (e.g. *Phellinus pini*). In a rot of this type the lignin is broken down unevenly so that numerous small, elongated cavities are formed in which the cellulose persists. The red rot

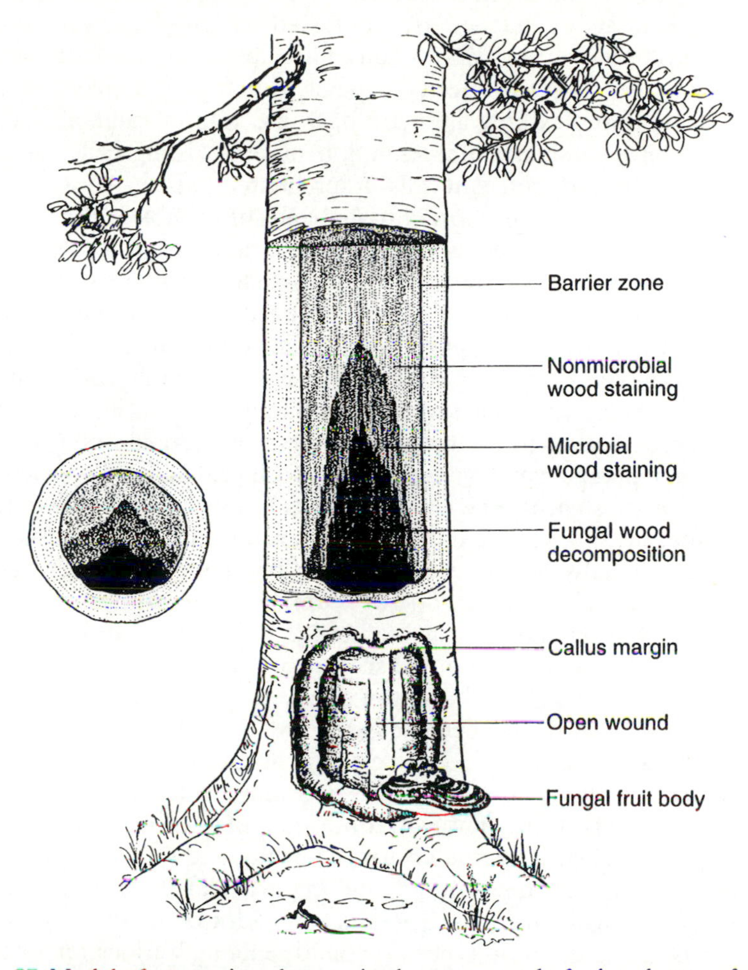

Fig. 87 Model of successive changes in the stem wood of a beech tree after prior injury to the bark (after Shigo 1979)

('Rotfäule') of Continental authors is also a white rot according to its chemistry, but the initially pale coloration is concealed by other color reactions.

There is another type of wood decomposition, known as soft rot, that is caused by some ascomycetes and deuteromycetes. It can be distinguished from the classical white and brown rots, although the appearance of the affected wood is very typical of a brown rot. A distinctive feature is that the rot takes place at or near the surface of the wood, where it causes a marked softening. The microscopic characteristics of this kind of decay are tunnel-like, oval, or diamond-shaped cavities in the secondary walls of the cells of the late wood. Soft rots are thought to be of only minor importance in the stems of standing trees, but they are economically significant in stored and structural timber.

Trees can resist microbial invasion of wood through a variety of defensive systems, most of which are non-specific and also protect against abiotic damaging agents. Some components of defence are conferred by pre-existing features of the wood, contributing to its natural durability; an attribute that varies with the species of tree. These are termed 'passive' defences, unlike 'active' defences which come into play only after a fungal infection or cambial injury. Active defences involve either the impregnation of pre-existing cells, or the production of new, particularly resistant cells. Impregnated cells can be seen on a cross-section of the stem, where they form brown bands of variable width. These are physical and biochemical barriers which, like static defences, can prevent or retard microbial invasion from injured or decayed wood into the surrounding sound tissues. However, both kinds of defensive barrier can be enzymically detoxified and penetrated by certain phenoloxidase-secreting fungi.

The most effective type of defensive boundary is the 'barrier zone' which develops within the annual ring that forms immediately after an injury to the cambium [191]. In broadleaved trees, this layer differs from normal woody tissue in consisting largely of parenchymatous cells loaded with the chemical precursors of antifungal phenolic substances. The walls of such cells may also become partly suberized. In coniferous species, the barrier zone is rather different, consisting of wood that is exceptionally rich in resin canals—a phenomenon that can also be observed even in tree species which do not normally have resin canals, such as the firs. The term 'barrier zone' was coined in the literature as an indication of its ability in most cases to confine any discoloration and decay to the wood present at the time of injury [210]. This removes any direct threat to the life of the tree, but decay can nevertheless become extensive in the pre-existing wood, leading to a risk of mechanical failure and consequent safety hazards.

Arboricultural treatments are nowadays being ever more widely practised, particularly on urban trees. There are, however, some areas of uncertainty about the long-term consequences of such treatments for the health and safety of the tree. This is especially the case with tree surgery, and in particular the use of wound treatments to prevent fungal infections. There is a need for better evaluation of such treatments, including different types of wound sealant. Various textbooks provide information on the present state of knowledge regarding practical tree care [13,189].

DECAY AGENTS

The principal agents of decay in the stem wood of living trees are fungi, nearly all of which belong to the Orders Aphyllophorales and Agaricales of the Basidiomycotina [102]. The few exceptions are ascomycete fungi. There are three main groups within the Aphyllophorales, which differ in the structure of their fruit bodies (basidiocarps). Members of the Stereaceae have thin fruit bodies, appressed to the substrate, and often leathery and persistent. Their spore-bearing surface (hymenium) comprises a simple external layer. The basidiocarps of the Sparassidaceae are stipitate, while those of the Polyporaceae are mainly bracket- or hoof-like, with the hymenium lining the surfaces of vertical tubes which emerge on the underside as rounded, daedaleoid, or lamellate pores. The Order Agaricales are distinguished by fleshy, mostly stipitate basidiocarps bearing hymenium-covered lamellae on the undersides.

The characters of the fruit bodies are the most important features for the identification of genera and species [28,29,31,70,85,110]. If fruit bodies are absent, decay fungi can sometimes be isolated from the wood and identified by cultivation of fruit bodies in the laboratory [209], or by carrying out enzyme tests on cultures [116], sometimes in combination with morphological characters of the mycelium and growth rate determination [201].

One way of categorizing decay fungi is to divide them into root rotting and stem rotting species, according to the location of the initial decay. Decay fungi that initiate their attack in the roots can sometimes invade large volumes of wood and some of them (e.g. *Heterobasidion annosum*) can spread into the stem, although others are confined to the roots (e.g. *Rhizina undulata*). The stem rots begin in aerial stem wounds or zones of dieback. Differentiating rots in this way is not without relevance to their control since this is more feasible for stem rotting fungi, whose attacks can often be avoided by preventive maintenance. Failing this, existing attacks on branches can sometimes be arrested by surgery. In contrast, direct treatment of root decay is possible only with difficulty and should preferably be restricted to methods which reinforce the defences of the tree (e.g. specific fertilizer applications, irrigation etc.). These methods are of course feasible only for trees in streets and parks.

Root Decay Fungi

Rhizina undulata Fr.

This fungus, a member of the Pezizales, is a specific root parasite which is capable of attacking both seedlings (in the case of pine) and also older trees, particularly 20–30 year-old stands of *Picea sitchensis* and *Pinus nigra*. Less often, *R. undulata* occurs in young stands of *Larix decidua*, *Picea abies,* and *Pinus sylvestris*. Affected trees become apparent by the failure of their newly formed shoots to extend fully and the yellow discoloration of their needles. Death eventually follows. The typical pattern of attack in a stand takes the form of a more or less circular group of dying trees which increases in size as time goes on. As such a phenomenon can have other causes, diagnosis requires the detection of the fruit bodies of the fungus, or its yellowish mycelial strands which can be found in the rooting area of diseased trees and on the surface of the killed roots [155].

The fruit bodies, which are inconspicuous and hemispherically domed, appear on the soil surface in summer, always at the periphery of an affected area. They are chestnut to blackish brown, 3–8 cm across, and edged with a white border. Their connection with the subterranean mycelium is by whitish, later brownish, strands that extend from their hollow undersides. The upper surface is occupied almost entirely by the hymenium, which is composed of paraphyses and closely packed asci, each with 8 hyaline, spindle-shaped ascospores, $30-40 \times 8-10$ μm in size (Fig. **88**).

The ecological peculiarity of the fungus lies in its close association with

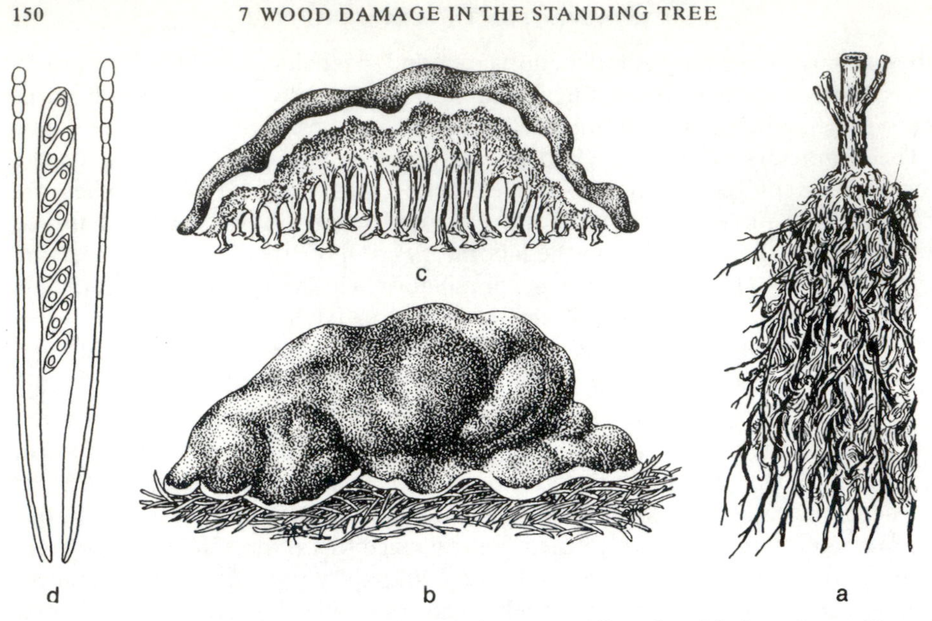

Fig. 88 *Rhizina undulata*. **a** infected root of a young Silver fir with fungal mycelium, **b** side view of fruit body, **c** cross-section of fruit body, **d** asci with ascospores and paraphyses (a after Hartig 1900; b, c, d from Butin and Kappich 1980)

fire sites from which it spreads, earning it the name 'pine fire fungus.' The spores lie dormant in the ground until they are heated for a short period to a temperature of between 38 and 45°C [111]. Providing that the soil is acidic, the resulting mycelium can spread saprotrophically on coniferous roots and may thus encounter living roots of susceptible host plants. Initially, the mycelium makes intimate contact with the roots and subsequently infects them, causing root rot. One to two years elapse between the germination of spores and the appearance of the first disease symptoms. The fungus requires the same length of time to form its fruit bodies.

The surest way of preventing 'group dying' follows from the biology of the causal fungus: fires of any sort on the ground should be strictly avoided, particularly in stands of Sitka spruce and *Pinus nigra*. This also applies to areas which are to be replanted with conifers and where there is often a desire to clear the site by burning 'lop and top.' Once the fungus has gained a foothold, further spread can be prevented by excavating ditches. However, this is only likely to be effective in the early stages of disease development, and when the dying groups are still small.

Heterobasidion annosum (Fr.) Bref.

Syn: *Fomes annosus* (Fr.) Cooke

The disease caused by this fungus, in forestry the most important of the fungal pathogens, is commonly referred to as Fomes root rot. It can cause considerable damage, principally in coniferous stands [85]. This damage takes two forms; the killing of roots and the decay of the wood in the form of a white rot extending up into the stem (butt rot). The resulting loss of timber value is economically even more serious than the root rot, but it should be noted that only about 70% of butt rots are caused by this fungus, the remainder being brought about by other basidiomycetes with a similar life style. These include *Armillaria* species, *Coniophora puteana* (Schum.:Fr.) Karsten, *Resinicium bicolor* (Alb. & Schwein.) Parm. and *Serpula himantioides* (Fr.:Fr.) Karsten.

Heterobasidion annosum is actively parasitic in the early stages of its attack, causing a root rot which may be severe, and which, in pines, leads to the death of a large proportion of the roots. In the spruces, the activity of the fungus soon becomes confined to the interior of the roots without very much affecting the vigour of the tree.

The fungus enters a largely saprotrophic phase as it penetrates the heartwood or the relatively inactive oldest zones of sapwood. The effect on the tree can vary with the tree species. In the very resinous Scots pine, the spread of the fungus into the stem is insignificant, although the trees may die as a result of the parasitic attack on the roots. In larch the mycelium develops in the heartwood/sapwood zone, and it has been reported from some forests that the stem decay may reach a considerable height, although in other instances its extent is comparable with that in Scots pine. In the spruces, the fungus can eventually spread in the mature wood up into the crown. Finally, it has been noted that 'Fomes' decay in the stem wood of Douglas fir can also be extensive [196], but this tends not to occur in trees under 80 years of age. The infected wood at first shows a grey to violet streaking, then reddish brown patches of decay containing small, scattered, white, spindle-shaped pockets in which black flecks later develop. Later, the wood disintegrates into a fibrous decay and is later completely destroyed by other organisms so that the interior of the stem becomes hollow.

The fruit bodies are the surest sign of an attack by *H. annosum*. They are found at the base of the stem or on superficial roots, sometimes covered with needle litter. The perennial brackets are mainly 10–20 cm across, though sometimes ranging between 2 and 40 cm, with a brown, corrugated, often zonate, leathery upper surface and a creamy white undersurface which is pierced with fine, closely set pores. In addition to the basidiospores which are abstricted from the basidia, conidia of the form genus *Oedocephalum* are produced by the fungus on moist wood. These can also contribute to the spread of the fungus (Fig. **89**).

In culture, isolates of *H. annosum* show considerable differences in mycelial growth, spore size, and in certain physiological characters. In addition to this, various ecotypes have also been identified in the field, linked to certain tree species [119]. However, little is yet known of the pathological significance of this variation.

Infection of the tree takes place almost exclusively in the roots, either from spores which are washed into the upper layer of the soil and germinate on the root, or by root contacts with a diseased tree or thinning stump. Thinning stumps may be infected directly by spores landing on the cut surfaces and

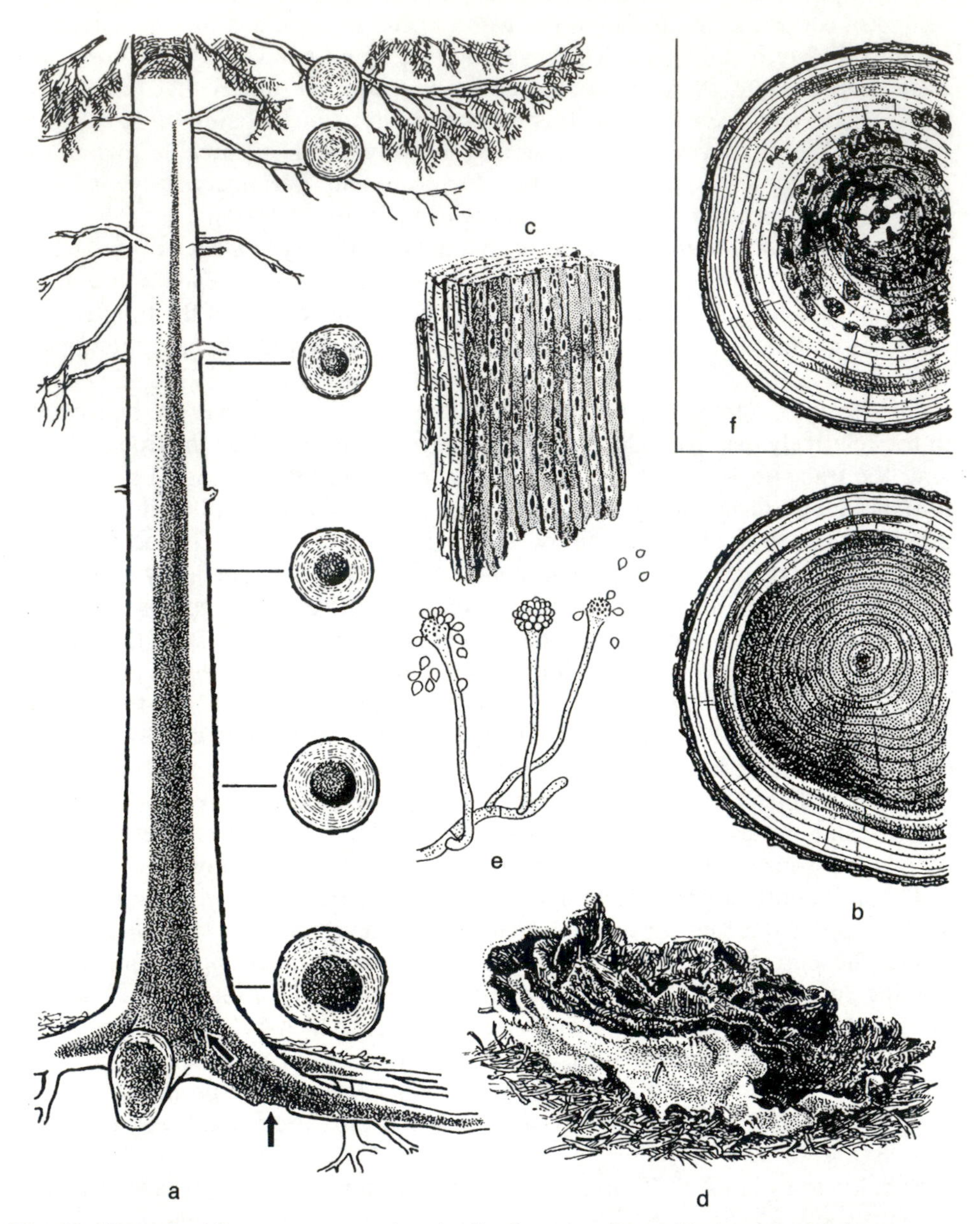

Fig. 89 *Heterobasidion annosum*. **a** longitudinal section through a spruce with heart rot, with stem cross-sections, **b** cross-section through a diseased spruce stem at an early stage of the disease, **c** a late stage in the decomposition of the wood, **d** fruit bodies, **e** conidiophores with conidia, **f** a heart rot caused by honey fungus

are the most important source of root infection in some areas. Infections rarely occur on the stem and are confined to the root collar zone. Infection is apparently favoured by wounds on the roots or at the root collar, although thin barked roots can also be infected directly by the fungus without being wounded.

First rotation crops on previous agricultural land are particularly at risk from infection. Also conifers on base-rich and compacted ground, and on sites with very variable moisture content suffer more from Fomes root rot than those on acidic, more open soils with a more uniform water supply. On many 'Fomes' sites, however, the growth of spruce is sufficient to compensate for the lost value represented by the unsaleable butt ends.

Prophylactic treatments against the fungus are not yet possible, but there are preventive measures that can provide significant control. In particular, the introduction of broadleaves into coniferous stands at the time of their establishment can reduce the eventual incidence of decay considerably. Another valuable technique is the treatment of thinning stumps to prevent them from becoming foci of infection, using sodium nitrite, urea or borax; chemicals which inhibit infection by *H. annosum* but not the decomposition of the stumps by saprotrophs. It is similarly possible to exclude the fungus from stumps by applying a spore suspension of an antagonistic fungus, and this is an established method in the case of pine, for which preparations of *Phlebiopsis (=Peniophora) gigantea* (Fr.) Jülich are available in Great Britain [164]. If a stand is already thoroughly infected by *H. annosum*, these preventive treatments are of little value. There is, however, some prospect of producing more resistant spruces by breeding and provenance selection.

Sparassis crispa (Wulfen:Fr.) Fr.

This species, sometimes called the cauliflower fungus, lives primarily as a root parasite, mainly in older pines, but it can spread upwards as far as 3 metres into the stem. This results in a brown rot, confined largely to the heartwood. The infected wood shows a yellow to dark reddish brown discoloration, darkening finally almost to black, and breaks up into cubical pieces. The fungus causes appreciable loss of timber value, chiefly in Scots pine. It has been discovered fairly recently [195] that younger Douglas fir stands can be quite badly damaged. Other hosts include spruce, Silver fir, and larch, although the fruit bodies are less often seen on these species.

The creamy white to ochre fruit bodies, which are up to 30 cm across and 20 cm high, arise at the base of living trees or on the cut surfaces of freshly felled stems. In their size and structure they resemble heads of cauliflower. They arise from thick, more or less deeply rooted, fleshy stalks which branch above into numerous, flattened, tape-like, twisted ends with sinuously lobed margins. The flat branches are covered with the hymenial layer which contains closely packed basidia from which the basidiospores are shed in autumn (Fig. **90**). The fungus is edible when young and a profitable

Fig. 90 *Sparassis crispa*. **a** Douglas fir stump with central rot and a fruit body to the side, **b** fruit body (part), **c** basidium with basidiospores

culinary species; older fruit bodies taste bitter. Fruiting is from August to November.

Phaeolus schweinitzii (Fr.:Fr.) Pat.

Syn. *Polyporus schweinitzii* Fr.

Along with *Sparassis crispa*, this fungus is the most important cause of stem rot in pine, Douglas fir, Sitka spruce, and larches in Great Britain [9]. It occurs initially on the roots, but the precise circumstances of infection have not yet been clarified. Possibly it requires honey fungus to pave the way [9]. Later a brown rot is found in the heartwood, extending only a short way up into the stem in pine, but up to 6–8 m in Douglas fir [85]. This can be recognized both by the typical cubical cracking and a marked smell of turpentine. Another characteristic sign of this fungus are the yellowish or creamy white, chalky-fluffy mycelial remains which coat the walls of the shrinkage cracks. Where the decay is confined to the lower part of the stem, it is referred to as a butt rot.

The short-lived fruit bodies of the fungus are found on the ground near roots and sometimes on the stem bases of infected trees. On pine, they quite often develop later on the stump surface after felling. The fruit bodies are up to 30 cm across with stalks thickening upwards; at first they are top-shaped, and later form several imbricate pilei. The colour varies—depending on the conditions in which they develop—from yellow-brown to chestnut brown. The upper surface is covered with a rusty red woolly felt, while the underside bears the strikingly

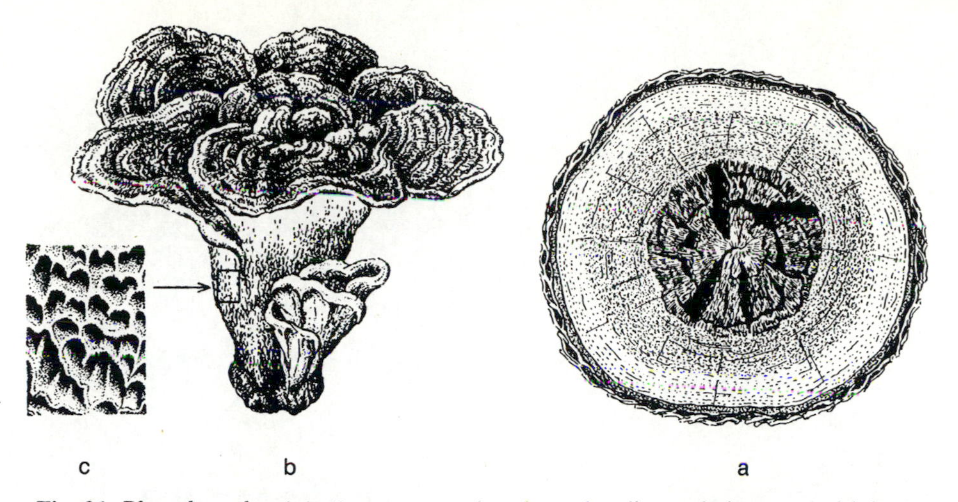

Fig. 91 *Phaeolus schweinitzii*. **a** cross-section through a diseased pine stem with heart rot, **b** fruit bodies, **c** surface (detail) of the pore layer on the underside of the fruit body

yellowish green pore layer which turns dark brown when touched. This consists of pores which are at first rounded, and later oblong-labyrinthine (Fig. **91**). Fruiting is from May to October, but the persistent, dead fruit bodies can be also seen at other times.

Other Root Rotting Pine Fungi:

- *Onnia tomentosa* (Fr.:Fr.) Karsten is easily confused with *Phaeolus schweinitzii.* Fruit bodies are thick and fleshy, yellowish cinnamon red, softly hairy, 6–12 cm, usually with an eccentric stipe. The fungus causes a root and butt rot.
- *Calocera viscosa* (Pers.:Fr.) Fr. Fruit bodies are rather like a goat's beard with several small, dichotomously branched arms, 3–7 cm high, golden yellow, and sticky in damp weather. It lives mostly saprotrophically on coniferous stumps, and is also parasitic in pine, Douglas fir, and spruce causing a root and butt rot (heart rot).

***Meripilus giganteus* (Pers.) Karsten**

The 'Giant polypore' colonizes the roots of mainly older trees and extends only a short distance into the stem where it causes a white rot. We still know relatively little about its life history. It seems to grow predominantly as a saprotroph on dying or already dead roots, while also extending into the live bark as a parasite.

In the initial stages of an attack, the mycelium occurs only in the central part of the root system, especially in the more deeply penetrating roots. Later, the fungus also penetrates the outer part of the bark thereby disturbing the uptake and transport of water. The result is that the leaves partially wilt, grow smaller

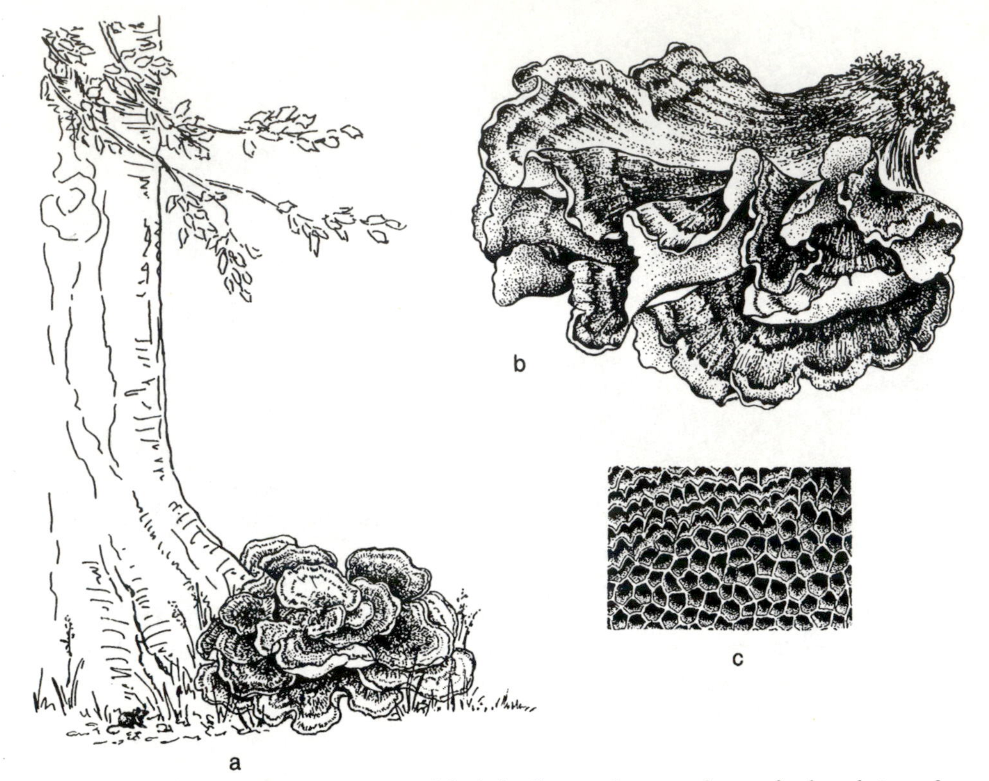

Fig. 92 *Meripilus giganteus*. **a** group of fruit bodies at the stem base of a beech tree, **b** fruit bodies, **c** under-surface of a fruit body (detail)

than normal, and the crown becomes thin. As the white rot progresses, the stability of the tree is reduced; perhaps with unpleasantly surprising results, particularly in urban areas.

The premature death and consequent infection of roots can be caused, for example, by the asphalting of a path near to mature broadleaves. Similar situations can arise from building operations which indirectly damage roots. In all cases the appearance of this fungus is always a sign of a severely decayed root system and a limited safe retention time for the tree which is usually an old specimen. Attacks are almost confined to broadleaved species, in particular beech and—very much less commonly—lime, oak, whitebeam, and plane.

The fruit bodies of this fungus can hardly be missed. They occur on stumps of recently felled trees or around the stem bases or major roots of standing trees, often in parks and other non-woodland locations. They grow up to 1 metre across and consist of numerous, often imbricate, tiered, fan-shaped, yellowish brown pilei. The base of the fruit body forms a stout, tuberous stalk which often emerges from the root system at some distance from the stem (Fig. **92**). Fruiting is from July to October.

A Similar Root Parasite on Broadleaved Trees:

- *Grifola frondosa* (Dickson:Fr.) S.F. Gray is a long-lived parasite and the cause of a white rot on roots and the lower parts of the stem of oak [86]. In southern Europe it occurs also on Sweet chestnut. Fruit bodies are at the base of older trees, 20–40 cm across, bushy, with many brownish, laminar, elongated pilei arising from a common, fleshy stalk.

Armillaria mellea (Vahl) Kummer and Related Species

Honey fungus—prized by fungus eaters but alarming to foresters—is among the most important of tree pathogens, having a worldwide distribution in both forest and non-forest habitats, and being able to attack almost all species of broadleaves and conifers [87]. As a saprotroph it lives in the ground on dead woody remains and on dead stumps of broadleaved and coniferous trees. Its mycelium is also to be found among the roots of healthy trees without initially causing any damage. Some members of the species complex may be primary pathogens, but in many cases the transition to the parasitic phase occurs only if the tree is weakened by certain stress factors (e.g. planting shock, attack by pests, waterlogging, lack of water and nutrients, or pollution by chemicals).

The fungus can enter the tree through wounds or directly through the bark of the roots and can then spread out in the cambial region between the bark and wood if the tree's structural or chemical barriers fail to check its progress. Eventually, perhaps after several years, the stem may become girdled, so that the infected tree dies. This mode of sustained attack on the cambium represents the main danger of honey fungus, and for this reason continental foresters sometimes give it the name of 'Kambiumkiller.'

As well attacking the roots as a parasite, the fungus is able to spread within the stem without attacking the cambium, eventually causing a white rot. As the decay is confined to the roots and stem base, it is another example of a butt rot. Among the more commonly attacked broadleaves are maple, poplar, and Red oak; in conifers, this type of decay occurs especially in spruce (Fig. **89f**). During the invasion of the wood, pseudosclerotial layers are quite often formed between the decayed and sound wood, and can be seen in sections as blackish 'zone lines.'

One of the typical external signs of an attack is the appearance of chlorotic foliage; for example, young spruces show a yellowish green discoloration of the needles which eventually die and go brown. Although such symptoms can be attributable to other damaging agents (e.g. insect attack or nutrient deficiency), unequivocal evidence of the presence of honey fungus is provided by the white, fan-like, radiating mycelial sheets which occur at the stem base between bark and wood. Also, heavy resin bleeding at the base of young conifers indicates a honey fungus attack. Another sign are the rhizomorphs under the bark and in the vicinity of the roots. They are typically dark-coloured elastic strands, 1–2 mm across, but their morphology and abundance vary between members of the species-complex. The rhizomorphs can grow through the soil from tree to tree and carry nutrients between different parts of the fungal thallus. In another capacity, they play a role as organs of infection.

A very conspicuous sign that the fungus is present are its fruit bodies which can be found in late autumn on dead stumps or at root buttresses. The fruit body consists of a ringed stem surmounted by a usually domed cap, 5–15 cm in diameter. It can be any kind of honey colour from yellow or brownish to olive. The top of the cap bears dark, appressed scales at its centre, which vary in prominence between members of the species-complex. On the underside are radially arranged gills whose surfaces are covered in minute basidia, each of which bears 4 basidiospores (Fig. **93**).

The *Armillaria mellea*-complex now embraces five intersterile species in Europe [170] which are separated on the basis of both morphological and ecological characters [120]. Their differing characteristics and requirements explain why the original aggregate species seemed to occur so widely on a very broad range of substrates and tree species. The pathogenic characteristics of the individual species are of particular interest here. Thus, *Armillaria mellea sensu stricto* occurs mainly on living broadleaved and coniferous trees; as a weak parasite on forest trees but as an aggressive primary parasite in orchards, parks, and gardens. Similarly, *A. obscura* (= *A. ostoyae*), which is common in central Europe, also has parasitic ability and can cause considerable damage in conifer stands. In contrast, *A. bulbosa* (= *A.gallica*), which occurs on broadleaves and

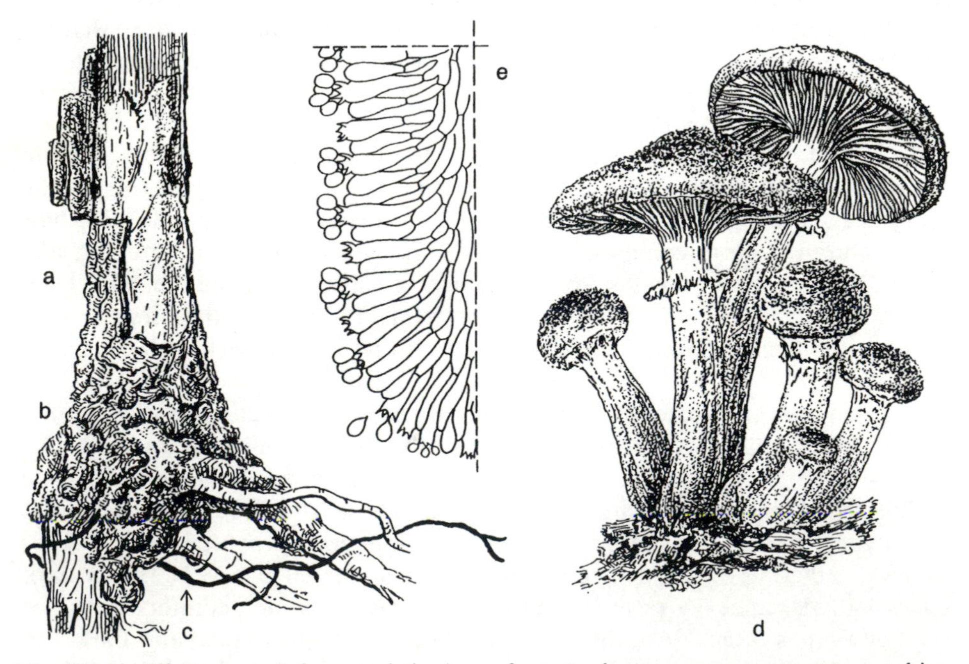

Fig. 93 *Armillaria* sp. **a-d** characteristic signs of an attack on a young spruce tree: **a** white mycelial sheet beneath the killed bark, **b** resinosis, **c** rhizomorphs, **d** fruit bodies; **e** cross-section (detail) through a spore-bearing gill

conifers, is predominantly saprotrophic. The same applies to *A. borealis*, which has the most northerly distribution. The remaining species currently recognized is *A. cepistipes*, of which two forms are known; a saprotroph and a weak parasite.

The control of honey fungus is difficult, as it is ubiquitous in the soil and inaccessible within the tree. The best course of action is to reduce the vigour of the fungus, which is heavily dependent on readily assimilated food sources, by grubbing out stumps and roots that are known to be infected with any of the more pathogenic forms of the species-complex. This measure should be undertaken in parks and gardens where trees and shrubs have become infected. It should also be done on sites where conifer plantations are to be established, but the routine grubbing out of all stumps and roots on such sites, though occasionally practised, could be counter-productive (as well as removing dead-wood habitats for wildlife) because it disrupts the microbial ecosystem of the forest floor and soil. An important role in this system is played by cord-forming decay fungi which may act as a natural control of pathogens such as *A. mellea*.

Fungicidal treatments can certainly kill the fungus in soil, but eradication from infected stumps cannot be effected without specialized injection techniques. Also, the costs of adopting chemical control widely would be prohibitive, and, in any case, this would be fundamentally opposed from the point of view of environmental protection. On the other hand, the use of biologically effective antagonists (e.g. *Trichoderma*) could be recommended, although it has so far proved effective only in small trials and laboratory experiments [185].

Other Root Rotting Gill Fungi:

- *Collybia fusipes* (Bull.:Fr.) Quélet causes damage to roots and bark and a butt rot in oak. It occurs only in the temperate zone [90]. Fruit bodies are in clumps at the base of stems and dark reddish brown with a furrowed or grooved and often twisted stem.
- *Pholiota destruens* (Brond.) Quélet is a parasite on living and freshly felled stems of poplar that causes a white rot. Fruit bodies are tough, wood-coloured, with the edge of the cap persistently rolled. Cap and stem have white, evanescent scales (Fig. **94b**).
- *Pholiota squarrosa* (Müller:Fr.) Kummer is a root parasite and cause of a white rot in the lower stem of broadleaves, e.g. maple, birch, beech, lime, poplar, willow, and apple, and also on spruce. Fruit bodies are at the stem base in clumps, rusty yellow, and cap and stem with shaggy, squarrose, recurved, reddish brown scales (Fig. **94a**).

Ustulina deusta (Fr.) Petrak

This fungus, also known by the name of *Hypoxylon deustum* (Hoffm.:Fr.) Jülich, is one of the few ascomycetes which can damage the wood of standing stems. In lime, for example, it can invade the wood through injuries to the roots, causing a severe white rot with black 'zone lines.' The infection can spread upwards within the woody cylinder of the stem and via the root buttresses into the wood of

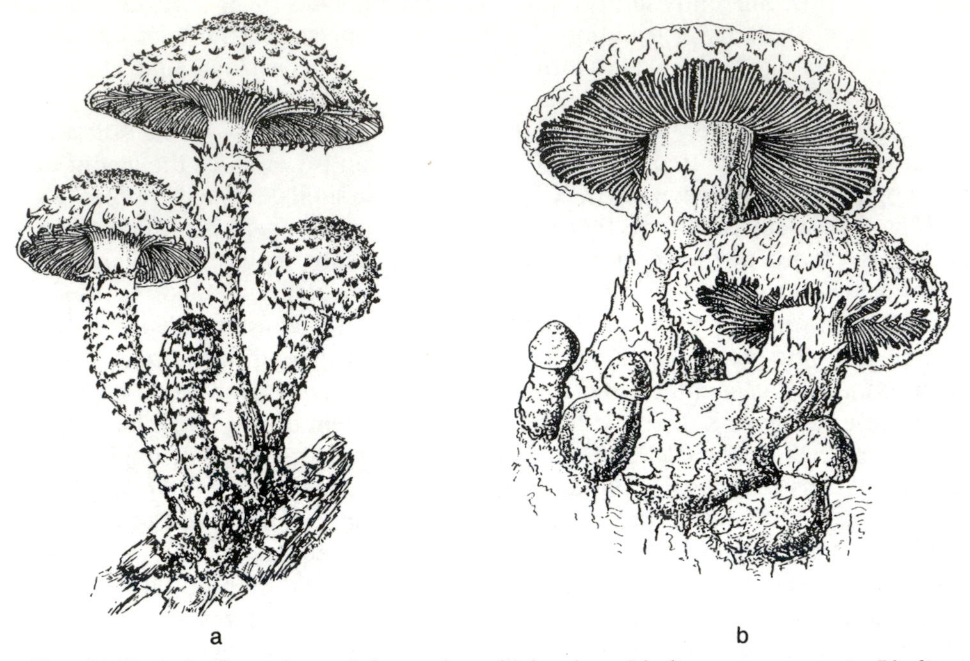

Fig. 94 Fruit bodies of wood-destroying gill-fungi. **a** *Pholiota squarrosa*, **b** *Pholiota destruens*

the roots. Often an attack escapes notice until the sudden collapse of the tree [225]. Beech, Horse chestnut, elm, and plane can be similarly attacked. On the other hand, the fungus can develop entirely saprotrophically in the stumps of broadleaves, especially beech, where it is one of the regular inhabitants.

The fruit bodies of *Ustulina deusta* are flat, crust-like stromata, on the upper surfaces of which conidia are formed in spring, appearing as a flour-like, whitish grey coating. In summer, the stromatal mass, which reaches up to 20 cm across, changes into a black, charcoal-like, very dry, lumpy structure with numerous perithecia buried in the upper surface. The perithecia are 1 mm across and contain ascospores 30–40 × 8–12 μm in size. As the fungus usually infects through wounds, injuries to the bark, especially at the stem base, should be avoided. Street trees are particularly at risk.

Stem Decay Fungi

Stereum sanguinolentum (Alb. & Schwein.:Fr.) Fr.

The 'Bleeding Stereum' is the most important of the fungi involved in the disease of spruce known in German as 'Wundfäule'—wound rot. It becomes established in wounds larger than 10 × 10 cm within a few days of their infliction, by the germination of spores on the wound surface. The resulting mycelium soon invades the wood above and below the wound—mainly in the younger annual rings—and stains it a reddish colour. If the wound calluses over, the white rot is halted or spreads only slowly; otherwise the fungus can extend several metres up the stem, in which case almost all parts of the stem cross section are damaged. However, the wood formed after wounding remains free of infection because of the formation of the antifungal, resinous barrier zone. Development of decay here would be possible only following a new infection from the outside. Less often, the fungus invades a central column within the stem, and it can also colonize felled wood of spruce, pine, and Silver fir in which it causes a red streaking (Fig. **103a**).

The fruit bodies of the fungus may form around the edges of wounds on standing trees, but only rarely. They appear much more frequently on felled timber, mainly on the end grain where their mussel-shaped pilei often form a covering like overlapping tiles. Quite often the greyish-ochre hymenium becomes resupinate, spreading out widely over the surface of the substrate. The hymenium contains translocatory hyphae which, if wounded, exude a liquid that turns red in the air. It is this which has given the fungus its name (Fig. **95**).

This wound rot is a relatively recent problem which seems to be a result

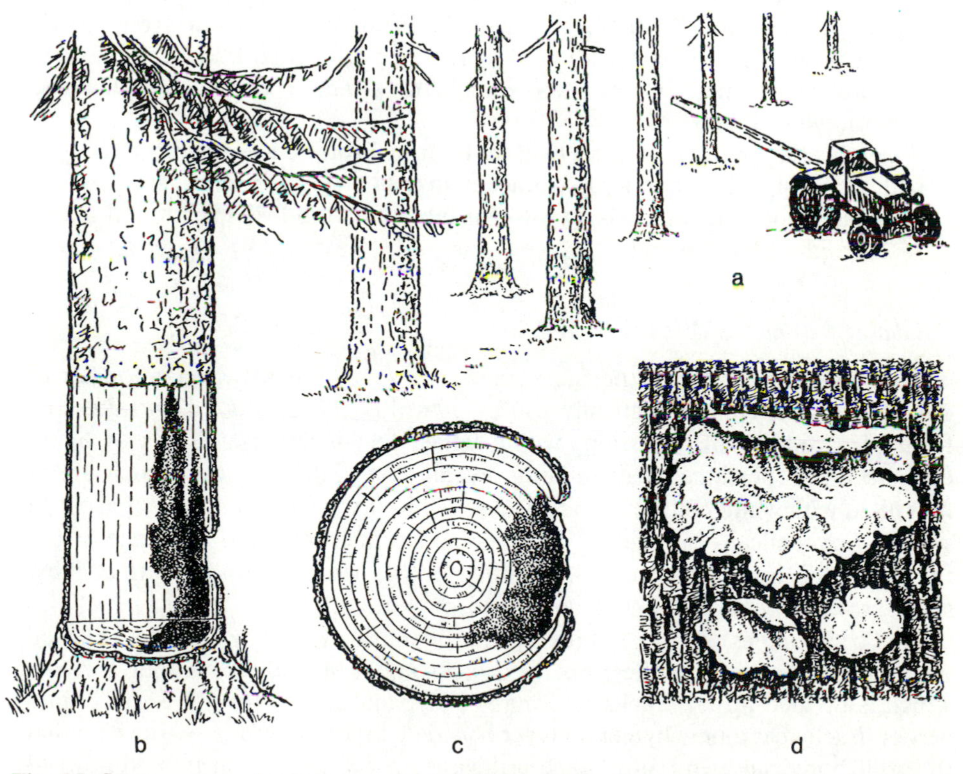

Fig. 95 *Stereum sanguinolentum*. **a** bark injuries arising in a spruce stand, **b** the extent of decay in a spruce stem two years after wounding (semi-diagrammatic), **c** stem cross-section with decay and onset of occlusion, **d** fruit bodies on dead bark

of the increasing mechanization of forestry. With the employment of heavy machinery for the extraction of thinnings, large bark wounds occur more and more frequently, bringing more or less extensive decay in their wake. The principal recommended means of control is to carry out thinning operations more carefully. Apart from this, the danger of infection can be reduced by the use of wound sealants.

Other Wound Rot Fungi [29,70]:

- *Chondrostereum purpureum* (Pers.:Fr.) Pouzar. Fruit bodies with shallow pilei are broadly effused across the substrate, pale purple or lilac, older specimens are purplish brown or leached to a dirty white. The fungus lives in the wood and bark of broadleaved trees (for example, maple and plane), where it is a wound parasite, killing bark and cambium. It is more common as a saprotroph on the end grain of felled stems of beech, poplar, and birch and causes silver-leaf disease of fruit trees.
- *Trametes hirsuta* (Wulfen:Fr.) Pilát. Fruit bodies are thin, protruding from the substrate, upper surface with grey or yellowish, stiff hairs, and underside have roundish, yellowish white pores. It occurs as a weak parasite, for example on sun-scorched beech but is more common as a saprotroph on stacked, debarked stems of various broadleaves. It also causes a white rot (Fig. **64c**).
- *Tyromyces caesius* (Schrader:Fr.) Murrill. Fruit bodies are in groups, pale bluish or an intense blue with a whitish edge, otherwise like *T. stipticus* in size and consistency. It occurs on wounds of extraction-damaged spruce and also on lying stems.
- *Tyromyces stipticus* (Pers.:Fr.) Kotl. & Pouzar. Fruit bodies are mostly single, soft and fleshy, 2–4 cm across, bracket-shaped, whitish with a sharp, bitter taste. It is the cause of a brown rot in the stem wood of spruce and also a saprotroph in lying wood.

Xylobolus frustulatus (Pers.:Fr.) Boidin

This fungus, a member of the Stereaceae, is a specific heartwood colonizer of oak in which it characteristically causes a 'white pocket rot.' This takes the form of a reddish brown stain, within which locally intense points of lignin decomposition produce small, oblong, evenly distributed cavities which are at first filled with white cellulose. Wood attacked in this way is known as 'partridge wood.' Economically significant damage can result from the occurrence of the fungus in the heartwood of old, standing oaks. Attacks can also affect lying stems, sawn remains, and even converted timber.

The fruit bodies of the fungus appear in bark cracks or on the exposed heartwood. They are in the form of a crust and white on the upper surface, which is divided up by cracks into many polygonal pieces from 0.5 to 1.5 mm across. Each year a new hymenial layer is laid down on the upper surface so that the fruit body can eventually reach a thickness of 8 mm. Each individual fruit body is narrowed towards its base in the form of a stem and contains numerous, thick-walled, brownish cystidia with short bristles (Fig. **96**).

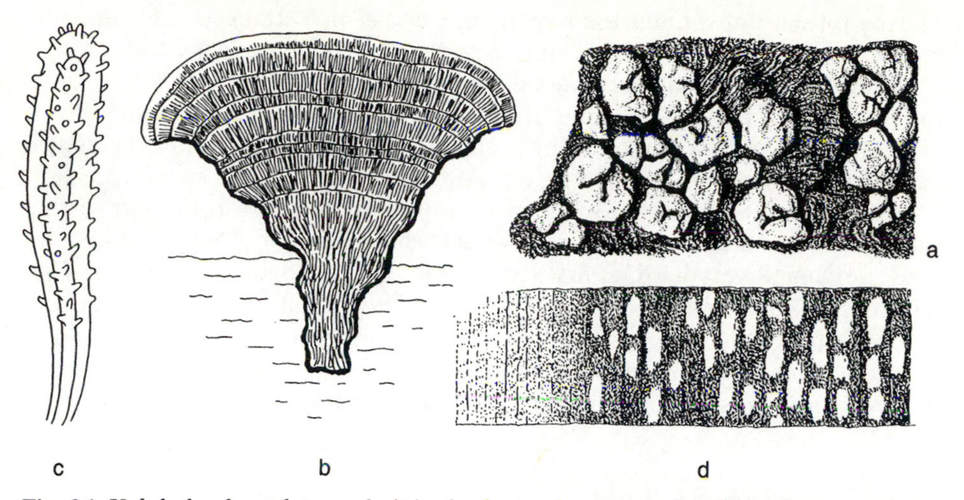

Fig. 96 *Xylobolus frustulatus*. **a** fruit body cluster, **b** cross-section through a several-year-old fruit body with multiple hymenial layers, **c** cystidia from the hymenium, **d** longitudinal section through infected oak wood with white pocket rot (b,c after Jahn 1971)

Fistulina hepatica (Schaeffer:Fr.) Fr.

This fungus, known as 'Beefsteak fungus' or 'Ox-tongue' is the cause of 'brown oak.' It occurs mainly in the stem wood of older oak trees, initially causing the development of irregular, brownish red stains in the heartwood of the lower stem, which it may invade extensively in bands or tongues. The staining may also be uniformly dark. At this stage there is little reduction in the strength of the wood so that it can still be used for various purposes. When infection is well advanced, a brown cubical rot with cracks develops. The fungus also grows as a saprotroph on stumps and stems lying on the ground.

The fruit body of this unique fungus has the shape of a sinuate tongue or liver and is fleshy, 10–30 cm across, 3–6 cm thick, and strikingly blood red or brownish red. When cut, the 'flesh' exudes a reddish juice. The underside consists of closely packed, separate tubes, each about 1 cm long, 'reminiscent of macaroni' [110].

Phellinus pini (Brot.:Fr.) Ames

Syn. *Trametes pini* Thore: Fr.

This is a specialized stem-rotting fungus which only infects living trees via dead branch stubs, for which reason the infection courts are mostly quite high on the stem. The fungus spreads deeply into the stem via the vascular traces of the branch stubs, from which it extends upward and downward within the zone between the sapwood and heartwood [191]. As a result of the tree's periodic defence reactions, the fungus is restricted to particular ring-shaped zones and so a ring rot develops. Later the remaining wood is also attacked. This initially becomes stained reddish brown before spindle-shaped white flecks and a white pocket rot form. The fungally invaded wood remains quite firm for a long time so that, although reduced in value, it can still be used for many purposes.

The fruit bodies are brackets which appear at various heights on the stems of older trees. They only begin to develop 10–20 years after an attack and, as the fungus cannot grow through living sapwood, they are restricted to the areas around knotholes or beneath dead branches (Fig. **97**). They are 5–12 cm across, perennial, very hard, at first red-brown, and becoming dark brown. The concentrically furrowed upper surface is at first felty rough, often finely cracked when old; the yellow to grey-brown pores are roundish or oblong and 0.2–0.4 mm in diameter. It is not usually possible to make out layers of tubes.

Phellinus pini occurs predominantly in northeastern Europe, where it is

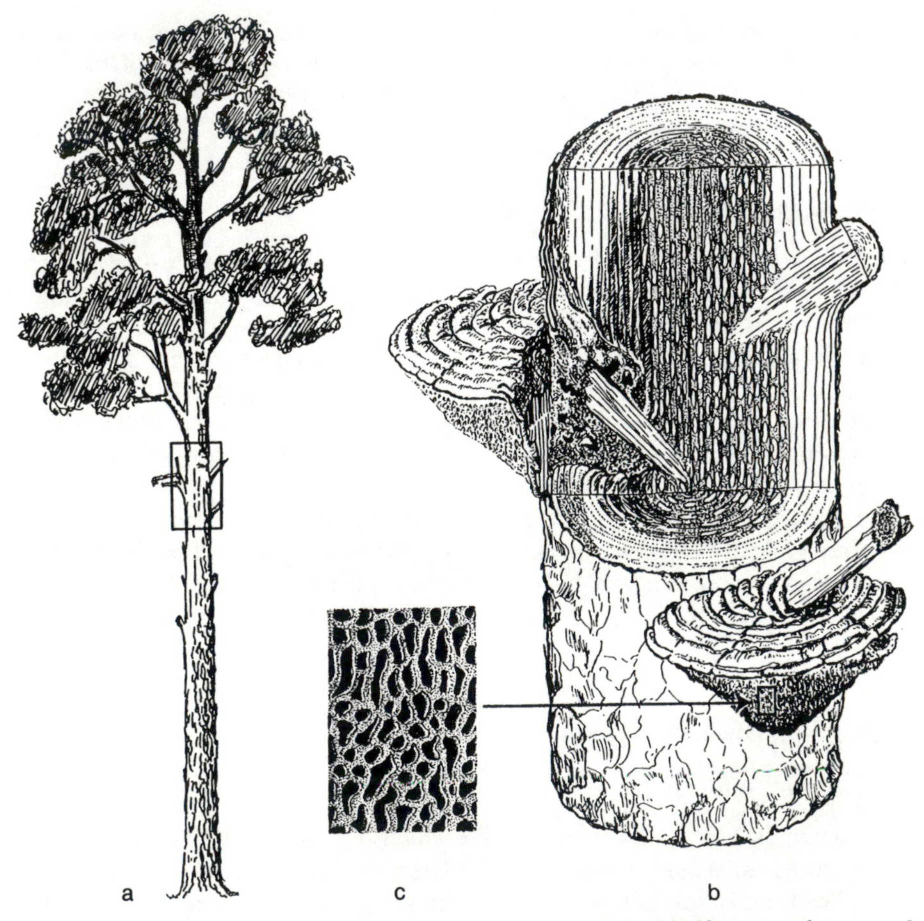

Fig. 97 *Phellinus pini*. **a** general view of an older pine infected halfway up the stem, **b** stem section with 'honeycomb rot' (white pocket rot) of the heartwood together with fruit bodies growing on the exterior, **c** surface view of the pore layer (detail)

economically significant on pines. It is certainly able to attack other coniferous species, but the damage it causes on spruce, larch, and Douglas fir is relatively slight.

*Other **Phellinus** Species* [70,176]:

- *Phellinus chrysoloma* (Fr.) Donk is very similar to *Ph. pini* but with thinner, elongated brackets and narrower pores (2–4 per mm). It occurs in spruce forests in mountain areas and also on pine and cedar (*Cedrus* spp.). It causes a white rot.
- *Phellinus hartigii* (All. & Schnabl) Pat. is very similar to *Ph. robustus*, with large, very hard brackets, with a grey-brown upper surface, and cinnamon yellow pore layer. Annual zones are not clearly demarcated. It is parasitic on old Silver firs.
- *Phellinus igniarius* (L.:Fr.) Quélet, known as the false tinder fungus, forms brackets which are 10–25 cm across, 4–10 cm thick, blackish brown and very hard with thick, rounded margins. It causes a white rot and occurs mainly on willow as well as on apple trees. It occurs in a number of varieties [176].
- *Phellinus pomaceus* (Pers.:Fr.) Maire. Fruit bodies are 2–7 cm across; tube layer is grey-brown to yellow-brown, and continues down on to the substrate. It is mainly parasitic on *Prunus* species.
- *Phellinus robustus* (Karsten) Bourd. & Galz. is a slow growing parasite that occurs mostly on older oak, less often on robinia and Sweet chestnut. It causes a white rot in the upper stem. Fruit bodies are about the size of a fist, very hard, with cinnamon-yellow pores; upper surface is often green with algae. It occurs mostly in groups around wounds left by fallen branches and occasionally also on the bark if the cambium has been killed.

***Fomitopsis pinicola* (Swartz:Fr.) Karsten**

This is a powerful destroyer of cellulose and thus a good example of a brown rot fungus. In continental Europe (it is apparently absent from the British Isles) the fungus occupies the same niche as *Fomes fomentarius* (tinder fungus) on beech and the two species often occur together. Nevertheless, the areas decayed by the two are sharply demarcated. The fungus attacks severely damaged or dying trees, especially in the lower part of the stem. It can continue growth and fructification on felled or blown stems. In the Alpine and Mittelgebirge regions it occurs more commonly on Silver fir and spruce. Wood attacked by the fungus eventually becomes brown, crumbly, and friable. As a result of volume loss, transverse and longitudinal cracks develop as it dries so that the final stage is a typical, crumbly, cubical rot (Fig. **98**) which can be reduced to a brown powder by rubbing between the fingers. Another feature of wood attacked by this fungus is the persistence of white mycelial shreds in the shrinkage cracks.

The perennial fruit bodies are bracket-shaped, 10–30 cm across, the upper surface fairly smooth and with a resinous crust, at first orange-red to rusty red with a yellowish margin, later blackish with a dark red margin. The pores are yellowish to pale brown with a yellowish white spore powder. 'To distinguish this

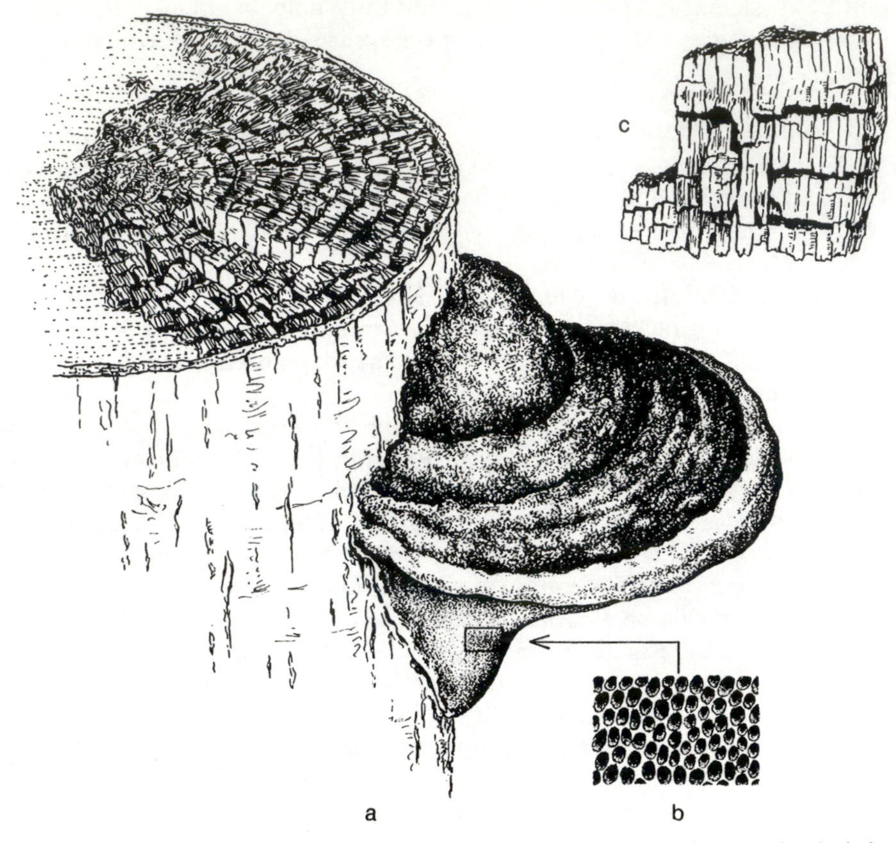

Fig. 98 *Fomitopsis pinicola*. **a** stem section of a beech with brown rot and a fruit body on the exterior, **b** underside of a fruit body (detail) with pores, **c** cuboidal disintegration of the decomposed wood

from the sometimes similar tinder fungus it is sufficient to hold a match to the crust: in the case of *F. pinicola*, it becomes sticky' [110].

Fomes fomentarius (L.:Fr.) Fr.

The 'Tinder fungus' is among the most striking of large polypores, occurring mainly on beech in continental Europe, and also in England where it is rare. Generally, it occurs less often on birch, alder, and hornbeam, but is frequent on birch in Scotland. It is a stem decay fungus, invading the wood of weakened trees via bark wounds or broken branches and causing a white rot. Typically, the wood decayed by the tinder fungus contains black boundaries ('zone lines'), which divide individual colonies from other mycelia or from the still healthy wood. They are produced by intensified phenoloxidase activity which converts fungal or host substances into melanin. If a large volume of wood is decayed, the strength of the stem is reduced so that it may snap in stormy or wet weather.

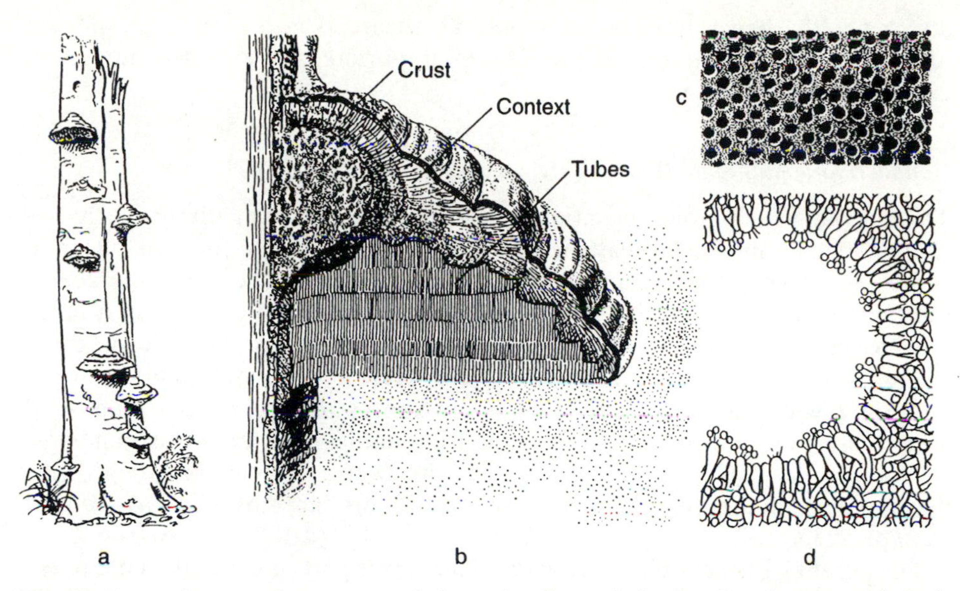

Fig. 99 *Fomes fomentarius*. **a** general view of a diseased, broken beech stem with fruit bodies, **b** cross-section through a bracket releasing a cloud of spores from the underside, **c** detail of underside of fruit body, **d** cross-section through a pore (detail) with basidia on the inner wall of the minute tube

This often happens to trees that have been damaged by the beech bark disease complex, perhaps many years earlier, following entry of the fungus through areas of necrotic bark. In England, this so-called 'beech-snap' following bark disease is usually caused by basidiomycetes other than *F. fomentarius*.

The fungus is usually conspicuous in any beech wood where it is present because of its hoof-shaped fruit bodies, which can continue growing for many years. They occur on wounds and also directly on the bark of overmature or dead trees. They can also occur in large numbers on fallen stems, in which the fungus can continue to live saprotrophically for many years after the death of the tree.

The fruit bodies are 10–50 cm across, pale brown, or grey to blackish or, often brownish when young, and the domed upper surface has raised zones. The crust is very hard, 1–2 mm thick and covers a yellow-brown, tough, fibrous trama, below which layers of brown tubes are attached. From spring to summer, vast numbers of basidiospores are released and carried away in air currents. In dry weather these can be seen as a white dusty deposit on the upper surface of the fruit bodies or close by (Fig. **99**).

In forestry, the fungus reduces the timber value of trees that are already weakened or diseased. It is therefore important to salvage the timber of affected trees as quickly as possible in order to avoid further economic loss. As mentioned above, *Fomes fomentarius* is one of various decay fungi that can destroy the timber value of stems affected by beech bark disease. However, despite the economic significance of the fungus, it is too weakly parasitic to attack healthy beech trees and it should be regarded as a natural component of the beech woodland ecosystem, with an important role as a decomposer of unusable timber residues.

Ganoderma applanatum (Pers.) Pat.

This fungus, which grows on various broadleaved trees, particularly beech, has characteristic semicircular, flattened fruit bodies which range from 10–50 cm in width but reach a thickness of only a few centimetres. The upper surface of the lumpy wrinkled brackets are concentrically zoned and covered in a hard, grey to brown crust (Fig. **100a,b**). On the underside are the fine pores of the brownish tube layer which can eject several thousand million spores daily in favourable conditions in summer. Spores are quite often deposited as a red-brown powder on top of the fruit bodies and their surroundings. In some areas, fruit bodies are found with warty or peg-like outgrowths developing from their undersides. These are galls produced by a fly, *Agathomyia wankowiczi*.

The perennial fruit bodies are found on the lower part of the stem of old trees, frequently those already colonized by other fungi; but this fungus lives principally

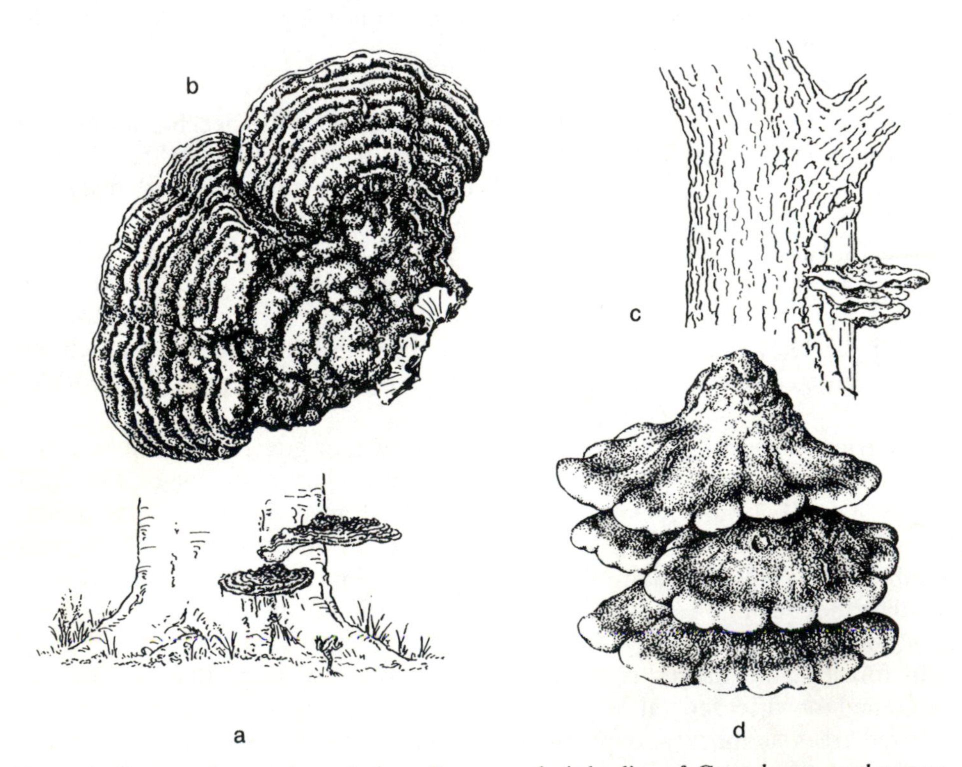

Fig. 100 Causes of stem decay in broadleaves. **a** fruit bodies of *Ganoderma applanatum* on the stem base of a beech tree, **b** upper surface of a fruit body; **c** fruit bodies of *Laetiporus sulphureus* on an oak stem, **d** fruit bodies

as a pure saprotroph on old, already dead stumps where it brings about an intense white rot.

*Other **Ganoderma** Species:*

- *Ganoderma adspersum* (S. Schulzer) Donk, in contrast to the similar *G. applanatum*, is a weak parasite with thicker, swollen, often overlapping brackets and a crust 1–2 mm thick. It causes a white rot and occurs most commonly in gardens and parks, but also on street trees. It occurs on oak, lime, Horse chestnut, beech, and whitebeam.
- *Ganoderma lucidum* (Curtis:Fr.) Karsten. The fruit body is pale yellow-brown to reddish on the upper surface and stalk, as if varnished. It causes a white rot in broadleaves also on spruce, pine, and larch.

Laetiporus sulphureus (Bull.:Fr.) Murrill

The large and brightly coloured fruit bodies of the 'Sulphur polypore' are visible from some distance and make it one of the most striking of the tree polypores. These 15–40 cm-broad brackets are orange-yellow above and sulphur yellow below. They can be found from May onwards, both at stem bases and high up in the tree, growing like overlapping tiles (Fig. **100c,d**). The fungus penetrates to the interior of the stem via wounds, where it continues to grow for many years in the heartwood region, bringing about a severe brown rot which can continue until the wood is completely decomposed. At the same time, the mycelium grows along the fissures which develop in the wood, lining them with broad, ribbon-like, yellowish strips. The sulphur polypore is able to grow in the heartwood because it can detoxify the heartwood phenols with tyrosinase, one of the phenoloxidases. On the other hand, it cannot usually attack the sapwood, so that the infected tree remains alive for a long time if it is not brought down by a storm. The question of wind-firmness arises principally with trees in streets and parks.

The fungus preferentially attacks broadleaves with a coloured heartwood, particularly oak, robinia, *Prunus* species, and can also be common on yew. Other fairly frequent hosts include trees with soft wood like rowan, Horse chestnut, and willow. In the mountains it also occurs on larch. It is also one of the very few polypores which is good to eat, for which reason one of its English names is 'Chicken of the woods.'

Polyporus squamosus (Hudson:Fr.) Fr.

This widely distributed polypore has a particularly large fruit body, reaching up to 50 cm across. This is kidney- or fan-shaped, pale yellow to ochre, with an eccentric stalk, blackish at the base. A particular distinguishing feature are the concentrically arranged brown scales on the upper surface of the pileus. The tube layer on the underside of the pileus is characterized by the pale yellow pores, 1–2 mm across, which are joined in a net-like fashion.

The fungus occurs mainly on trees in streets and parks where it can cause

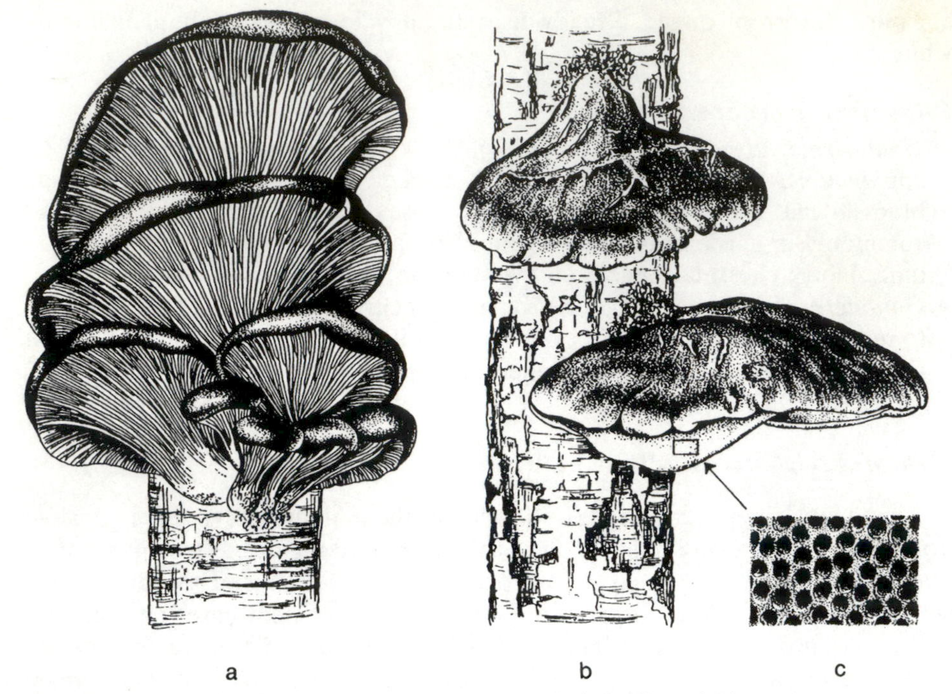

Fig. 101 Causes of stem decay in broadleaves. **a** fruit bodies of *Pleurotus ostreatus* on beech; **b** fruit bodies of *Piptoporus betulinus* on birch, **c** pore openings (detail)

an extensive white rot in the stem. Less often it occurs as a saprotroph on stumps of various species, including maple, lime, poplar, Horse chestnut, elm, and sycamore.

Piptoporus betulinus (Bull.:Fr.) Karsten

The 'Birch polypore' is a strongly host-specific stem rotter which to date has been found only on birch. It can be termed a weak parasite for it attacks only older trees or those suffering from a lack of light or weakened in some other way. It invades the upper part of the stem through branch stubs, then spreads upwards in the stem, causing a severe brown rot, dark red to brownish in colour. The annual fruit bodies appear singly or in groups, often several metres up the stem. In their young state they are domed in the shape of a bell with a bare and smooth upper surface. Later they often become cracked and covered in a thin, leathery, skin-like, pale grey-brown to brownish crust. The lower part of the fruit body consists of a grey-white tube layer. The tubes are short and open at very small, roundish pores. A short stalk arising close to the edge joins the fruit body to the substrate (Fig. **101b,c**). Birch trees in the parks or gardens on which the fruit bodies are observed should be felled as soon as possible as there is an increased danger of their snapping.

Inonotus hispidus (Bull.:Fr.) Karsten

This fungus occurs only on broadleaves, particularly on ash, walnut, plane, and apple. It is one of the most active parasites among the polypores. It penetrates via broken branches or pruning wounds to the centre of the tree, where it causes a spongy white rot with fine, whitish longitudinal and radial streaks on a dark ground. Infected stems and branches often retain their strength for a long time, due to the maintenance of a sound, outer shell of wood, especially in the case of plane which has very good mechanical properties. However, in the case of ash, weakening of the wood beyond the infected zone has been detected and the fungus may also cause a 'canker rot', attacking the cambium and the usually resistant new wood formed later than the initiating wound.

The absence of breakage in many cases can mean that an attack by *I. hispidus* is not noticed until fruit bodies appear. These occur 2 to 10 m up on the stem or on thicker branches in the crown of the tree. They are annual, bracket-shaped, 20–30 cm across and covered in a shaggy felt on the upper surface. In colour, they are at first a bright yellowish rusty red, later rusty brown to dark brown and finally—after they have died in the winter—blackish. The soft, watery flesh (trama) is similarly coloured, as is the 1–3 cm-thick tube layer (hymenium), though the pores darken when touched. Copious watery exudations collect as droplets in 2–4 mm wide canals on the undersides of fresh, actively growing fruit bodies.

The fungus occurs mainly on street trees, as well as in apple orchards, and is associated particularly with pruning or other wounds. As the decay remains largely confined to the central portion of the stem, except in the canker rot of ash, it is seldom necessary to fell infected trees.

***Pleurotus ostreatus* (Jacq.) Kummer**

The 'Oyster mushroom' grows as a wound parasite on living trees, preferring Horse chestnut, beech, lime, poplar, and willow; as a saprotroph it colonizes stumps and lying stems where it causes a white rot in the heartwood. The fruit bodies, which appear in clusters on the wood, are pale-gilled fungi with semicircular or kidney-shaped caps and laterally placed stalks. The cap colour is slate grey, blackish blue, or even olive green. Systematically, even though it looks like a true gill fungus, it is placed in the Polyporaceae (Fig. **101a**).

The fruit bodies of the Oyster mushroom are prized and sought after as culinary species; they are found relatively late in the year, persisting even after frosty nights. Anyone who would like to be independent of chance finds can cultivate the fungus in the garden or in the forest.

8 *Damage to Fallen and Felled Timber*

If a tree is felled, or snaps, it undergoes a series of changes which start with the progressive death of its living tissues and may continue for many years after death. As the tissues lose vitality, their active defence systems collapse, allowing colonization by a characteristic succession of micro-organisms and animals. This succession continues for many years, if the timber is not first utilized by man, and eventually results in the complete breakdown of the tree. Logs generally need to be utilized promptly if their commercial value is not to be seriously reduced.

Two main kinds of deterioration can occur in timber before it is utilized. Of these, discoloration (staining) starts to develop first in lying timber within a few weeks of felling. Staining does not usually affect the strength of the timber but helps to reduce its value. It may be abiotically induced, or can involve wood-inhabiting fungi. The second kind of deterioration is fungal decay, which becomes a significant problem only after the log has been lying for a considerable time. Many fungi may be involved in decay; most commonly basidiomycetes, which eventually produce brown or white rots. Decay can also be caused by ascomycetes and by deuteromycetes (Fungi Imperfecti), but then often takes a less severe form, known as 'dote.' Bacteria also colonize dead wood, but only in exceptional circumstances do they cause significant breakdown of cell wall materials—for example in wet stores—and even then they cause little loss of wood strength.

WOOD DISCOLORATIONS

Oxidation Stain

This type of discoloration, which occurs in the wood of broadleaved trees, starts at the exposed end grain of lying stems and extends along them for varying distances. The colour of the stain may be brownish (as in beech) or greyish brown (as in oak). The staining develops when oxygen enters the wood and comes into contact with the phenolic contents of the cells (oxidation stain). Fungi or other organisms are not involved in this process.

If a beech stem, for example, remains lying in the forest for several weeks or months, the wood begins to dry out slowly. The xylem parenchyma cells remain viable but, when the moisture content drops to about 50% of the dry weight, they begin to form balloon-like protrusions through the pits of neighbouring vessels, blocking the vessel lumens. These structures, known as tyloses, contain part of the cytoplasm together with the cell nuclei (Fig. **102**). Soon afterwards, the parenchyma cells die and their contents become oxidized, imparting a yellowish brown discoloration to the wood. Tylosis formation does not impair the mechanical properties of the timber, but it prevents satisfactory impregnation

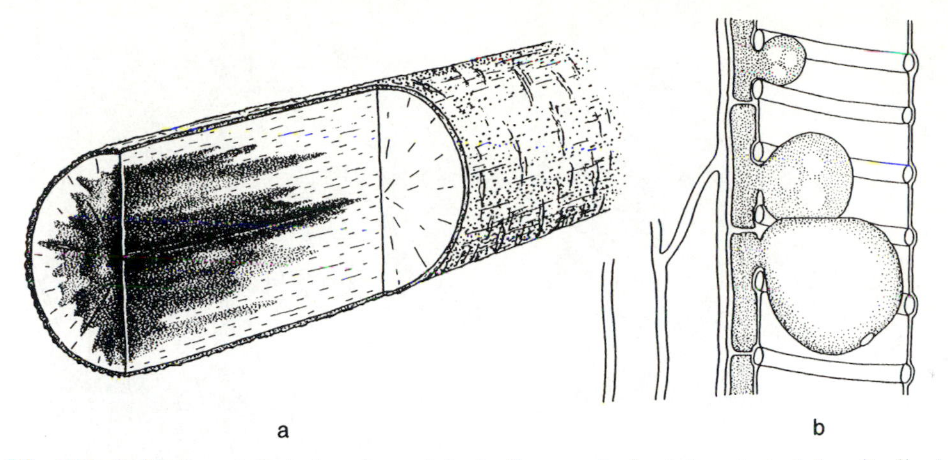

Fig. 102 Oxidation stain in beech. **a** stain in the wood of a lying stem, **b** longitudinal microscope section with tylosed vessels

with preservatives. Moreover, the staining can depress the value of the timber by rendering it unsuitable for veneer production. If logs continue to lie in the forest, they are later affected by white rot fungi, which weaken the timber.

Oxidative staining can be prevented by leaving the bark on roundwood and storing it as wet as possible. Excessive drying can be inhibited by applying a protective coating to the exposed end grain. The ends of particularly valuable veneer logs are protected in this way by painting them with hermetic sealants such as whitening, opaque resin varnish, or paraffin wax.

Other Types of Abiotic Wood Staining:

- *Grey wood stain*; superficial weathering caused by ultraviolet light and the subsequent leaching of the freed lignin by rain. What remains is a silvery white, cellulose-rich surface. It occurs on timber of various conifers used in building and is principally found at high altitude.
- *Charring* or *browning* of wood surfaces is caused by intense insolation in dry weather. Lignin is broken down as in grey-stain, but the products (humic acids) are retained as dark brown, amorphous encrustations. It occurs mostly at even higher altitudes and can be confused with sooty moulds.

Red Streaking

Red streaking is a type of fungal discoloration, occurring only in conifer wood, that is recognized as a distinct condition in continental Europe, and is known as 'Rotstreifigkeit' in German. The discoloration is reddish brown, or sometimes reddish to yellow, extending into logs both from their bark-covered faces and from their cut ends. The causal agents are white rot fungi.

The fungal invasion responsible for red streaking develops quite rapidly in logs of moderate moisture content, spreading axially from the cut ends or

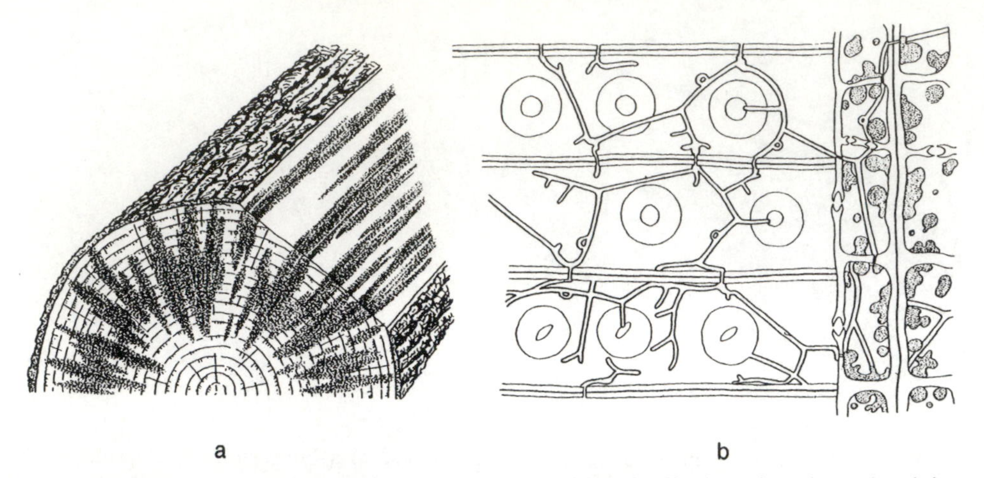

Fig. 103 Red streaking in spruce. **a** transverse and longitudinal section through a lying spruce stem with discoloured streaks, **b** radial-longitudinal section through red-streaked sapwood with fungal hyphae in the tracheids and rays

radially from the bark. The staining is mainly an oxidative effect and can be seen as streaks of varying width in a longitudinal section of the wood. Under the microscope, numerous fungal hyphae can be seen in the red-streaked wood; these are fine and hyaline with frequent anastomoses. (Fig. **103**). Although the fungi involved are white rotters that can break down the lignin, there is little such breakdown in the early weeks or months of colonization, and the affected timber can thus still be used for various purposes.

In spruce, *Stereum sanguinolentum* is the most important cause of red streaking; it occurs also as a wound rot fungus in standing stems (Fig. **95**). In pine, red streaking is due mainly to *Trichaptum abietinum* whose violet, white-fringed fruit bodies are either resupinate or stand out from the wood as cap-like brackets. Its most important feature are the violet-coloured pores.

Red streaking develops especially if the wood remains in a semi-moist state over a long period, particularly in the warmer part of the year. This means that winter felling is a useful first-stage preventive measure. In high altitude forests, felling can safely commence in the autumn. After felling, the development of red streaking can be further inhibited by rapid drying of the debarked stems in airy timber yards.

Blue Stain

This is another kind of staining which affects coniferous wood and, as the name implies, it has a bluish or bluish grey tint. There are various fungi that cause it; the so-called blue stain fungi, which attack only the sapwood. It occurs both on the surface of the wood in the form of dark spots or streaks, and also within it. A prerequisite for its development is a moderate wood moisture content, in the range 30% to 120% of the dry weight. The main economic losses from blue stain occur on pine.

The cause of the staining can be traced to the presence of dark-coloured hyphae whose growth is mostly radial, along the nutrient-rich xylem rays (Fig. **104**). Growth from cell to cell takes place either via the pits or by mechanical perforation of the cell walls by thin penetration hyphae.

Blue staining of coniferous timber can occur during various stages in storage and utilization. It becomes most extensive in un-debarked pine stems which are allowed to dry out slowly over weeks or months while lying in the forest. The fungi that cause blue stain in unsawn logs, (primary blue stain) belong either to the Ascomycotina or to the Fungi Imperfecti. The Ascomycotina are represented by numerous species of the genera *Ceratocystis* and *Ophiostoma* of which the most frequent on coniferous wood is *O. piceae* (Münch) H. Sydow & Sydow. This forms long-beaked perithecia and brush-like coremia whose spores are dispersed by bark beetles (Fig. **104**). Of the imperfect fungi, the form genus *Leptographium* and *Discula pinicola* (Naum.) Petrak are worth mentioning. The latter is the main cause of a distinct condition called 'internal blue stain' which is characterized by a central discoloration of the wood without any external damage.

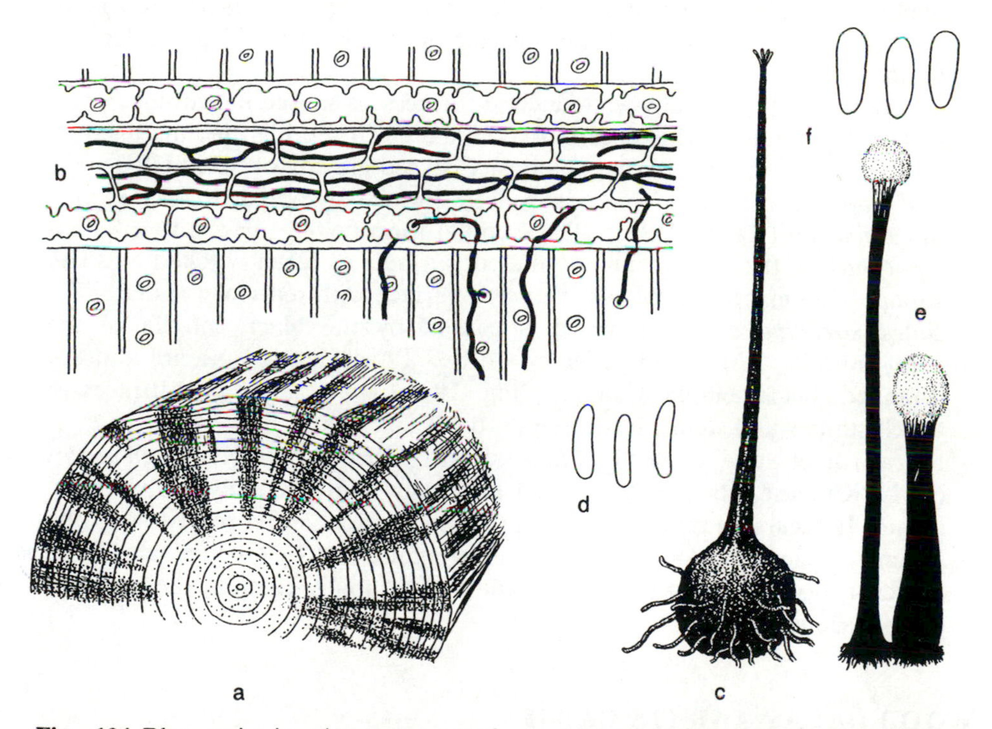

Fig. 104 Blue stain in pine stem wood. **a** sectored staining of the sapwood, **b** radial-longitudinal section through blue-stained sapwood with hyphae of blue stain fungi in the cells of a ray, **c** perithecium of *Ophiostoma piceae* with ascospores (**d**), **e** *Graphium* state with conidia (**f**)

Blue stain can also occur secondarily in sawn timber that is not completely air dry; this is caused by fungi that infect the cut surfaces. The main species involved are members of the genus *Cladosporium*, but *Strasseria geniculata* also occurs frequently; this fungus is one of the causes of the shoot tip disease of pine seedlings (Fig. **12e**).

Finally, there is a form of blue stain (tertiary blue stain) that occurs after the timber has been converted into products and painted, having ceased to be the concern of the forester. It affects products that tend to re-imbibe moisture while in service; for example, garage doors and clear-varnished window frames. The causal agents are the imperfect fungi, *Aureobasidium pullulans* and *Sclerophoma pithyophila* (Fig. **23c**).

The surest protection against blue stain in logs is to fell the trees at the appropriate time—from October to January—and to utilize them quickly. In cases of delay in getting the timber away, storage in water or beneath water sprinklers has proved worthwhile. Blue stain in sawn timber can be avoided both by storing in airy conditions and by the use of aqueous fungicides. To prevent blue stain on painted wood, a fungicidal priming coat can be used.

Other Wood Stains Caused by Fungi:

- A *grey stain* of poplar wood, caused by the numerous, finely divided, brown hyphae of the imperfect fungus, *Phialophora fastigiata* (Lagerb. & Melin) Conant, occurs in stems of various poplar species left lying for lengthy periods.
- *Blackening* of periodically wetted wood surfaces, is caused by various imperfect 'sooty moulds.' This occurs predominantly in the high mountains, and mostly in association with the abiotically caused greying of wood [39].
- *Red spotting* of beech wood is caused by the ascomycete *Melanomma sanguinarium* (Karsten) Sacc. The conidial form belongs to the form-genus *Aposphaeria.* It occurs mainly on the cut surfaces of recently felled logs and stumps. The mycelium colours the wood surface with red-violet spots.
- *Black streaking* of beech wood is caused by the black conidia of the hyphomycete, *Bispora monilioides* Corda. Typical symptoms are radially arranged, black, elliptical streaks (Fig. **109a,b**) on freshly cut surfaces of beech stumps and stems, less often on birch and oak.
- *'Green rot'* of wood, caused by *Chlorociboria* (*Chlorosplenium*) *aeruginascens* (Nyl.) Kanouse, stains rotten wood internally and externally a blue-green colour. It occurs on thicker branches of various broadleaves left lying on the ground. A similar colour effect can be brought about by green algae which, however, occur only on the surface and preferentially colonize the cut ends of coniferous stems.

WOOD DECAY AND ITS CAUSES

The felling of a tree creates new ecological conditions for wood-inhabiting saprotrophic fungi and other organisms, which may either gain a foothold in the wood as it dries out or may already be established in the living tree. Among the wood decomposing species that may already be active, there are some, such as *Fomes fomentarius*, that can continue to grow and even to fructify for some time. However, the environment soon becomes so extreme that existing colonists, if present, are succeeded by others that can thrive better in drying wood. This microflora includes many species of Aphyllophorales as well as organisms that are not active in standing stems; these include specialized ascomycetes, imperfect fungi, and bacteria that are responsible for the later stages of breakdown of the wood or wood residues.

The intensity and speed of decomposition of the wood depends only partly on the species of fungi involved. Other factors that determine this include humidity, temperature, and also the inherent wood durability of the tree species concerned. So, for example, the heartwood of larch, pine, and Douglas fir is relatively resistant to fungal attack, whereas their sapwood is easily decayed. In the case of spruce and Silver fir, the whole of the stem wood has little decay resistance. Among broadleaved trees, the heartwood of robinia, oak and elm is very durable, while the wood of beech, maple, birch, and Horse chestnut can be readily decomposed by fungi.

There is a great range of wood-decomposing fungi which occur on fallen and felled timber, but only a few that are noticed for their size, colour or frequency will be described here. For further study of the fungal groups concerned, more comprehensive mycological books must be consulted [29,110,175,176].

Gloeophyllum sepiarium (Wulfen:Fr.) Karsten

This is among the most active of the wood-rotting fungi on felled coniferous timber. It occurs regularly on conifer stumps, and it also colonizes converted coniferous timber such as posts, fencing stakes, poles, or other wooden structures if they have re-absorbed moisture and have not been treated with an appropriate timber preservative. The fungus transforms the wood into a disintegrating, cubical brown rot. Although it likes moisture, it can survive long periods of dryness, then resuming its destructive activities. As the decay at first develops only in the interior of the wood, an attack may go unrecognized for a long time.

The fruit bodies of the fungus normally occur on older stumps where they grow mostly in rosette-like clusters; on vertical surfaces they develop pilei which stand out clearly from the wood and which may coalesce. These are about 3 cm across, semicircular or kidney-shaped and with an initially rusty yellow to rusty brown upper surface which later darkens. They are characterized by a gilled hymenium on the underside, with crowded gills (about 14–24 per cm) which are occasionally fused together (Fig. **105**).

*Other **Gloeophyllum** Species on Coniferous Wood:*

- *G. abietinum* (Bull.:Fr.) Karsten; fruit bodies are pale ochre to cinnamon brown and often in the form of narrow strips up to 30 cm long, emerging from drying cracks; gills are less crowded than in the previous species

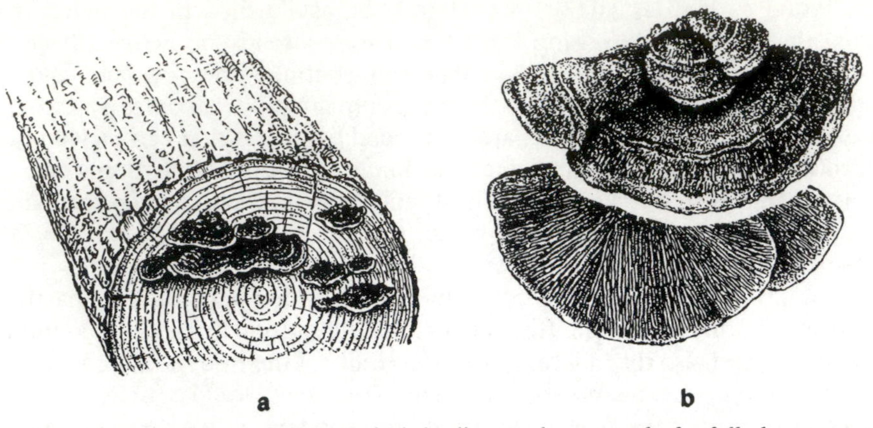

Fig. 105 *Gloeophyllum sepiarium*. **a** fruit bodies on the cut end of a felled spruce stem, **b** upper surface (above) and lower surface (below) of a fruit body

(only 9–11 per cm). It causes an active brown rot, especially on converted coniferous timber.

- *G. odoratum* (Wulfen:Fr.) Imaz.; fruit bodies are often appressed to the substrate, are yellowish orange when young, and later with a black boss. It typically smells of fennel or aniseed and causes a brown rot. It frequently occurs on the cut surface of older conifer stumps.

Schizophyllum commune Fr.:Fr.

The 'Split gill' can be found all year round. It is the only species of its genus and occurs preferentially on lying broadleaved stems where it is among the first colonizers, quite often in company with *Trametes hirsuta*. It often appears on beech logs left lying outside the forest where it can survive strong insolation and temporary desiccation without harm. However, its fruit bodies are also to be found on standing beech trees which have suffered sunscorch. Its remaining habitat is converted and painted timber. The fungus causes a white rot.

The fruit bodies are mostly clustered, 2–5 cm wide, mussel- or fan-shaped, sessile, tough leathery, and often have scalloped margins. Their upper surfaces are pale grey and felty-woolly; on the undersides are the reddish brown 'gills,' arranged like a fan and split lengthwise along their edges (Fig. **64**).

Daedalea quercina (L.:Fr.) Fr.

The 'Maze-gill,' which occurs principally on oaks but also on Sweet chestnut in southern and some other parts of Europe, finds a habitat as a saprotroph both on stumps, and on converted timber where it causes an intensive brown rot. It can, however, also transfer to living trees as a wound parasite. The greatest risk applies to Red oaks planted along roads and in parks where attacks quite often raise questions of traffic safety.

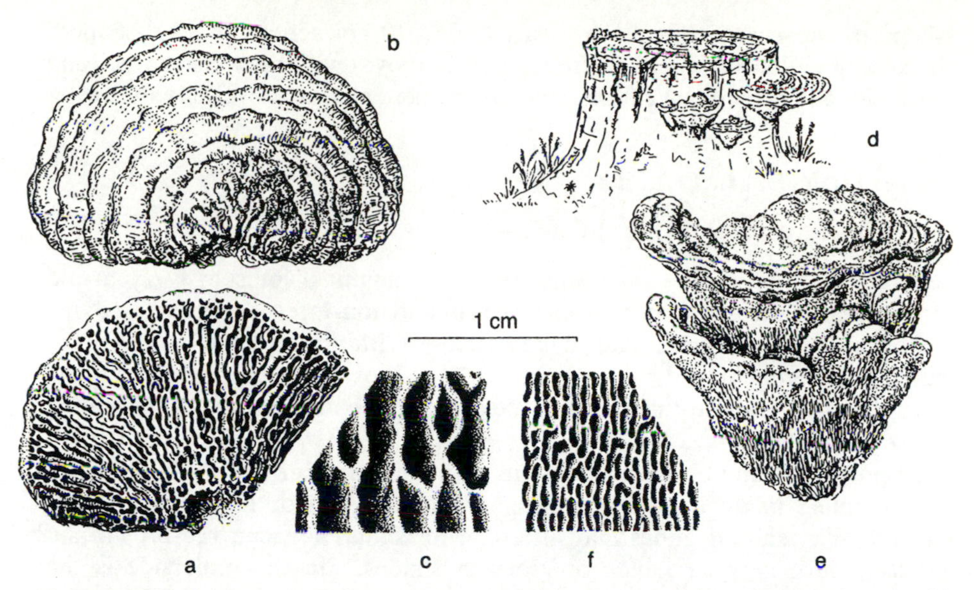

Fig. 106 Fruit bodies of polypores on broadleaves. **a-c** *Daedalea quercina*: **a** underside, **b** upper surface, **c** enlarged portion of the underside; **d-f** *Trametes gibbosa*: **d** fruit bodies on a beech stump, **e** view of fruit bodies, **f** enlarged portion of the underside

This easily recognized fungus is characterized by its bracket-shaped fruit bodies, 5–20 cm across, which can develop on standing stems both near ground level and up to a height of several metres on branch wounds or dead branches. The relatively flat and smooth upper surface of the pileus is pale brownish to greyish brown, with more or less distinct concentric zonation (Fig. **106a,b**). Its undersurface is characterized by millimetre-wide, widely separated tubes, broken up in a labyrinthine, gill-like pattern. This appearance has earned the fungus its generic name, which comes from Daedalus who, in Greek mythology, is said to have constructed the labyrinth for the Minotaur on Crete. Its tough-leathery fruit bodies can persist for several years, during which time the tubes continue to grow without visible dividing lines. If, for the purpose of identification, exclusive reliance is placed on the tube layer, confusion could arise with *Trametes gibbosa* though this has more closely spaced, radially elongated, white pores, and, in addition, grows preferentially on beech (Fig. **106d–f**).

Similar Species with Elongated Pores on Broadleaves:

- *Trametes (Pseudotrametes) gibbosa* (Pers.:Fr.) Fr. is frequent on beech stumps and lying wood or other broadleaved species. Fruit bodies are 8–15 cm across, often with a characteristic boss on the top surface; upper surface is grey-white, often greenish when old due to light algal growth, with elongated pores beneath. It causes a white rot. Brackets are readily eaten away by beetle larvae (Fig. **106d,f**).
- *Daedaleopsis confragosa* (Bolton:Fr.) J. Schröter occur predominantly on still-standing but dead stems or branches of birch, alder, or willow. The

shape of the fruit bodies is very variable, 3–10 cm across, bracket-shaped, becoming violet-red where bruised, smooth above, with brownish zones, and with elongated pores or gill-like hymenium beneath. It causes a white rot.

Trametes versicolor (L.:Fr.) Pilát

Syn. *Coriolus versicolor* (L.:Fr.) Quél.

The 'Many-zoned polypore' is a regular and wide-ranging colonist of 4–6 year-old beech stumps in the forest, but can also be found in non-forest areas and on dying trees or on stumps of any broadleaved species. Within the fungal communities occurring on broadleaved tree stumps, it is one of the most vigorous wood decomposers. A white rot develops in the affected wood, occasionally assuming a straw yellow tint.

This polypore can be recognized by its thin, semicircular or fan-shaped, 5–8 cm broad pilei, its upper surface is often colourfully zoned. The concentrically arranged, silky shining zones may alternate in colour between red-brown and yellowish, dark grey, and blue or greenish shades. On the underside is the yellowish white tube layer, which becomes pale orange on drying. The pores appear indented and can only be seen with difficulty by the naked eye (Fig. **107**).

Similar Species with Rounded Pores on Wood of Broadleaves:

- *Trametes (Coriolus) hirsuta* (Wulfen:Fr.) Pilát is a saprotroph on lying stems, also on sun-scorched trees (Fig. **64c**) and on converted timber. Fruit bodies have grey or yellowish, stiff hairs above and roundish, yellowish white pores beneath. It causes a white rot.
- *Pycnoporus cinnabarinus* (Jacq.:Fr.) Karsten lives saprotrophically on dead wood of broadleaves, particularly on birch, beech, and rowan. It colours the wood red. Fruit bodies are 3–8 cm across, flat, and bright vermilion. It causes a white rot.
- *Bjerkandera adusta* (Wild.:Fr.) Karsten. is a saprotroph on stumps of many broadleaves, also occasionally on dying branches of living trees. The undersides of fruit bodies are characteristically silver-grey with very small pores, becoming black where bruised. It causes a white rot.

Stereum hirsutum (Willd.:Fr.) S.F.Gray

The 'Hairy Stereum' lives predominantly as a saprotroph on felled stems and lying wood of various broadleaved species, occurring as a primary colonist and capable of causing a definite white rot after a few months.

This member of the fungal genus *Stereum* is characterized by its smooth, poreless hymenium and the yellowish upper surface which, in the fresh state, is orange with felted hairs. As a rule, the fruit bodies form carpets of imbricate brackets. These are often laterally joined, then forming horizontally arranged undulating rows. In the woods or in the timber depot, this fungus is often to

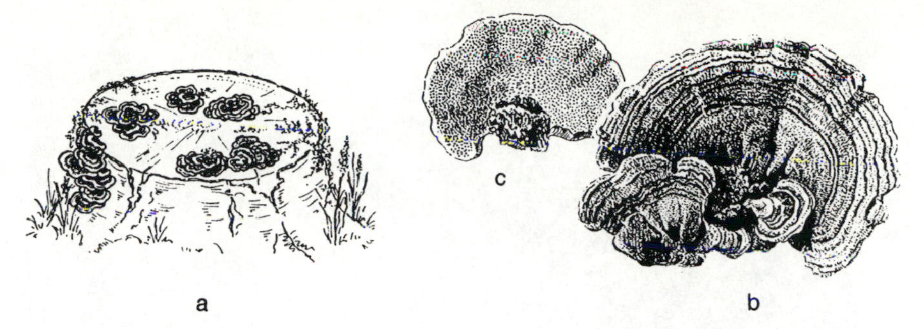

Fig. 107 *Trametes versicolor*. **a** beech stump with fruit bodies, **b** upper surface of a fruit body, **c** underside with pore layer

be found on the cut ends of stacked oak logs, its fruit bodies colonizing the sapwood at the outer edge of the stem. If it is allowed to continue developing for a longer period, its fruit bodies will also be found on the bark all over the stem (Fig. **108a**).

Similar Species with Smooth Hymenia, on Wood of Broadleaves:

- *Stereum rugosum* (Pers.:Fr.) Fr. Fruit bodies are closely appressed to the substrate with a white, free margin. Hymenium are whitish to ochre, reddening when wounded. It is a common saprotroph and cause of 'Weißstreifigkeit' (white streaking) in the wood. It also causes a canker disease on Red oak.
- *Chondrostereum purpureum* (Pers.:Fr.) Pouzar. Fruit bodies are violet on the hymenial surface and attached broadly to the substrate. The upper margin runs out into extended pilei. A primary colonist of cut surfaces on broadleaved

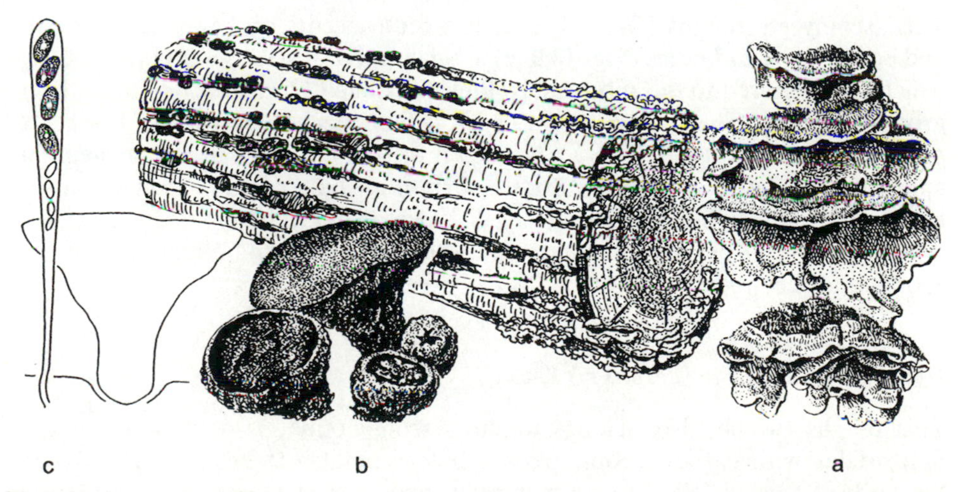

Fig. 108 Storage rot fungi on oak. **a** fruit bodies of *Stereum hirsutum*; **b** fruit bodies of *Bulgaria inquinans*, **c** ascus with ascospores

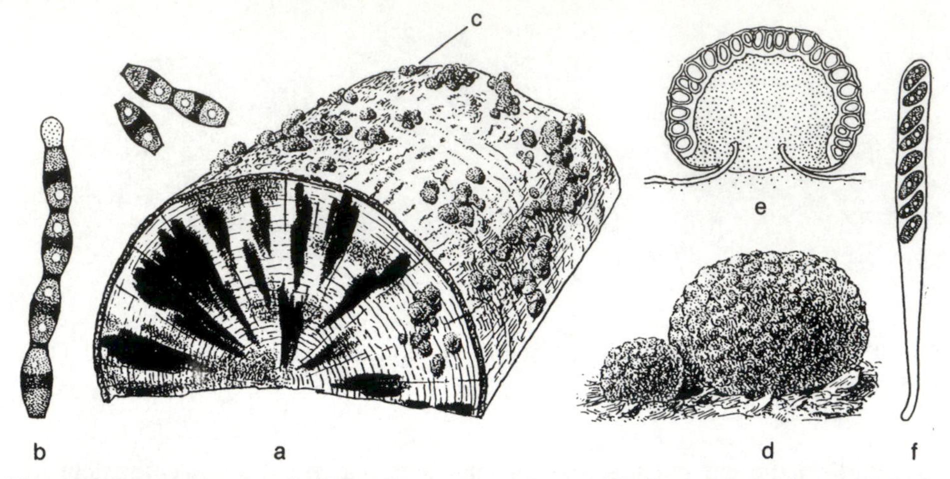

Fig. 109 Storage rot fungi on beech. **a** spore mats of *Bispora monilioides* on the cut surface of a beech butt, **b** conidial chains; **c-f** *Hypoxylon fragiforme*: **c** fruit bodies on a piece of beech stem left lying, **d** view of fruit bodies, **e** cross-section through a mature fruit body with immersed perithecia, **f** ascus with ascospores

timber, particularly beech, it causes silver-leaf disease in standing fruit trees (cf. p. 162).

- *Cylindrobasidium evolvens* (Fr.:Fr.) Jülich. Fruit bodies are in the form of small, white patches, soon running together, with a narrow, white margin. Older fruit bodies cream ochre in colour and cracking. It occurs on fresh cut surfaces, especially on beech.

Bulgaria inquinans Pers.:Fr.

This ascomycete fungus (Order Helotiales) often occurs on lying stems of oak and beech. Its fruit bodies (Fig. **108b,c**) are gelatinous discs, 1–3 cm in diameter, which are anchored to the substrate by their top-shaped basal portion and which grow in groups. Their upper surfaces are shiny black to brownish black. A peculiarity are the asci which are arranged in parallel rows in the hymenium and which contain two different forms of spore: 4 larger, dark, kidney-shaped and 4 smaller, hyaline ascospores. Both types are germinable. Any observer touching the fruit bodies will be struck by the blackness and abundance of the spores that they eject.

Hypoxylon fragiforme (Pers.:Fr.) Kickx

This fungus, which also belongs to the Ascomycotina, is a member of the Sphaeriales with carbonaceous, perithecoid ascomata. It occurs regularly on the bark of freshly felled stems or broken branches of beech causing a white rot, although this does not penetrate very deeply into the wood. The decayed wood contains fine, black 'zone lines'. The spherical fruit bodies, about 1 cm across, arise mostly in groups on the bark-covered surface. They are reddish brown outside, black within, and have the consistency of charcoal. Immersed in the outer surface of the fruit body are similarly spherical perithecia in which elongated asci are formed, each containing 8 ascospores (Fig. **109d–f**). Older groups of fruit bodies stand out because the bark in their immediate vicinity appears to be dusted with a black powder—spores.

Xylaria Species

The members of the ascomycete genus *Xylaria* are characterized by the particularly large and striking fruit bodies they produce, sometimes of fantastic shapes. They are predominantly saprotrophic, occurring on wood, bark or substrates containing cellulose. Only a few species are capable of gaining a foothold on living trees. Among the more common species are:

- *Xylaria hypoxylon* (L.:Fr.) Grev. occurs from November to March, usually in groups on rotten, decaying broadleaved tree stumps and lying wood, mostly beech. Development begins with the formation of small, black-stalked clubs a few centimetres high which, in the course of the winter, become differentiated into structures up to 6 or 8 cm in height with their tips branched like antlers (Fig. **110d**). The white tips, which look as if they are dusted with flour, are in fact covered in conidia. The perithecia develop on the swollen, club-like base, but not until the spring. Inside these, the black ascospores are formed, unevenly spindle-shaped and measuring 10–14 × 5–6 μm.
- *Xylaria polymorpha* (Pers.:Fr.) Grev. similarly colonizes broadleaved stumps, especially those of beech. Such stumps are usually dead and rotting, but there is a suspicion that the fungus can also attack living trees, giving rise to a

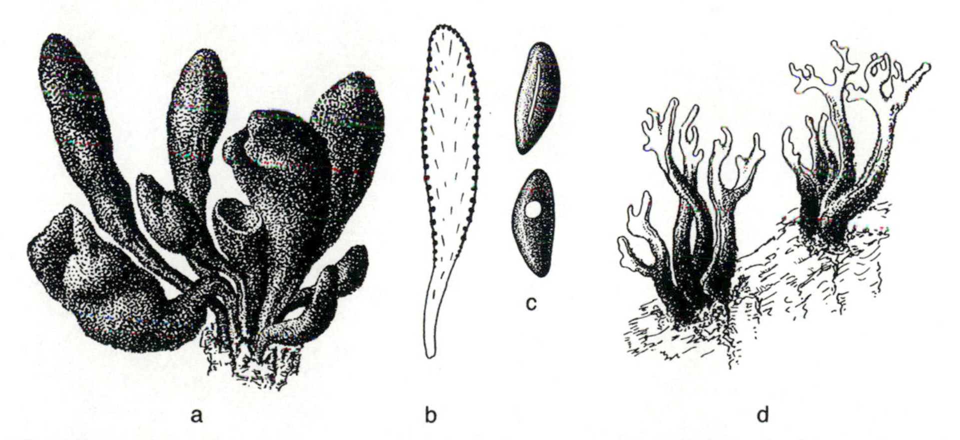

Fig. 110 Fruit bodies of ascomycetes on broadleaves. **a-c** *Xylaria polymorpha*: **a** view of fruit bodies, **b** cross-section through a fruit body, **c** ascospores; **d** fruit bodies of *Xylaria hypoxylon*

black root rot [197]. The fungus forms clusters of black fruit bodies which are finger-like to irregularly club-shaped and originate on the butt or on major roots. These structures are the basis for the common name 'Dead man's fingers' (Fig. **110a–c**). The ascospores are black, unevenly almond-shaped, and measure 20–30 × 6–9 μm.

9 *Epiphytes, Symbionts, and Parasitic Flowering Plants*

EPIPHYTES

Epiphytes are organisms which use trees or other plants exclusively as support, without parasitizing them. Some are capable of synthesizing their own organic nutrients (e.g. algae and lichens), while others (mainly fungi) live saprotrophically on dead or non-living material accumulated on the bark. Although epiphytes are not parasitic on trees, there are some (e.g. *Athelia, Trichonectria*) which are nevertheless 'parasitic,' attacking other organisms that live on the tree bark.

Among the epiphytes which occur in Europe, there are a few which attract the attention of the forester, or of others with an interest in trees because they are either economically important or have a particularly striking appearance.

Algae

Algae are autotrophic organisms, most of which live in water; only a few species are capable of thriving out of water—which they can do by absorbing water vapour from the air. In contrast to the filamentous, marine forms, the aerial algae form more or less rounded cells of the *Pleurococcus* type (Fig. **112b**) which join together in parcel-like complexes of cells. They colonize the bark of many tree species, forming extensive green coatings. Their development is favoured in areas of high air humidity, (e.g. near the coast or in moist, sheltered sites), and high nitrogen input (e.g. from anthropogenic emissions) also seems a favourable factor for certain green algae. Algae on the bark of trees are of no significance to the forester. A few aerial algae can, however, also colonize twigs and needles of conifers, sometimes enveloping them in a thick crust. This can have a detrimental effect on the quality of decorative Douglas fir and Silver fir foliage, as the alga-covered needles lose their sheen and become unsightly. For the tree itself, a thick algal coating results in a slight loss of photosynthetic efficiency. It is possible to prevent the formation of algal crusts through silvicultural methods; for example, by using more suitable sites and preventing the development of dense stands.

Fungi

Among the epiphytes that occur on trees there are numerous fungal species. Since they use the tree only for support, they are of no pathological significance but some, such as the sooty moulds, can spoil the aesthetic value of needles or leaves. Also, there are fungi which are falsely blamed for causing disease, the

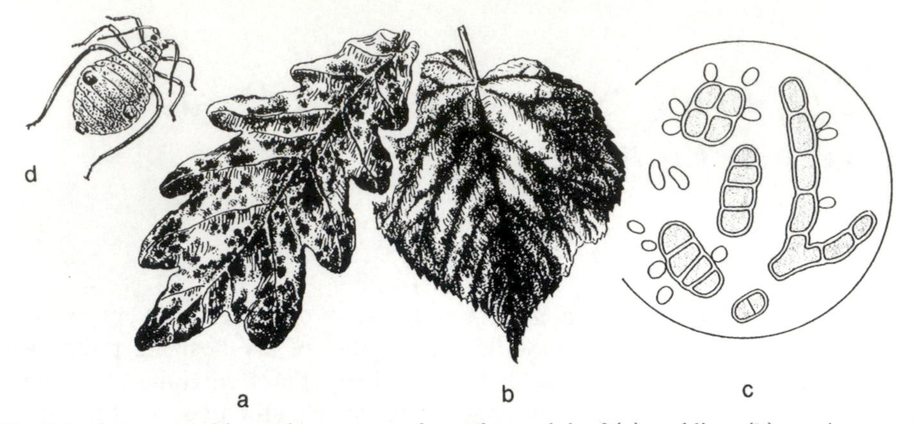

Fig. 111 Sooty moulds on the upper surface of an oak leaf (**a**) and lime (**b**), **c** microscopic view of sooty mould fungi, **d** oak bark aphid,' a honeydew producer

more so because they occur in striking and apparently threatening form. One example is *Athelia epiphylla*.

- The term 'sooty mould' is applied to a number of systematically disparate fungal species, characterized by their epiphytic mode of life, which form black coatings on leaves, needles or twigs of various tree species. Some are known as 'black yeasts'. They obtain their nourishment principally from sugary excretions (honeydew) of various insects or from other organic materials secreted by the plant itself. They tend to be noticed as a dark, albeit thin, coating on the upper surfaces of smooth leaves such as those of oak, lime, willow, and plum. Quite often, members of the lay public suppose this to be a deposit of dirt from the street (Fig. **111**).

  If a sooty mould is examined under the microscope, numerous brown, mostly thick-walled, roundish cells can be seen which either form multi-celled spore packets or short chains of cells. This appearance is typical of most of the diverse range of fungal species that occur as sooty moulds, all of them being suited to the high osmotic value of the sugary substrate. Among them are several species of the imperfect genera *Hormiscium, Triposporium,* and *Sarcinomyces* together with better known black fungi such as *Aureobasidium pullulans* and *Cladosporium herbarum*.

  For the tree, a heavy coating of sooty moulds can mean some, if only a little, reduction in assimilation. For the forester, sooty moulds are undesirable where Douglas fir or Silver fir branches are to be harvested for decorative foliage. These black fungi can also cause a nuisance in the urban setting if they are deposited in rain and leave behind dark spots on the clothing of passers-by. Occasionally, as a precaution against such eventualities, the use of insecticides against the insects which produce the honeydew is recommended. Application is by drenching, spraying, or by the use of injector capsules.
- *Athelia epiphylla* Pers:Fr. is a basidiomycete in the family Corticiaceae. Owing

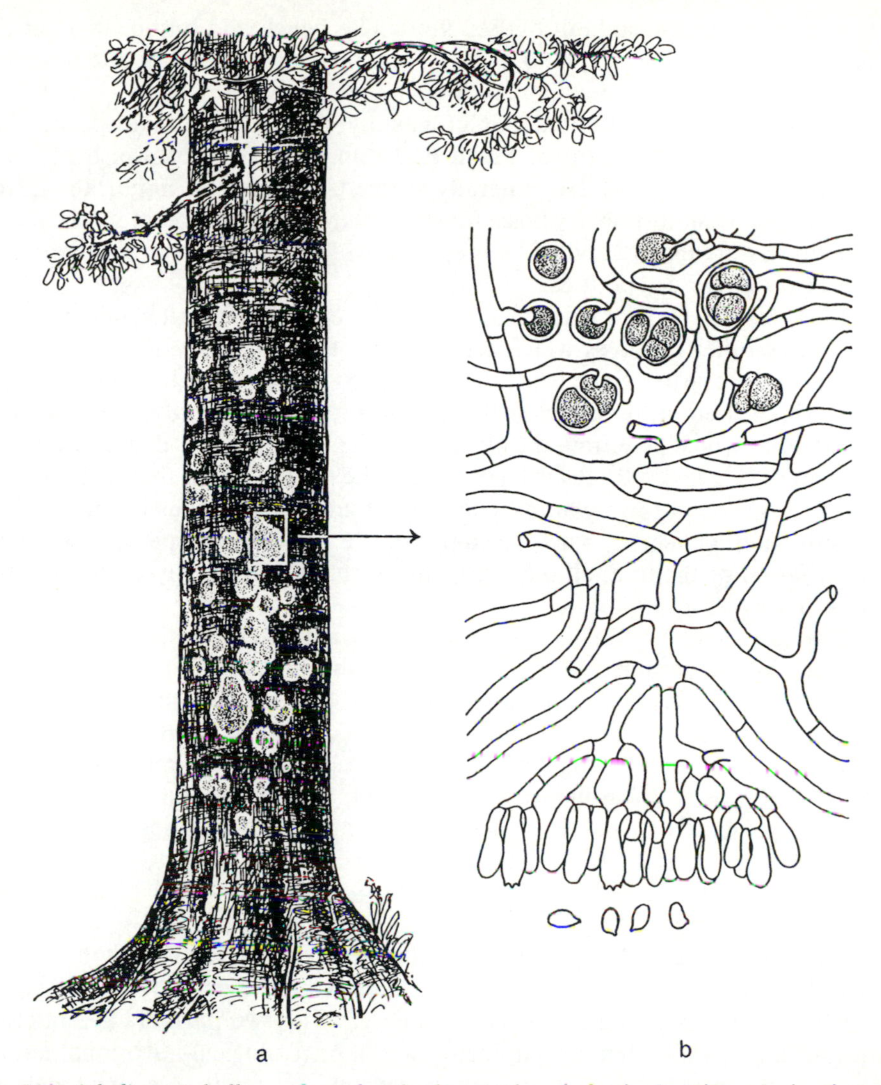

Fig. 112 *Athelia epiphylla*. **a** fungal colonies on beech bark, **b** microscopic view of a mycelial mat with algae (top), mycelium (middle), and basidia (bottom) (b after Oberwinkler 1977)

to its striking appearance, it can be erroneously taken to be the cause of disease. The 'symptoms' are the formation of white, often coalescent patches and rings on the bark of various tree species, clearly visible at a distance (Fig. **112**). The fungal colonies, which are closely appressed to the bark surface, often reach the size of the palm of a hand where forest conditions are optimal for their development (high air humidity). In urban situations, the patches are mostly smaller and then coloured brownish grey in the centre. In winter, the white patches disappear and the natural colour of the bark returns. Favoured tree species are the smooth barked kinds like beech and hornbeam, but the fungus also occurs on maple, oak, lime, and poplar; even spruce and larch are used as a substrate in certain conditions.

The fungus is distributed sometimes by means of basidiospores, but these are not always formed. Small (0.2 mm diameter), brownish mycelial aggregations (sclerotia) are generally formed and may be important in the absence of basidiospores whose function they seem to assume. They also function as organs of survival by means of which the fungus overwinters and can also survive dry periods.

With regard to its mode of life, *Athelia epiphylla*—which, incidentally, has many other relatives which are not easy to distinguish from each other [115]—is a parasite, since it attacks the green algae which live on the bark, enmeshing them in its mycelial threads and forming fine absorptive hyphae (haustoria) which penetrate the algal cells (Fig. **112a,b**). The damage that this does to the algae can be easily be seen with the naked eye, as their colonies die and turn greyish brown wherever the mycelium of the fungus has been. Any bark-inhabiting lichens that are also present (e.g. *Lecanora* species) suffer the same fate. For the tree, all these organisms are merely epiphytes that use the bark purely as a support.

Similar patches on the bark, though smaller and greyish violet in colour, are formed by the ascomycete *Trichonectria hirta* (Bloxam) Petch. This fungus also obtains its sustenance parasitically from green algae and uses the tree merely as a support. To distinguish this from *A. epiphylla*, it is necessary to look in summer for its orange fruit bodies, containing several-celled, cylindrical conidia, 25–38 × 7 μm in size (Table **II,19**); and in autumn for the perfect state, represented by reddish perithecia surrounded by bristles.

Lichens

Numerous lichens can be counted among the epiphytes on trees. Lichens occupy a special position in the Plant Kingdom, insofar as each species has a dual structure which comprises both a fungus and an alga. The two partners live in close symbiosis and form both a morphological and a physiological-nutritional entity. Lichens occur in many morphological forms; they grow either as encrustations ('crustose' lichens) or as foliose or frondose thalli. The tree-living forms occur principally in places of high air humidity or high rainfall, growing on dead or living branches, on the bark, but also on the ground, on rocks, or wooden posts.

Resistant though the lichens are to desiccation, heat, and cold, most species react very sensitively to anthropogenic pollutants, especially to SO_2. On the basis of their differing sensitivity to sulphur dioxide [101], certain species of lichen can be used as direct bioindicators for gauging air quality. Among the very sensitive species are members of the genera *Usnea* and *Ramalina*, which are nowadays rarely found in heavily populated areas. In contrast, *Hypogymnia physodes* is an acid-tolerant species which still occurs in areas with an SO_2 concentration of 60 μg m^3 (Fig. **113a**). *Lecanora conizaeoides* can tolerate even higher concentrations (Fig. **113b,c**); this species is far more common now than

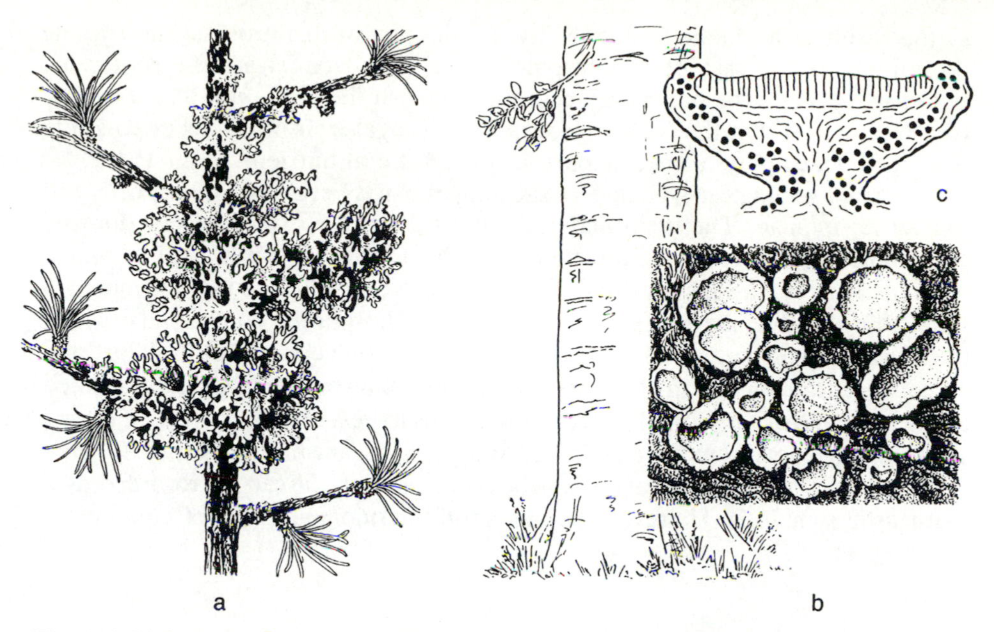

Fig. 113 Lichens. **a** *Hypogymnia physodes* on a larch twig; **b** *Lecanora conizaeoides* on beech bark, **c** cross-section of fruit body

previously, as it is able to penetrate even into the cities—which otherwise would be counted as 'lichen deserts.'

In forestry, the lichens have no undesirable significance, although they impair gas exchange in plants where they are exceedingly prolific. In commercial fruit growing, lichen growth is not welcome, as harmful insects are able to overwinter more successfully beneath the lichen cushions.

SYMBIONTS

Mycorrhizas

The term mycorrhiza denotes an intimate coexistence of a fungus with the roots of a higher plant. In this way of life, there is considerable exchange of substances which in most cases is mutually beneficial.

Since the time of Frank [78], two forms of mycorrhiza have been distinguished on the basis of morphological and anatomical features: the endotrophic, vesicular-arbuscular mycorrhizas, (VA mycorrhizas), and the ectotrophic mycorrhizas. The VA mycorrhizas occur with many herbaceous plants and also with tropical forest tree species. On the other hand, nearly all our woodland and parkland trees form associations with ectotrophic mycorrhizal fungi [94].

Ectomycorrhizal roots are morphologically different from the thread-like absorptive roots of the host in its non-mycorrhizal state, becoming dichotomously branched or coralloid, shortened, and thickly cylindrical—presumably as the result of hormone secretions by the fungus. In this process, the hyphae do not—as in the case of endotrophic mycorrhizas—penetrate the root cells, but grow between the outer bark cells and enmesh them in a net-like structure (the Hartig net). On the root surface, a loose fungal weft or a thicker covering is formed which plays a specific part in the uptake of nutrients (Fig. **114**).

A close physiological exchange of substances exists between the root cells and the fungal hyphae. The main materials that the fungus extracts from the tree are simple, soluble sugars and certain growth substances, e.g. thiamine, which it is unable to synthesize for itself. The tree receives water and various nutritive salts from the fungus; principally nitrogen and phosphorus compounds, which the much-branched mycelial net takes up from the soil very efficiently. Thus, the mycorrhiza functions as an efficient nutrient-absorbing organ for the tree. Linked to the presence of a mycorrhiza, there also seems to be a certain protective effect, shielding the roots from the attacks of pathogenic organisms in the rhizosphere.

As the mycorrhizal association benefits both partners, it can be regarded as a mutualistic symbiosis. However, the apparently harmonious state of equilibrium

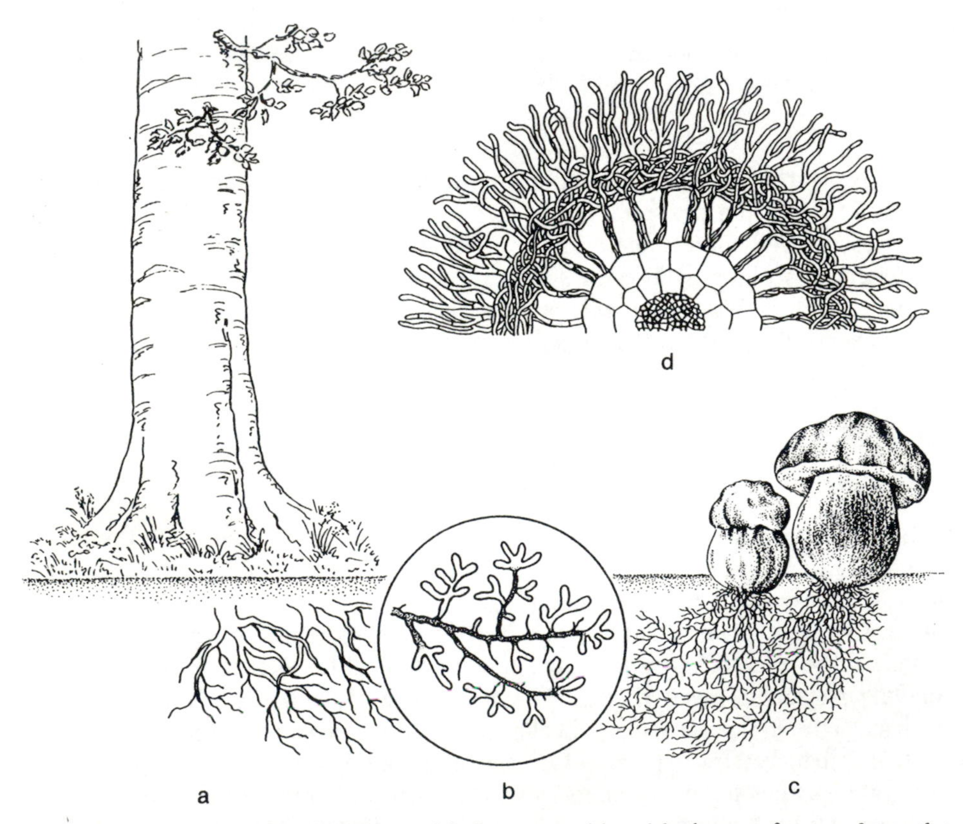

Fig. 114 Mycorrhizas. **a** beech in symbiotic partnership with the cep fungus, **b** rootlets distorted by fungal mycelium (view through a hand lens), **c** fruit bodies of the cep (*Boletus edulis*), **d** cross-section (detail) through a root tip with an ectotrophic mycorrhiza

can be tilted by environmental influences in favour of one partner or the other, triggering clearly aggressive or defensive reactions. If one considers this latent aggressive character of the mycorrhizal partner, the reciprocal relationship between tree and fungus is best illuminated by Melin's statement, '. . . that fundamentally the mutualistic symbiosis is a state of dual parasitism exploited by both participants.'

The mycorrhiza-forming fungi include many members of the Agaricales and Boletales. Some of these are strongly host specific (e.g. *Suillus grevillei* on *Larix* species); others form mycorrhizas with various species of tree (e.g. *Boletus edulis*). Further examples of mycorrhizal fungi are the agarics *Amanita, Lactarius*, and *Russula* spp., various poroid fungi (boletes) of the genera *Boletus, Suillus, Xerocomus*, and *Leccinum*, together with other basidiomycetes, such as *Scleroderma* spp. Among the ascomycetes, there are only a few isolated examples of species which enter into an ectotrophic symbiosis with trees. Among these, the genus *Tuber* can be mentioned. Finally, there are also some deuteromycetes (Fungi Imperfecti) which are known to form mycorrhizal partnerships.

In a healthily structured forest, mycorrhizas occur naturally and require no special management. Special measures are required, however, where afforestation is to take place on land that contains no suitable mycorrhizal fungi, e.g. on steppes or prairies, at high elevations, or where soils are thin and impoverished. Here, care must be taken to see that the appropriate mycorrhizal fungi are added to the soil by incorporating woodland humus so that the seedling roots will come into sufficient contact with the mycelium.

Particular interest has recently been focused on mycorrhizas in connection with the 'new type' of forest damage. This damage involves not only the well known symptoms on needles and leaves, but also a number of quantitative and qualitative changes that have been discovered in the fine root systems of various tree species. The most striking feature is a reduction in the density of the fine mycorrhizal roots, linked with a diminution in the production of fungal fruit bodies and in the range of fungal species involved.

At the microscopic level, ectomycorrhizas from trees in air-polluted stands take on some of the appearance of endomycorrhizas, with hyphal penetration of cells in the bark parenchyma leading to a shift in the 'symbiotic' balance in favour of the fungus. If the amount of damage increases, the fine roots rot and the supply of nutrients and water is impaired. It is very probable that some of the striking symptoms of the present 'Waldsterben' (forest decline)—crown thinning, acute needle yellowing—are directly related to the loss of mycorrhizas [93].

PARASITIC FLOWERING PLANTS

Common Mistletoe

Viscum album L.

The Common mistletoe is a hemiparasite of the family Loranthaceae. Hemiparasites possess green leaves and therefore can secure a proportion of their nutritional requirements for themselves. However, they extract water and minerals from the host plant, to whose water-conducting vessels they are

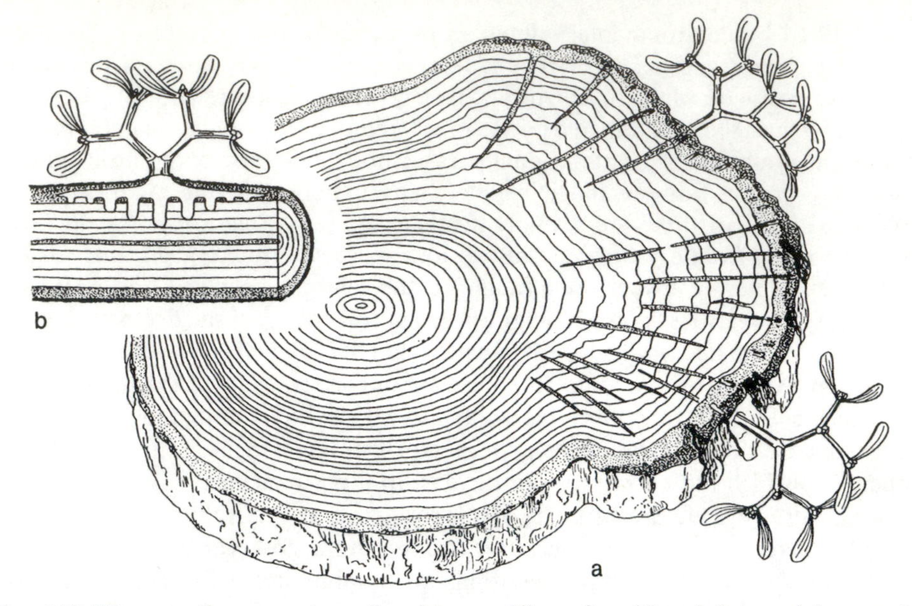

Fig. 115 *Viscum album.* **a** stem disc from a Silver fir with mistletoe sinker roots penetrating deep into the wood, **b** diagrammatic longitudinal section through a stem section from a poplar bearing mistletoe, showing 'bark roots' extending horizontally and 'sinker roots' extending vertically

attached. Many broadleaved and coniferous genera are known to act as hosts [221].

Mistletoe forms spherical, bush-like growths reaching as much as 1 m in diameter which are particularly eye-catching in winter when exposed by the loss of the leaves of the host tree. The small green stems of the parasite, which arise in clusters from a segment of the branch, form forked twigs, each bearing two leathery, green, opposite leaves. The plants are dioecious; in autumn, white, berry-like, pea-sized fruits with tough, slimy flesh and containing oval or three-cornered seeds arise from the female flowers.

The seeds, which are distributed by birds, germinate if they reach the bark of a suitable host plant, each initially forming a sticky adhesive disc at the tip of the radicle. From the middle of this disc, a fine primary sinker root grows and is able to penetrate the bark tissues. This main root penetrates as far as the woody cylinder, but does not grow into it. Each year, perpendicular, dowel-like, so-called bark roots arise from the zone of the main root that lies within the host bark. From these roots, two to three secondary sinkers (haustoria) reach out as far as the wood. As the tree continues to grow in thickness, the sinkers are overgrown by the wood so that they appear to be forcing their way into it. So that the connection with the green shoot is maintained, the sinkers must subsequently be able to lengthen themselves. This is achieved by intercalary growth of a meristematic zone of tissue situated in the cambial region of the infected branch. After 10–60 years, the ability to grow in length is lost; the sinkers die and leave channels behind in the wood which appear in boards as holes (Fig. **115**).

The Common mistletoe occurs as three subspecies in Europe; these are distinguished less by their particular morphological features than by their biological behaviour. Thus, the broadleaf tree mistletoe (ssp. *album*) occurs on about 36 broadleaved woody plants; the Silver fir mistletoe (ssp. *abietis*) lives almost entirely on species of Silver fir; and the pine mistletoe (ssp. *austriacum*) has various species of pine as its principal host plants.

In pine forest, mistletoe occurs mainly in the crown, and this can cause stunted growth. In Silver fir forest, mistletoe is not welcome, on account of the reduction in the value of the stem wood it causes. Damage to fruit trees can also be serious; mainly in the case of apple which is the most common host for mistletoe in Europe.

Control measures consist entirely of cutting out the infected branches below the point of attack—a procedure which can reap an additional commercial benefit, as mistletoe boughs are much sought after as room decoration at Christmas time.

Related Species:

- *Loranthus europaeus* Jacquin attacks various oak species, less often Sweet chestnut; its 20–40 cm high, dark brown branches bear deciduous leaves and yellowish green, berry-like false fruit (Fig. **116c**). It causes branch swellings in the host plant; heavier attacks lead to serious growth reduction and occasionally to the death of oak trees before they are ready for harvesting. The false fruits are distributed by birds, particularly thrushes. Control is by mechanical removal of the mistletoe 'bushes,' by crown reduction and by cultural measures from the age of 40 years. Distribution is mainly in the northern, subpannonic oak region.
- *Arceuthobium oxycedri* M. Bieb. occurs in southern Europe on various species of *Juniperus*; it forms yellowish green, leafless shoots which break out of the bark of the branches like pegs. In contrast to several North American *Arceuthobium* species, this is of no economic importance.

Dodders

Cuscuta species

Cuscuta species are full parasites (holoparasites) in the family Convolvulaceae, which have almost no leaves and no chlorophyll. In the forest, they attack mainly maple, hazel, poplar, and willow (Fig. **116a**). On these hosts damage is insignificant, although *Cuscuta* species are more important on agricultural crop plants.

Toothwort

Lathraea squamaria L.

Toothwort, which similarly is a chlorophyll-free holoparasite, colonizes the roots of various broadleaved trees and shrubs, particularly hazel, forming knot-like

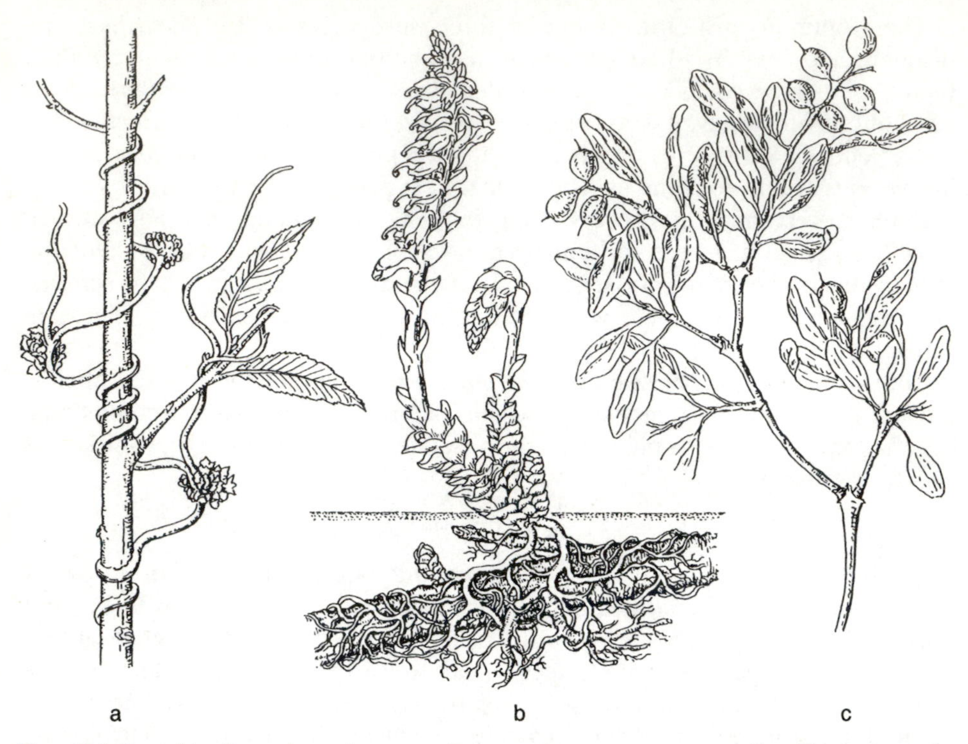

Fig. 116 Parasitic flowering plants. **a** *Cuscuta europaea* on a small willow stem; **b** *Lathraea squamaria* on hazel roots; **c** twig of *Loranthus europaeus* with ripe berries

haustoria by means of which it withdraws all its necessary nutrition from the host plant. Pink or pale violet-coloured inflorescences covered in scale-like leaves arise above ground in spring. There are no accounts of damage to the host plant (Fig. **116b**).

10 *Changes in Habit and Growth Abnormalities*

Every plant species varies within certain limits in the expression of its specific growth habit and the form of its component parts. The extremes of this variation occasionally attract attention but are not regarded as pathological changes. Such deviations are caused by various abiotic or biotic factors. Thus, for example, a light-demanding tree species grows with a bent stem if the light source is one-sided. Similar responses can be caused by the wind blowing predominantly from one direction. Animals can bring about various changes in form, such as the bushy growth which follows browsing by deer. Finally, plants too can influence the growth habit of trees; e.g. the stem constrictions and twisting caused by *Lonicera periclymenum* (honeysuckle). All these examples are induced by external factors and represent morphological adaptation within the basic pattern which is characteristic of that particular tree.

On the other hand, there are many morphological changes which tend to go beyond the intrinsic growth pattern of the tree species and which can be regarded as deformities. Examples of these include witches'-brooms on branches or stems, knobbly formations, tumours, burrs, or fasciations. In the case of leaves, the result can be curling or dwarfing, while abnormalities of inflorescences include excessively abundant cone production. Such deformations are caused in some cases by abnormalities in the genetic mechanisms that control development. In other cases, the developmental control can be disrupted due to infections by viruses and parasitic microorganisms, sometimes giving rise to strange and monstrous formations.

Witches'-brooms

Witches'-brooms consist of localized proliferations of bunched and mostly short twigs on morphologically normal branches. This results from the flushing of large numbers of dormant or adventitious buds. These branch in various directions to form a structure that may be round, flattened, or broom-shaped and which can reach up to a metre in diameter. Witches'-brooms evade the control over growth exercised by the tree; their shoots, unlike those of other twigs, are not plagiotropically arranged, but grow erect and to a certain extent take the form of miniature, independent trees on otherwise healthy branches. They arise predominantly on lateral branches (Fig. **117b**), but one may sometimes occur at the top of a coniferous tree whose upper crown can then be transformed into a single gigantic witch's-broom (Fig. **117a**).

Witches'-brooms can be seen on a very wide range of tree species. They quite often occur on birch, pine, Silver fir, and spruce. They are only occasionally

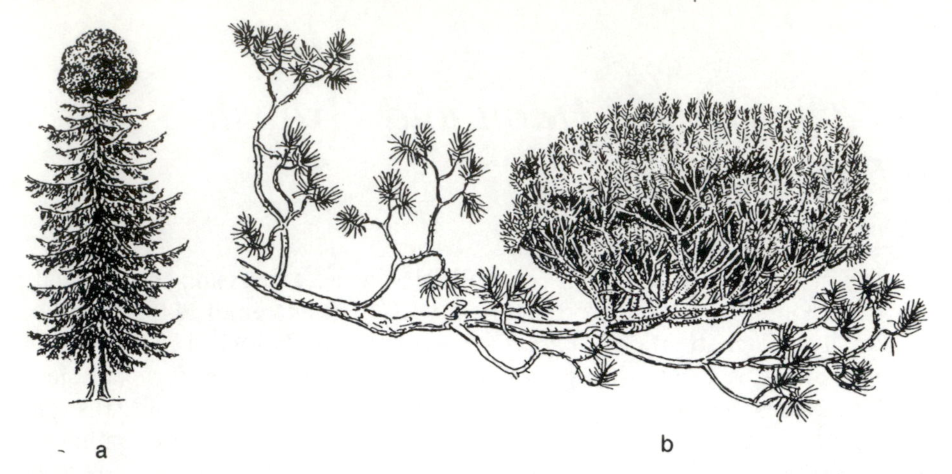

Fig. 117 Witches'-brooms. **a** witch's-broom formed on the leading shoot of a spruce tree; **b** 6-year-old witch's-broom on a lateral branch of Scots pine

found on beech, hornbeam, lime, and robinia and remarkably seldom on maple, Douglas fir, and larch.

Although witches'-brooms on different tree species have a fairly similar growth habit, they may be produced by different classes of causal agent. Thus, on birch and Silver fir, fungi are responsible for abnormal twig production; in the case of a witch's-broom on aspen, mycoplasma-like organisms (MLOs) have been shown to be the cause [186], while rickettsia-like organisms are responsible for the witch's-broom-like shoot abnormality which occurs on declining larch trees [143]. Finally, witch's-broom formation can also be induced by mites, as in birch, for example. Of the witches'-brooms that form due to fungal infection, the following merit further more detailed discussion:

- **Witches'-brooms on Silver fir** are induced by the rust fungus, *Melampsorella caryophyllacearum* (Link) Schröter. This heteroecious fungus lives out its haplophase on the fir and its dikaryophase on various members of the Caryophyllaceae. On the fir, the fungal mycelium first grows in the bark of young twigs where it causes small, knot-like outgrowths to appear. If a bud is penetrated by the fungus, it flushes in the following spring to form a miniature, erect, heavily branched 'tree'. The needles of these abnormal twigs are yellowish green and smaller than normal, and they bear the yellow fruit bodies (aecidia) of the fungus in summer. The aecidiospores from these must transfer to the dikaryotic-state host (Mouse-ear chickweed or stitchwort species) to complete the development of the life cycle.

  After the aecidiospores have been released, the needles are shed, so that there are never more than one year's needles on the twigs of the broom. Witches'-brooms of this kind can be many years old and can reach a considerable size. As most brooms occur on lateral branches, hardly any direct damage is detectable. Losses only arise if the fungus attacks the

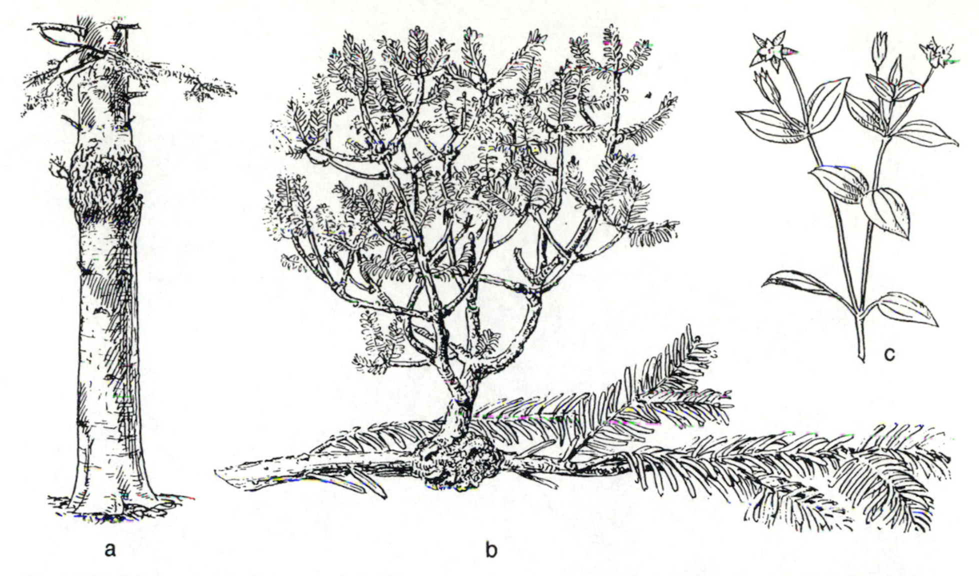

Fig. 118 *Melampsorella caryophyllacearum*. **a** symptoms on the stem of a European Silver fir ('Rädertanne'), **b** Silver fir witch's-broom; **c** *Moehringia trinervia* as a dikaryotic-phase host

main stem where it can give rise to the formation of goitre-like swellings in the wood. This malformation of Silver fir wood is known in German as 'Rädertanne.'

- **Witches'-brooms on birch** are the result of a persistent growth disturbance caused by the ascomycete *Taphrina betulina* Rostrup (Fig. **119**). Stimulated by specific chemicals produced by the fungus, large numbers of dormant buds flush, eventually forming densely branched, spherical bushes. The twigs infected by this fungus, unlike those occurring in mite-induced witches'-brooms, have onion-shaped swellings at the bases of their side shoots. Another feature of these shoots is that their tips point upwards as the result of negative geotropism. However, the most certain indication of fungal brooms are the asci which occur in early summer on the undersides of the leaves, and whose hyaline spores begin to germinate even while they are still in the ascus. As the fungus overwinters in the buds and in the bark, birch brooms can live for many years. No economic significance is attached to witches'-brooms on birch, although branches distal to them occasionally die.
- **Witches'-brooms on hornbeam** are initiated by the ascomycete *Taphrina carpini* (Rostrup) Johannsson. The bushes, which do not get very large, are not so common as the witches'-brooms on birch; also, the shoots usually die prematurely so that they can be mistaken for birds' nests.
- **Witches'-brooms on cherry** are similarly caused by a species of *Taphrina, T. cerasi* (Fuckel) Sadeb. The orthotropically growing, broom-like clusters

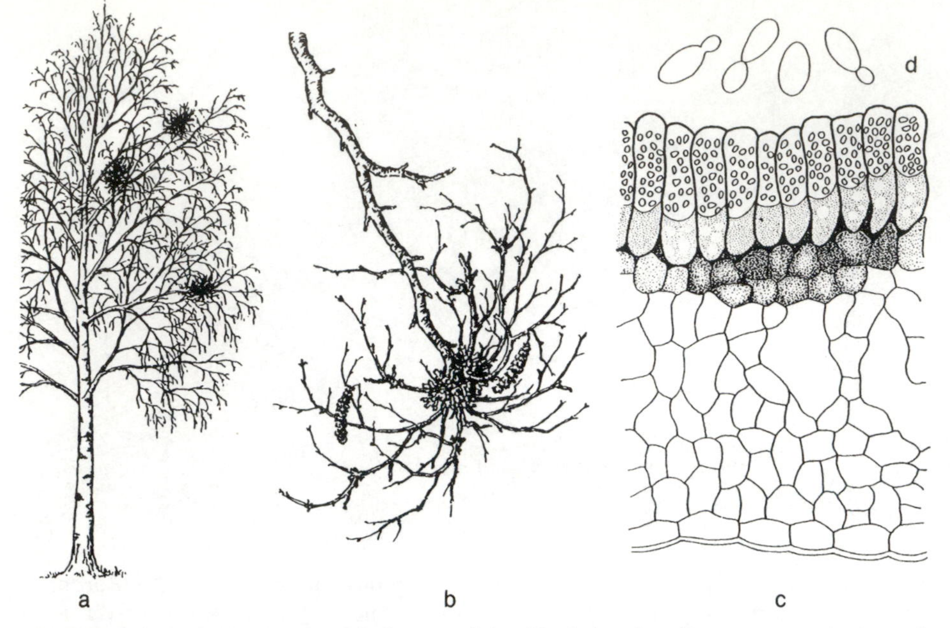

Fig. 119 Witches'-brooms on birch caused by *Taphrina betulina*. **a** general view of a downy birch with several witches'-brooms, **b** a young witch's-broom, **c** cross-section through an infected leaf, with asci, **d** budding ascospores (c after Ferdinandsen and Jørgensen 1938/39)

of shoots are particularly noticeable in summer due to their dense foliage. They are not welcome in orchards, as the fungus inhibits flowering and thus fruit production on the attacked shoots.

- **Witches'-brooms on pine** probably form as a result of bud mutations (Fig. **117b**). By taking cuttings, it has been possible to raise densely branched individuals with compressed growth and these are traded by nurseries.

Blastomania and Burls

Blastomania represents a phenomenon related to witches'-brooms, similarly based on an increase in bud formation, although flushing is weak and there is no abnormal branching. The abnormality involves the flushing of a mass of adventitious or dormant buds very close to one another. Many of these soon perish, and this results in the production of new buds which flush to form a dense mass of delicate twigs. These form a bushy structure which is sometimes recognized as a distinct condition (known as 'Stammhexenbesen' in German - Stem witch's-broom), although this is sometimes regarded as merely a form of epicormic shoot growth.

Although in some cases the cause of blastomania is thought to be a mycoplasma-like organism [179], witch's-broom formation on the stem can be induced artificially by the mechanical removal of normal 'water sprouts.' In this

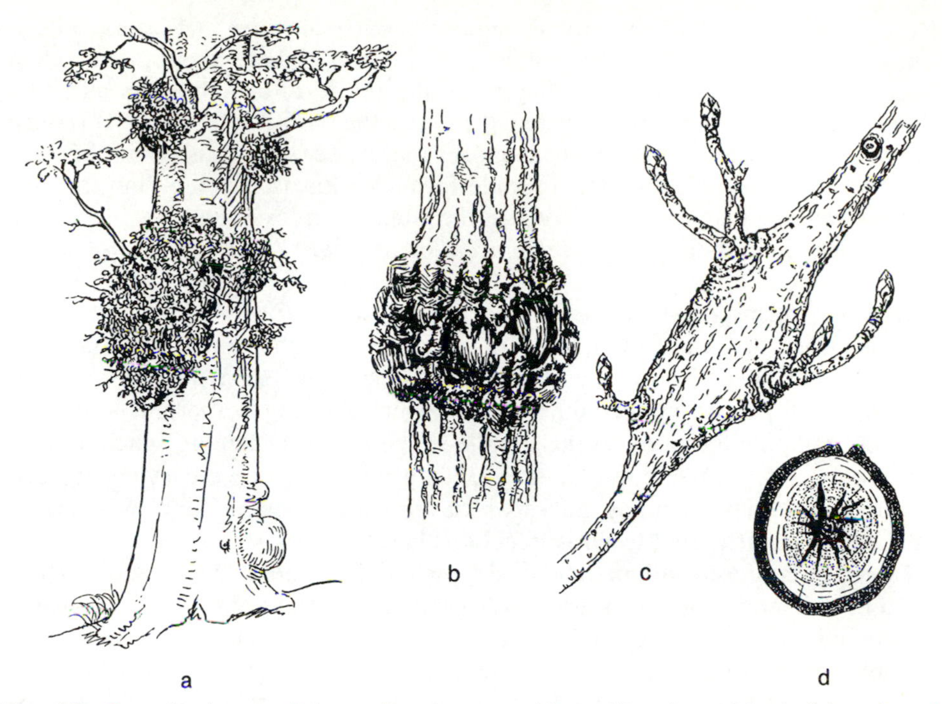

Fig. 120 Growth abnormalities. **a** beech stem with proliferation of buds (above) and sphaeroblast formation (below); **b** gall formation on oak; **c** spindle-shaped swelling on a horse chestnut twig, **d** cross-section through the same

case the tree responds to the loss of the twigs by flushing from adventitious buds. This phenomenon, which can be seen in particular on lime trees, is especially encouraged if the 'water sprouts' are removed each year. Other tree species with a tendency to blastomania are beech (Fig. **120a**), oak, maple, birch, Horse chestnut, and elm.

In blastomania, a woody excrescence on the main stem results from the formation of annual wood increments around the bases of repeatedly aborted buds. This is sometimes referred to as a burr, and the wood it contains is called burl-wood. Thus, blastomania and burl formation are very closely morphogenetically linked. A histological feature of burl-wood is the strong curvature of the xylem cells, which in places occur in tangled groups. Burl-wood in some species, such as maple and ash can yield veneers which are highly prized by the furniture industry.

Stem Swellings and Nodules

Tree stems sometimes produce swellings that appear rather similar to the burrs associated with blastomania, but are fundamentally different in their formation since they are not associated with bud proliferation and usually have a more or less smooth covering of bark. Such swellings are called 'Echte Kröpfe' (true swellings) in German, to distinguish them from burls ('Maserkröpfe'). They occur on beech, birch, and Horse chestnut, where they can reach a huge size. On oak (Fig. **120b**), they may take the form of numerous burr-like stem thickenings, and this is described in German as 'Kropfkrankheit' (goitre or swelling disease), although—as in most other cases—the cause is not known. Another term, 'Knollenbildung' (nodule formation) is used in German forestry if the swelling extends right around the stem. An example of this is the stem thickening induced by *Melampsorella caryophyllacearum* on Silver fir ('Rädertanne'—wheel fir).

In some forms of stem swelling, a ball of wood forms within the bark, isolated from the wood of the parent stem, so that the entire structure can easily be detached. The 'sphaeroblasts' of beech (Fig. **120a**) are an example of this, developing from individual buds in which flushing is aborted. A further variant of nodule formation is the spindle-shaped twig thickening which occurs, for example, on Horse chestnut (Fig. **120c,d**). In this case the xylem is dark coloured and contains many cracks which appear star-shaped in cross-section. This has been attributed to virus infections [179].

Not to be confused with pathological growth deformities is the local distortion of stem diameter which can arise as a result of grafting. This may involve a thickening only at the graft union, or there may be a difference in the overall diameter of the stock and the scion due to a mismatch in growth rates. In cases where the thickening is confined to the graft zone, it may reflect some degree of incompatibility between stock and scion. Stems showing graft-related deformities can quite often be seen in trees in parks and avenues, particularly on grafted Horse chestnuts.

Cankers

The term 'canker' in English is used to mean various bark and wood disorders which can persist for one or more years. If the disorder lasts for only one year (e.g. in the disease caused by *Phacidium coniferarum*), the term 'annual canker' is used. After this period, the actual disorder is over, and the process of occlusion begins. In contrast, a 'perennial canker' is the result of a struggle, often continuing for many years, between a pathogen and a host plant. The pathogen kills tissues, typically during the dormant season, and the plant responds by the 'compartmentalization' of the adjacent healthy tissue and by the process of 'callusing'. This results in an uneven pattern of increased wood production which takes the form of saucer-shaped or elliptically flattened deformities and can thus lead to a reduction in the value of the stem or to loss of timber.

Either fungi or bacteria can be involved in the formation of perennial cankers. A classic example of a perennial canker caused by fungi is *Lachnellula willkommii*, the cause of larch canker (Fig. **70**). In contrast, bacteria are responsible for canker formation in only a few cases. Some bacterial cankers are very gnarled, erumpent, and irregular due to hypertrophic growth induced by the pathogen, as in the example of ash canker, caused by *Pseudomonas syringae* ssp. *savastanoi* pv. *fraxini*, (Fig. **75c**). Indeed, due to the proliferation of the cells, this condition is sometimes referred to as 'ash tumour.'

Plant Tumours

Plant tumours can be defined as proliferating masses of tissue which are formed from cells that have undergone a particular transformation which causes their growth to become disorderly and unregulated by the plant. They exhibit deviant metabolic characteristics and to a large extent evade the morphogenetic control exercised by the plant. As they arise mostly from cambial cells or from bark parenchyma and show no further differentiation, they are—initially, at least—of a soft, fleshy consistency. The formation of tumours can be triggered by bacteria, viruses, or genetical events. In the field of forestry the following tumours are of interest:

- **Tumours on willow and poplar** caused by *Agrobacterium tumefaciens* (Smith & Towns.) Conn. Characteristic symptoms of the disease are the rounded or gall-like, warty-rough outgrowths which can occasionally reach the size of a small fist. On poplars and willows (Fig. **121a,b**), such proliferations occur predominantly on above-ground parts of the stem, less often on roots. In contrast to the tumours which occur on fruit trees and agricultural crop plants and which are known as 'crown galls,' these tumours are of no importance.
- **Tumours on Aleppo pine** caused by *Pseudomonas (Bacterium) pini* (Vuill.)

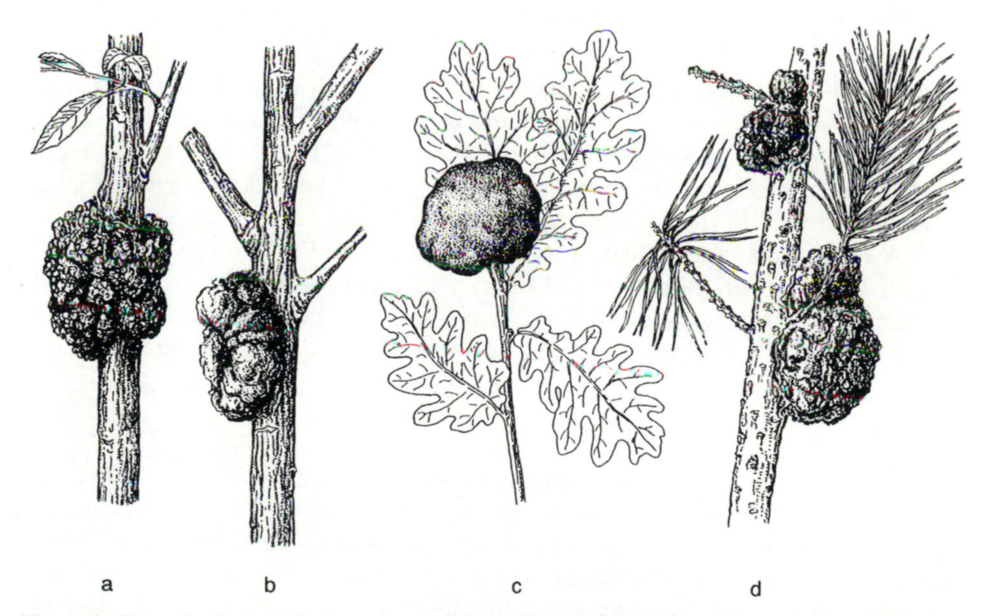

Fig. 121 Growth abnormalities. **a** tumour formation caused by *Agrobacterium tumefaciens* on willow (**a**) and poplar (**b**); **c** a gall (oak apple) caused by *Biorhiza pallida* (gall wasp); **d** twigs of Aleppo pine galled by *Pseudomonas pini*

Fig. 122 Fasciation on ash (**a**) and larch (**b**)

Petri. The characteristic symptom is the formation of tubercles on the twigs. These are spherical tumours with a core of wood which can reach a size of up to 5 cm across and an age of several years (Fig. **121d**). Twigs bearing large numbers of these can die after a number of years. They, like their host plant, are only found in the Mediterranean region.

A distinction can be made between tumours and the many types of plant gall. Galls are similarly soft and fleshy and can be very similar in outward form (Fig. **121c**); however, their growth is orderly, and differentiated tissues are formed. Their formation is induced predominantly by animals.

Fasciation

Fasciations are strap-like broadenings of shoots which arise when cell division within the growing points at the tip of the shoots becomes orientated in two opposite directions. In this way, the growing point is broadened into a line. In the simplest case, a terminal bud affected in this way grows into a flattened, upright, strap-like shoot. In the following year, the fasciation can continue to develop in the same way, but it is only the terminal bud that can do this. As a rule, the lateral buds still continue to grow into rounded, unfasciated side branches. If one corner of the fasciated shoot is favoured over the other, its growth curves in one direction, forming a scimitar-like shape or something similar.

Fasciations occur relatively infrequently in woodland. Among conifers, they are found most often on pine, larch (Fig. **122b**), and spruce; of the broadleaves, ash (Fig. **122a**), robinia, and willow seem particularly prone to the condition [224]. Little is still known about their genesis. In some cases they seem to be triggered by external stimuli, when growth is re-initiated following the normal completion of a phase of shoot development. Normal growth is then later resumed. In some tree species, the tendency to fasciation has become genetically stabilized to such an extent that most of the shoots that are produced are fasciated. An example of this is the fasciated willow *Salix sachalinensis* 'Sekka,' whose abnormally shaped shoots have been favoured by florists in recent times.

Cone Proliferation

The term 'Zapfensucht' in German (cone proliferation) denotes an abnormal condition in which large clusters of cones form on the branches of conifers. This phenomenon is most familiar in Scots pine, in which as many as 250 cones can occur in a cluster. These, though, are borne at the base of the long shoot where the male flowers normally develop. It is thought, therefore, that this cone proliferation is the result of a spontaneous alteration in the genome in which

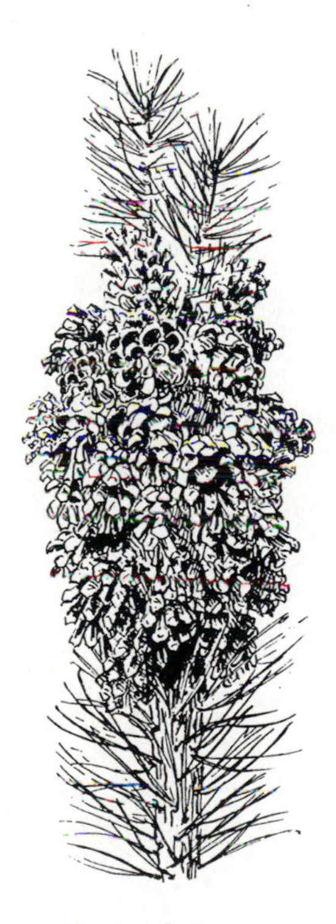

Fig. 123 Cone proliferation on Scots pine

the female and male genes are interchanged. Thus, all the positions on the shoot normally occupied by male cones instead bear female ones. The principal evidence for this theory is the heritability of the abnormality. The hope that the condition could be exploited for increasing seed production has, however, not been realized, as nearly all seeds from this source are inviable (Fig. **123**).

TABLE I

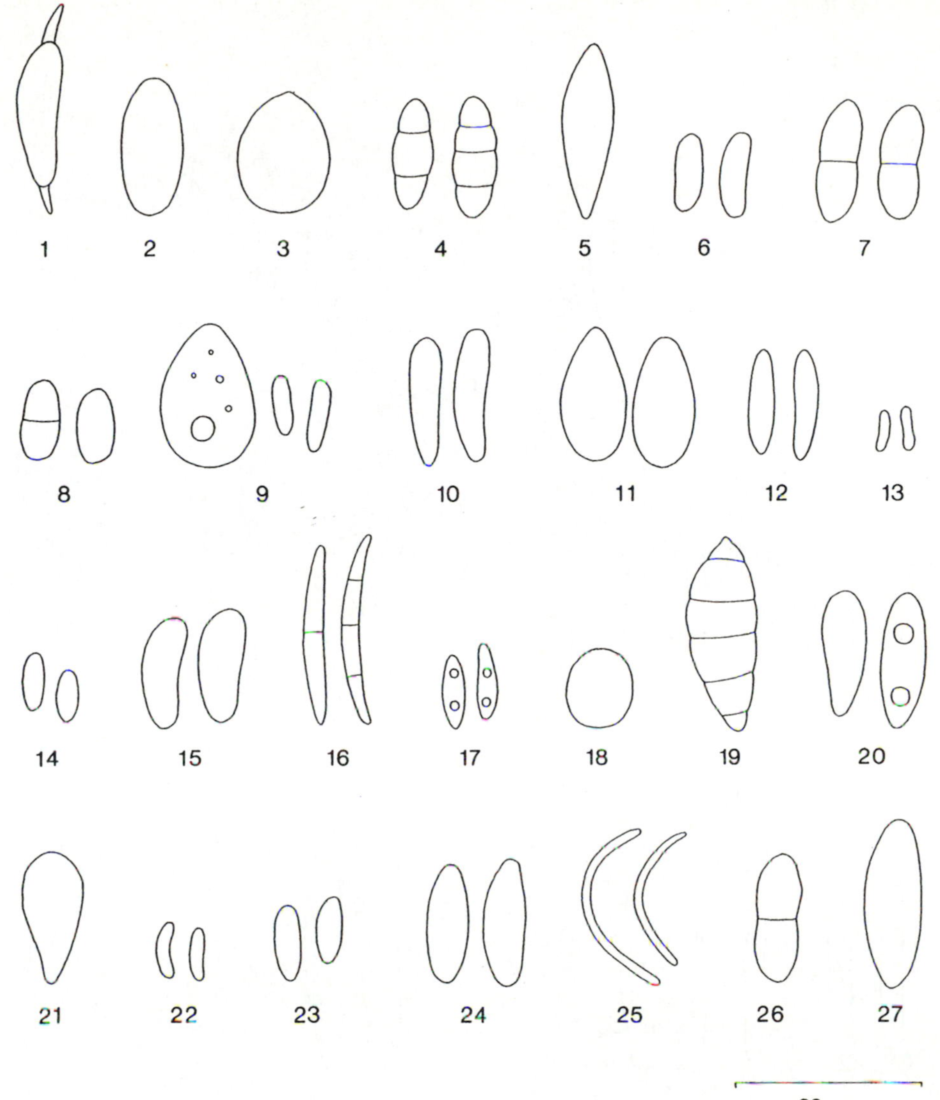

1 *Rosellinia thelena*
2 *Rhizosphaera pini*
3 *Rhizosphaera macrospora*
4 *Hendersonia acicola*
5 *Chloroscypha sabinae*
6 *Discula campestris*
7 *Diplodina acerina*
8 *Phyllosticta aceris*
9 *Phyllosticta minima*
10 *Discula betulina*
11 *Cryptocline cinerescens*
12 *Asteroma alneum*
13 *Phyllosticta osteospora*
14 *Phyllosticta populorum*
15 *Monostichella salicis*
16 *Brunchorstia laricina*
17 *Phomopsis occulta*
18 *Pithya cupressina*
19 *Seiridium cardinale*
20 *Myxosporium devastans*
21 *Melanconium betulinum*
22 *Cytospora salicis*
23 *Myxofusicoccum salicis f. microspora*
24 *Fusicoccum galericulatum*
25 *Libertella faginea*
26 *Nectria coccinea*
27 *Hypoxylon mammatum*

TABLE II

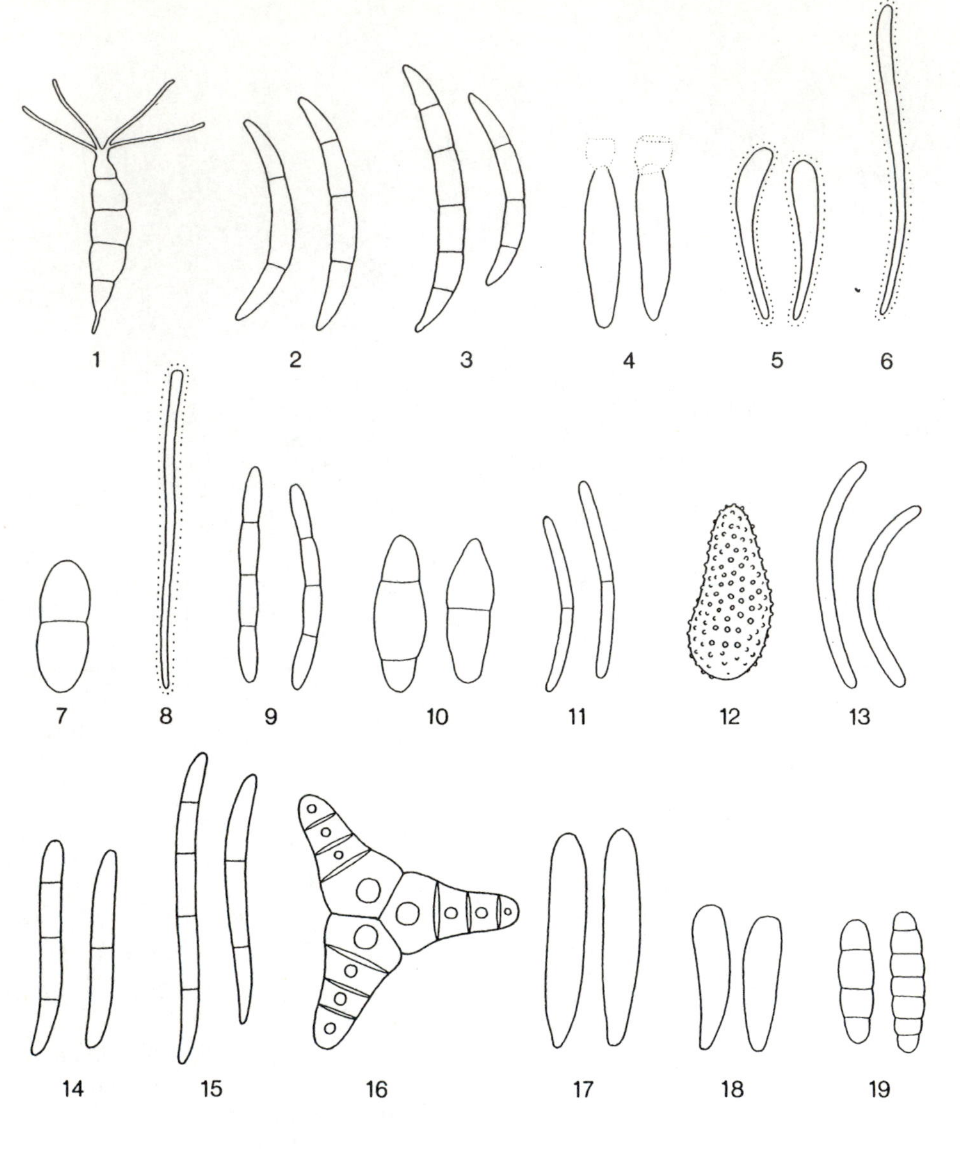

1 *Pestalotia funerea*
2 *Fusarium oxysporum*
3 *Fusarium culmorum*
4 *Tiarosporella parca*
5 *Lophodermella sulcigena*
6 *Lophodermella conjuncta*
7 *Herpotrichia coulteri*
8 *Lophodermium juniperinum*
9 *Phloeospora aceris*
10 *Pollaccia elegans*
11 *Septoria populi*
12 *Melampsora larici-populina*
13 *Cryptosporium betulinum*
14 *Cylindrocarpon cylindroides* var. *tenue*
15 *Cylindrocarpon willkommii*
16 *Asterosporium asterospermum*
17 *Fusicoccum macrosporum*
18 *Cryptosporiopsis abietina*
19 *Trichonectria hirta*

Classification of Fungi and Lichens

Classification of Fungi and Lichens Mentioned in this Book
(according to Ainsworth and Bisby's *Dictionary of the Fungi*, 1983)

Key to major taxonomic groups:
A = Ascomycotina M = Mastigomycotina
B = Basidiomycotina D = Deuteromycotina

Allantophoma	D, Coelomycetes
Alternaria	D, Hyphomycetes
Amanita	B, Agaricales
Apiognomonia	A, Diaporthales
Aposphaeria	D, Coelomycetes
Armillaria	B, Agaricales
Ascocalyx	A, Helotiales
Ascochyta	D, Coelomycetes
Ascodichaena	A, Rhytismatales
Asteroma	D, Coelomycetes
Asteromella	D, Coelomycetes
Asterosporium	D, Coelomycetes
Athelia	B, Aphyllophorales
Aureobasidium	D, Hyphomycetes
Bispora	D, Hyphomycetes
Biscogniauxia	A, Sphaeriales
Bjerkandera	B, Aphyllophorales
Boletus	B, Boletales
Botryodiplodia	D, Coelomycetes
Botryosphaeria	A, Dothideales
Botryotinia	A, Helotiales
Botrytis	D, Hyphomycetes
Brunchorstia	D, Coelomycetes
Bulgaria	A, Helotiales
Calocera	B, Dacrymycetales
Calyptospora	B, Uredinales
Cenangium	A, Helotiales
Cephalosporium	D, Hyphomycetes
Ceratocystis	A, Ophiostomatales
Cercospora	D, Hyphomycetes
Chalara	D, Hyphomycetes
Chalaropsis	D, Hyphomycetes
Chlorosplenium	A, Helotiales
Chlorociboria	A, Helotiales
Chloroscypha	A, Helotiales
Chondrostereum	B, Aphyllophorales
Chrysomyxa	B, Uredinales
Ciboria	A, Helotiales
Cladosporium	D, Hyphomycetes
Coleosporium	B, Uredinales
Colletotrichum	D, Coelomycetes
Collybia	B, Agaricales
Colpoma	A, Rhytismatales
Coniophora	B, Aphyllophorales
Conostroma	D, Coelomycetes
Coryneum	D, Coelomycetes
Cristulariella	D, Hyphomycetes
Cronartium	B, Uredinales
Crumenulopsis	A, Helotiales
Cryphonectria	A, Diaporthales
Cryptocline	D, Coelomycetes
Cryptodiaporthe	A, Diaporthales
Cryptosporiopsis	D, Coelomycetes
Cryptosporium	D, Coelomycetes
Cryptostroma	D, Hyphomycetes
Cucurbitaria	A, Dothideales
Cyclaneusma	A, Rhytismatales
Cylindrocarpon	D, Hyphomycetes
Cylindrosporella	D, Coelomycetes
Cytospora	D, Coelomycetes
Daedalea	B, Aphyllophorales
Daedaleopsis	B, Aphyllophorales
Delphinella	A, Dothideales
Diatrypella	A, Diatrypales
Didymascella	A, Rhytismatales
Didymosporina	D, Coelomycetes
Digitosporium	D, Coelomycetes
Diplodia	D, Coelomycetes
Diplodina	D, Coelomycetes
Discella	D, Coelomycetes
Discosporium	D, Coelomycetes
Discula	D, Coelomycetes
Dothistroma	D, Coelomycetes
Drepanopeziza	A, Helotiales
Endocronartium	B, Uredinales
Endothia	A, Diaporthales
Fistulina	B, Aphyllophorales
Fomes	B, Aphyllophorales
Fomitopsis	B, Aphyllophorales
Fusarium	D, Hyphomycetes
Fusicoccum	D, Coelomycetes
Ganoderma	D, Aphyllophorales
Gemmamyces	A, Dothideales
Gloeophyllum	B, Aphyllophorales
Gloeosporium	D, Coelomycetes
Glomerella	A, Polystigmatales
Gnomoniella	A, Diaporthales
Gremmeniella	A, Helotiales
Grifola	B, Aphyllophorales
Guignardia	A, Dothideales
Gymnosporangium	B, Uredinales
Helicobasidium	B, Auriculariales

Herpotrichia A, Dothideales
Heterobasidion B, Aphyllophorales
Hormiscium D, Hyphomycetes
Hormonema D, Hyphomycetes
Hypodermella A, Rhytismatales
Hypogymnia A, Lecanorales
Hypoxylon A, Sphaeriales
Inonotus B, Aphyllophorales
Kabatiella D, Hyphomycetes
Kabatina D, Coelomycetes
Keithia A, Rhytismatales
Lachnellula A, Helotiales
Lactarius B, Russulales
Laetiporus B, Aphyllophorales
Lasiodiplodia D, Coelomycetes
Lecanora A, Lecanorales
Lecanosticta A, Coelomycetes
Leccinum B, Boletales
Leptodothiorella D, Coelomycetes
Leptographium D, Hyphomycetes
Leucostoma A, Diaporthales
Libertella D, Coelomycetes
Lirula A, Rhytismatales
Lophodermella A, Rhytismatales
Lophodermium A, Rhytismatales
Macrophomina D, Coelomycetes
Marssonina D, Coelomycetes
Megaloseptoria D, Coelomycetes
Melampsora B, Uredinales
Melampsorella B, Uredinales
Melampsoridium B, Uredinales
Melanconium D, Coelomycetes
Melanomma A, Dothideales
Meria D, Hyphomycetes
Meripilus B, Aphyllophorales
Microsphaera A, Erysiphales
Monilia D, Hyphomycetes
Monostichella D, Coelomycetes
Mycosphaerella A, Dothideales
Myxofusicoccum D, Coelomycetes
Myxosporium D, Coelomycetes
Naemacyclus A, Rhytismatales
Nectria A, Hypocreales
Neofabraea A, Helotiales
Oedocephalum D, Hyphomycetes
Onnia B, Aphyllophorales
Ophiognomonia A, Diaporthales
Ophiostoma A, Ophiostomatales
Penicillium D, Hyphomycetes
Peniophora B, Aphyllophorales
Peridermium B, Uredinales
Pestalotia D, Coelomycetes
Pezicula A, Helotiales
Phacidiopycnis D, Coelomycetes
Phacidium A, Helotiales
Phaeocryptopus A, Dothideales
Phaeolus B, Aphyllophorales
Phellinus B, Aphyllophorales
Phialophora D, Hyphomycetes
Phloeospora D, Coelomycetes
Pholiota B, Agaricales
Phoma D, Coelomycetes
Phomopsis D, Coelomycetes
Phyllactinia A, Erysiphales
Phytophthora M, Peronosporales
Phyllosticta D, Coelomycetes
Piptoporus B, Aphyllophorales
Pithya A, Pezizales
Pleuroceras A, Diaporthales
Pleurotus B, Aphyllophorales
Pollaccia D, Hyphomycetes
Polymorphum D, Coelomycetes
Polyporus B, Aphyllophorales
Potebniamyces A, Rhytismatales
Pseudotrametes B, Aphyllophorales
Pucciniastrum B, Uredinales
Pycnoporus B, Aphyllophorales
Pythium M, Peronosporales
Ramalina A, Lecanorales
Rehmiellopsis A, Dothideales
Resinicium B, Aphyllophorales
Rhabdocline A, Rhytismatales
Rhacodiella D, Hyphomycetes
Rhizina A, Pezizales
Rhizoctonia D, Hyphomycetes
Rhizosphaera D, Coelomycetes
Rhytisma A, Rhytismatales
Rosellinia A, Sphaeriales
Russula B, Russulales
Sarcinomyces D, Hyphomycetes
Schizophyllum B, Aphyllophorales
Scirrhia A, Dothideales
Scleroderma B, Sclerodermatales
Scleroderris A, Helotiales
Sclerophoma D, Coelomycetes
Seiridium D, Coelomycetes
Septoria D, Coelomycetes
Septotinia A, Helotiales
Serpula B, Aphyllophorales
Sirococcus D, Coelomycetes
Sparassis B, Aphyllophorales
Sphaeropsis D, Coelomycetes
Stereum B, Aphyllophorales
Strasseria D, Coelomycetes
Stromatinia A, Helotiales
Suillus B, Boletales
Sydowia A, Dothideales
Taphrina A, Taphrinales
Thanatephorus B, Aphyllophorales
Thekopsora B, Uredinales
Thelephora B, Aphyllophorales
Tiarosporella D, Coelomycetes
Trametes B, Aphyllophorales
Trematosphaeria A, Dothideales
Trichaptum B, Aphyllophorales
Trichoderma D, Hyphomycetes
Trichonectria A, Hypocreales
Trichoscyphella A, Helotiales
Trichosphaeria A, Sphaeriales
Trichothecium D, Hyphomycetes
Trimmatostroma D, Hyphomycetes
Triposporium D, Hyphomycetes

Tubakia	D, Coelomycetes	*Valdensia*	D, Hyphomycetes
Tuber	A, Pezizales	*Valsa*	A, Diaporthales
Tubercularia	D, Hyphomycetes	*Venturia*	A, Dothideales
Tyromyces	B, Aphyllophorales	*Verticillium*	D, Hyphomycetes
Uncinula	A, Erysiphales	*Xerocomus*	B, Boletales
Usnea	A, Lecanorales	*Xylaria*	A, Sphaeriales
Ustulina	A, Aphyllophorales	*Xylobolus*	B, Aphyllophorales

Latin–English Names of Woody Plants Mentioned in this Book

(*after Elsevier's Dictionary of trees and shrubs, 1986, with amendment*)

Abies – fir
Abies alba – Silver fir
Abies nordmanniana – Caucasian fir
Abies procera – Noble fir
Abies veitchii – Veitch's silver fir
Acer – maple
Acer campestre – Field maple
Acer negundo – Box elder
Acer pseudoplatanus – sycamore
Acer platanoides – Norway maple
Acer rubrum – Red maple
Acer saccharum – Sugar maple
Aesculus – Horse chestnut
Aesculus hippocastanum – Common horse chestnut
Aesculus pavia – Red buckeye
Ailanthus – Tree of heaven
Alnus – alder
Alnus incana – Grey alder
Amelanchier – serviceberry

Betula – birch
Betula pendula – Silver birch
Betula pubescens – Downy birch

Carpinus – hornbeam
Carya – hickory
Castanea – chestnut
Castanea crenata – Japanese chestnut
Castanea dentata – American chestnut
Castanea sativa – Sweet chestnut
Catalpa – catalpa
Catalpa bignonioides – Indian bean tree
Cedrus – cedar
Cercis canadensis – redbud
Cercis siliquastrum – Judas-tree
Chamaecyparis – False cypress
Chamaecyparis lawsoniana – Lawson cypress

Corylus – hazel
Cotinus coggygria – Venetian sumach
Cotoneaster – cotoneaster
Cotoneaster integerrimus – Common cotoneaster
Crataegus – hawthorn
Crataegus laevigata – Midland hawthorn
Crataegus monogyna – Common hawthorn
Cryptomeria – cryptomeria
Cupressus – cypress
Cupressus macrocarpa – Monterey cypress
Cupressus sempervirens – Italian cypress
Cydonia – quince

Empetrum nigrum – Crowberry
Eucalyptus – eucalyptus
Euonymus – Spindle-tree

Fagus – beech
Fagus grandifolia – American beech
Fagus sylvatica – Common beech
Fraxinus – ash
Fraxinus excelsior – Common ash

Ginkgo biloba – ginkgo
Gleditsia – Honey locust

Juglans – walnut
Juglans regia – Common walnut
Juniperus – juniper
Juniperus chinensis – Chinese juniper
Juniperus communis – Common juniper
Juniperus nana – Dwarf juniper
Juniperus oxycedrus – Prickly juniper
Juniperus sabina – Common savin
Juniperus virginiana – Eastern redcedar

Larix – larch
Larix decidua – European larch
Larix gmelini – Dahurian larch
Larix laricina – American larch
Larix kaempferi – Japanese larch
Larix russica – Siberian larch
Ledum palustre – Labrador tea
Lonicera periclymenum – honeysuckle

Magnolia – magnolia

Malus – apple
Malus sylvestris – Wild crab

Picea – spruce
Picea abies – Norway spruce
Picea engelmannii – Engelmann spruce
Picea omorika – Serbian spruce
Picea pungens – Colorado spruce
Picea pungens forma *glauca* – Blue spruce
Picea sitchensis – Sitka spruce
Pinus – pine
Pinus cembra – Arolla pine
Pinus contorta – Shore pine
Pinus contorta var. *latifolia* – Lodgepole pine
Pinus flexilis – Limber pine
Pinus halepensis – Aleppo pine
Pinus lambertiana – Sugar pine
Pinus monticola – Western White pine
Pinus mugo – Dwarf mountain pine
Pinus nigra – Austrian pine
Pinus nigra var. *maritima* – Corsican pine
Pinus peuce – Macedonian pine
Pinus pinaster – Maritime pine
Pinus pinea – Stone pine
Pinus ponderosa – Western Yellow pine
Pinus radiata – Monterey pine
Pinus strobus – Weymouth pine
Pinus sylvestris – Scots pine
Pinus wallichiana – Buthan pine
Platanus – plane
Platanus hybrida – London plane
Platanus occidentalis – Western plane
Platanus orientalis – Oriental plane
Populus – poplar
Populus alba – White poplar
Populus balsamifera – Balsam poplar
Populus × *canescens* – Grey poplar
Populus × *euramericana* – Hybrid black poplar
Populus nigra – Black poplar
Populus nigra cv. *Italica* – Lombardy poplar
Populus tremula – aspen
Populus tremuloides – American aspen
Populus trichocarpa – Black cottonwood
Prunus – plums, cherries etc.
Prunus avium – Wild cherry
Prunus padus – Bird cherry

Prunus serotina – Black cherry
Prunus triloba – Flowering almond
Pseudotsuga menziesii – Douglas fir
Pyracantha – firethorn
Pyrus communis – Common pear

Quercus – oak
Quercus ilex – Holm oak,
Quercus petraea – Sessile oak
Quercus pubescens – Downy oak
Quercus robur – Pedunculate oak
Quercus rubra – Red oak
Quercus suber – Cork oak

Rhododendron ferrugineum – 'Alpenrose'
Rhododendron hirsutum – Hairy rhododendron
Rhus – sumach
Rhus typhina – Stag's Horn sumach
Ribes – currant
Robinia – robinia

Salix – willow
Salix alba – White willow
Salix alba var. *coerulea* – Cricket-bat willow
Salix caprea – Goat willow, Common sallow
Salix cinerea – Grey willow
Salix fragilis – Crack willow
Salix purpurea – Purple osier
Salix rigida – American willow
Salix sachalinensis – Sachalin willow
Salix triandra – Almond-leaved willow
Sequoia – redwood
Sequoia sempervirens – Coast redwood
Sequoiadendron giganteum – wellingtonia
Sophora japonica – Pagoda tree
Sorbus – whitebeams, rowans, service trees
Sorbus aria – whitebeam
Sorbus aucuparia – rowan
Sorbus intermedia – Swedish whitebeam
Syringa – lilac

Taxus – yew
Thuja – thuya
Thuja occidentalis – White cedar
Thuja plicata – Western Red cedar
Tilia – lime, linden

Ulmus – elm
Ulmus glabra – Wych elm
Ulmus × *hollandica* – Dutch elm
Ulmus minor – Lock elm
Ulmus procera – English elm

Glossary

abcission active shedding of redundant organs (leaves, flowers, twigs)
abiotic pertaining to non-living agents (e.g. weather)
abstriction the partitioning of a spore from its parent cell
acervulus (pl. acervuli) the sporiferous structure of the Melanconiales, subcuticular or subepidermal, applanate to thick-discoid; restricted to parasitic fungi
acrogenous borne at the end (e.g. of a specialized hypha)
aecidium (pl. aecidia) the basic fruiting structure (sorus) of the Uredinales, produced by the monokaryotic mycelium, bearing aecidiospores
aleuriospore a thick-walled conidium developed from the swollen end of a conidiogenous cell or hyphal branch, single or in chains
alternate host (pertaining to rust fungi that change hosts in their life cycles) one of the two host plants on which the fungus develops - the one other than that affected by the disease in question
anamorph asexual form of the imperfect state, characterized by conidiomata
anastomosis cross-linkage between gills or hyphae, through which cell contents can be transferred
antagonism a general term for any sort of combative interaction between organisms
anthracnose a disease on leaves (or perhaps other tissues in the case of an herbaceous plant) causing local, delimited dark lesions, soon drying out and becoming sunken
apothecium (pl. apothecia) a cup- or saucer-like ascoma in which the hymenium is exposed at maturity; sessile or stipitate
ascoma (pl. ascomata) a sporocarp producing asci
Ascomycotina, Ascomycetes group of fungi whose characteristic reproductive cell is the ascus
ascospore a spore borne in an ascus
ascus (pl. asci) the typical reproductive cell of the perfect state of the ascomycetes, in which ascospores are produced following karyogamy and meiosis
aseptate not partitioned, without cross-walls
asexual type of reproduction not involving nuclear fusion or meiosis
assimilates substances produced by biological processes (e.g. starch)
autotrophic having the nutritional capability to synthesize all organic cell constituents from inorganic compounds following the absorption of energy, usually from sunlight
Basidiomycotina, Basidiomycetes group of fungi in which the sexual reproductive cell is the basidium, on which basidiospores are borne
basidiospore a spore acrogenously budded from a basidium, containing one or two nuclei

basidium (pl. basidia) a terminal cell, diagnostic for basidiomycetes, from which basidiospores are produced, following karyogamy and meiosis

biotic pertaining to living agents

biotrophic parasite a parasite which feeds on host cells while they are still living

blastospore an asexual spore produced by budding

blight a common name for diseases causing rapid dieback of leaves, shoots, or branches

brown rot a type of wood decay, in which the cellulose and hemicellulose components are broken down by specific enzymes

callus a parenchymatous layer of new cells that forms over a cut or damaged plant surface; also used to describe the tissues that occlude injuries in which the cambium overlying wood is killed or removed

cambium a meristematic layer of tissue in higher plants which is responsible for the development of secondary thickening

canker collective term for a group of bark and wood disorders, which can persist for one year (annual cankers) or longer

canker rot a disease caused by a pathogen that both decays the wood of a tree and prevents occlusion of its infection court by killing the surrounding bark and cambium

canker stain a necrotic bark and wood disease, characterized by an intense discoloration of the affected tissues, in which the pathogen spreads axially within the host along xylem vessels and radially along the xylem rays

chlamydospore an asexual thick-walled spore originating intercalarily by modification of a hyphal segment

chlorosis a yellowish to whitish general discoloration, associated with deficient production or breakdown of chlorophyll

cleistothecium (pl. cleistothecia) a closed ascocarp, opening by rupture; characteristic of the Erisyphales

clone a set of individual organisms which have been propagated asexually from a single source plant

Coelomycetes a class of Deuteromycotina whose conidia are produced in pycnidia (Sphaeropsidales) or acervuli (Melanconiales)

conidioma (pl. conidiomata) a multicellular conidia-bearing structure

conidiophore a specialized hypha or sporophore which bears conidia

conidium (pl. conidia) a specialized, non-motile asexual spore, usually formed at the apex of the tapering end of a conidiophore and readily deciduous

coremium (pl. coremia) a stalk-like structure which consists of a group of united conidiophores producing asexual spores at their tips

cotyledon seed leaf

cystidium (pl. cystidia) a sterile cell of distinctive shape, typically found between the basidia or in the trama of fruit bodies of the Basidiomycetes

cytoplasm the living contents of a cell, including the organelles and excluding the cell nucleus

Deuteromycetes, Deuteromycotina (*see* Fungi imperfecti)

diagnosis the recognition and naming of a disease on the basis of symptoms

dichotomous branching in a bifurcated pattern

dieback the death, often progressive, of tissues so that the extent of the living crown, root system, or cambium around a wound is reduced

dikaryophase, dikaryotic-state the binucleate phase, following cytoplasmic fusion between two monokaryotic mycelia whose different nuclei, of complementary mating types, remain separate within the cells and functionally independent until the later stage of nuclear fusion

dikaryotization the process whereby a mycelium becomes dikaryotic, resulting from cytoplasmic fusion (plasmogamy) between two monokaryons

diploid having the $2n$ number of chromosomes in one nucleus

discomycetes (cup fungi) the traditional term for taxa whose ascomata are sessile, open, or cup shaped

dormancy the resting stage of an organism or a particular organ, in which metabolism is reduced; mainly under constitutive control, as in the germination dormancy of seeds

ecotype a group of individuals of a species, which show a distinctive genetic adaptation to their local environment

ectoparasite a parasite that lives externally on its host, absorbing nutrients via specialized organs, such as haustoria

ectotrophic gaining nutrition while developing externally or superficially (e.g. ectomycorrhizas)

emission airborne pollutant or chemical precursor of pollutants

endoconidium (pl. endoconidia) an asexual spore that develops internally from a conidiophore

endogenous originating from within an organism, or extending outwards from within

endoparasitic living within a host cell

endophytes microorganisms, especially fungi and bacteria, which live in higher plants, without immediately causing obvious disease symptoms; their further development may follow with either a saprotrophic or overtly parasitic phase

endotrophic (of mycorrhizas) growing inter- and intracellularly in the host roots

enzyme any of a large class of protein substances produced by living cells which catalyze biochemical changes, usually in the presence of other substances (coenzymes)

epidemic an infectious disease that occurs over an increasingly large geographic area

epidemiology the study of the origin and development of mass infection within a population

epidermis the outer layer of cells of a fruit body or any plant tissue

epiphytic a plant growing on the outside of another, but not as a parasite

erumpent (of fungal fruit bodies, usually on bark) bursting out through the host surface

facultative parasite a fungal or other organism that has the ability to live either parasitically or saprotrophically, and which can therefore be cultured

flagellate equipped with one or more whip-like appendages for propulsion

Fungi imperfecti (=Deuteromycotina) a diverse assortment of fungi whose common feature is the lack of a sexual stage (teleomorph)

fungicide a synthetic substance that kills fungi; may be used in plant protection
fungistatic inhibiting fungal growth and reproduction
fusiform spindle shaped
geotropism growth response of plant or fungal organs to gravity
haplontic-state host (of Uredinales) a host plant on which the haploid, monokaryotic mycelium develops
haplophase (=haplontic phase) a developmental stage of fungi (Uredinales) with a single, haploid type of cell nucleus
haustorium (pl. haustoria) a specialized hyphal branch within a living host cell, by which some fungal parasites absorb nutrients; also, a corresponding multicellular structure of parasitic flowering plants
heartwood wood, usually discoloured, that has been altered by a normal aging process; it contains no living cells, but may retain chemical reactivity in response to injury (cf. sapwood)
hemiparasite *see* semiparasite
heteroecious (of the Uredinales) requiring more than one host for completion of the life cycle
hyaline glassy-transparent; colourless
hymenium (pl. hymenia) the fruiting surface of ascomycetes and basidiomycetes; an aggregation of asci or basidia in a cohesive layer, mixed with paraphyses or other sterile cells
hyperparasite an organism that parasitizes a parasite
hypertrophy excessive and often abnormal growth of cells, organs, or organisms
hypha (pl. hyphae) a fungous filament
Hyphomycetes (syn. Hyphales) class of the Deuteromycotina characterized by mycelial forms which are sterile or which fruit only by forming conidia
hypocotyl part of the stalk between roots and cotyledons
hypophyllous present on the undersides of leaves
hypovirulence a state in which a pathogen has a weakened ability to attack a particular host
hysterothecium (pl. hysterothecia) the elongate fruit body of the Hysteriales, which is closed during dry conditions and opens by a long slit when moistened
imago (pl. imagines) fully developed, sexually mature insect
imperfect state (*see* anamorph)
infection the entry of a pathogen into the body of a host, with the establishment of a stable, irreversible parasitic relationship
inflorescence a flower-bearing structure, representing a modified vegetative shoot system
intercalary not apical, but between the apex and the base
intercostal in between the veins of a leaf
karyogamy fusion of nuclei within a cell; union of two nuclei of matching sexual mating types
Lammas shoot a shoot formed during summer, following the pause in growth at the end of the main period of annual flushing

lenticel a ventilation pore in a stem or root

lignin a high molecular-weight phenolic and lipid polymer in plants that, after cellulose, is the main constituent of wood

loculus (pl. loculi) a cavity in a stroma, especially the ascigerous cavity, without a perithecial wall

macroconidium (pl. macroconidia) a large conidium produced usually at a different period or on a different spore-bearing structure from the microconidia

meristematic pertaining to cells whose function is to divide and lay down new tissues, as found at growing points

microconidium (pl. microconidia) a small conidium produced usually at a different time from macroconidia; in some cases proved to be a spermatium

microcyclic (of the Uredinales) having a short cycle, with mycelium and spores only in the monokaryotic phase

mildew a mealy, loosely attached mat on the surface of green tissues of a plant, consisting of the mycelium or spores of a parasitic fungus

monokaryon *see* dikaryophase

morphogenesis genetically determined programme by which a organism attains its characteristic form

morphotype a group of individuals within a species which are morphologically distinct but have no taxonomic significance

mycelium (pl. mycelia) the collective term for a group or mass of hyphae

mycoplasma-like-organisms (MLOs) pleomorphic micro-organisms that parasitize phloem cells and cannot be cultured

mycoplasmas the smallest pleomorphic micro-organisms living in host cells; lacking a true cell nucleus or definite cell wall; can be cultured on nutrient media

mycorrhiza a mutualistic symbiosis of a fungus with the roots of a higher plant. The fungus grows either intracellularly in the cortex (endotrophic) or predominantly on the root surface and between the outer root cells (ectotrophic)

necrosis death of cells or tissues

necrotrophic parasite (*see* perthophyte)

needle-cast a premature and copious falling of needles, caused by fungal attack or abiotic factors

oogonium (pl. oogonia) a female sexual organ, which at maturity contains one or more oospores

Oomycetes a class of Mastigomycotina, typically aquatic, sometimes terrestrial, saprotrophic or parasitic; forming oospores, biflagellate zoospores, and conidia

oospore the resting spore from a fertilized oogonium, usually thick-walled

orthotropic upright; pertaining to the growth orientation of shoots

ostiole or **ostiolum** (pl. ostiola) the schizogenously formed canal in the tip of a true perithecium, lined with periphyses; also the opening of a pycnidium

paraphysis one of the sterile filaments that occur in the hymenium of the fructifications of fungi, especially in ascomycetes, arising in the ascogenous layer

parasite an organism living upon another and deriving food from it, with or without fatal effect on the host

parenchyma the soft tissue of higher plants, consisting of thin-walled, mostly isodiametric, cells, usually living

pathogen a parasite capable of causing disease

pathogenicity the genetically determined ability of an organism to cause disease

pathotype a group of individuals within a parasitic species, which differ from the rest of the population in their pathogenicity

perfect state (*see* teleomorph)

periderm secondary, protective covering tissue, mainly on wood plants

peridium (pl. peridia) the outer enveloping coat of a sporangium or fruit body

periphysis one of a number of hair-like projections inside the ostiole of a perithecium or pycnidium

perithecium (pl. perithecia) oval, round, or pyriform ascocarp which opens by a pore or a slit and within which asci are born

perthophyte pathogenic organism which, by secreting toxic substances, kills plant tissues and then colonizes them

phellem a layer of cork cells which is laid down by the phellogen

phellogen cork cambium; a secondary meristem which lays down cork cells on its outer side

pheromones substances which are produced by individuals of a species and which induce a specific reaction in others of the same species; especially in insects

phialide a flask-shaped or fusiform-beaked structure from the apex of which conidia are abstricted

phialospores spores formed in succession, endogenously or exogenously, producing chains or spore heads on phialides

Phycomycetes a name still widely used for a group of 'lower' fungi with mostly tube-like, aseptate, multinucleate mycelium, recently classified as Mastigomycotina and Zygomycotina

phytotoxic toxic to plants

pileus the cap of a basidiomycete fruit body

pit a perforation in the secondary wall of a plant cell, communicating across a semipermeable membrane with a corresponding perforation in an adjacent cell

plagiotropic orientated at an angle to the gravitational direction

plasmogamy cytoplasmic fusion, without there necessarily being an accompanying fusion of nuclei

pleomorphic, polymorphic having many forms

poroid pore-like; type of basidiomycete fruit body structure, in which the hymenium lines numerous vertical tubes which release spores through their external openings

provenance a term applied to plants, denoting place of origin

pseudosclerotium an irregularly formed, sclerotium-like structure, which consists of sclerotized hyphae enclosing part of the substrate

pycnidium (pl. pycnidia) a flask-shaped or globose ostiolate conidioma, the inner surface of which is lined by conidiogenous cells; typical in Sphaeropsidales

reaction zone a protective chemical boundary, formed within the existing wood of a tree in response to wounding, and separating healthy wood from the zone of damage and microbial colonization

resupinate reversed; bent backwards

rhizomorph a root-like aggregation of hyphae having an apical meristem and frequently differentiated into a rind of small, dark-coloured cells surrounding a central white core; functions include transport of water and nutrients; also significant in the infection process

rhizosphere the zone surrounding a root, in which mycorrhizas occur

rhytidome bark formed by a tertiary process of stem thickening, in which a rough corky surface is formed externally by a succession of secondary periderms that develop within the inner bark

rickettsias rod-shaped or spherical prokaryotic microorganisms with a cell wall; mainly vertebrate pathogens. Somewhat distinct from these are 'Rickettsia-like organisms' (RLOs) which are plant pathogens with corrugated or furrowed cell walls

saprotroph, saprophyte an organism using dead organic material as food

sapwood wood that contains living cells with different functions: transport, storage, mechanical support, and protection

schizogenous produced by splitting or fragmentation

sclerotium (pl. sclerotia) a firm, often black, mass of hyphae of variable size and morphology, often acting as a resting body

scutellum, thyriothecium an inverted flattened ascoma having a more or less radial structure and lacking a basal plate

semiparasite, hemiparasite facultative parasite which is partly able to manufacture its own essential foodstuffs

senescence natural ageing

septum (pl. septa) a cross-wall or partition between two cells

sorus a fruiting structure in certain fungi, especially the spore mass of the Uredinales

spermatium (pl. spermatia) a nonmotile, uninucleate structure, which acts as the male fertilization cell

spermogonium (pl. spermogonia) a walled, pycnidium-like structure in which spermatia are produced

sphaeroblast a ball-like excrescence of bark, with a woody core, which can develop from a previously dormant bud

sporangiophore a sporophore bearing a sporangium

sporangium (pl. sporangia) an organ producing endogenous asexual spores

spore reproductive body of a cryptogam, within one or more cells, of various shapes and colours, produced sexually or asexually

sporodochium (pl. sporodochia) a conidioma in which the spore mass is borne on a cushion-like aggregation of short conidiophores

sporophore a spore-producing or spore-bearing structure

stoma, stomate (pl. stomata) respiratory aperture of higher plants

stroma (pl. stromata) a dense mass of vegetative hyphae in or on which spores or fruit bodies are produced

subiculum a net- or crust-like growth of mycelium under fruit bodies

symbiosis the living together of dissimilar organisms with mutual benefits and mutual exploitation

symptom distinct sign of a disease, characterized by external or internal alterations in an organism following the activity of a damaging agent

syndrome a complex of symptoms; the concurrence of a number of individual symptoms which characterize a disease

synonym another name for a species or a higher taxon; a later or invalid name not currently employed for the taxon

systemic (of a parasite) spreading throughout the host;
(of a fungicide) translocated within the plant

taxonomic pertaining to the classification of organisms

teleomorph, perfect state sexual form of a fungus, characterized by ascomata or basidiomata

teleutosorus (pl. teleutosori, telia) sori producing teleutospores

teleutospore, teliospore resting spore of a rust fungus

thallus the vegetative body of a primitive plant (e.g. alga) or fungus, lacking a vascular system and without demarcation between roots and shoots

trama sterile tissue supporting a hymenium; the tissue between adjacent hymenia in the basidiomycetes

tylosis (pl. tyloses) a cellular intrusion into a vessel, entering via a pit in the wall

tyrosinase an enzyme of the phenoloxidase group, one effect of which is to convert phenolic substances into a toxic form

uredosorus, uredinium (pl. uredosori, uredina) a sorus producing uredospores

uredospore, urediniospore the summer spore of a rust fungus

vector an organism which carries the living agent of disease from one host to another

virulence (adj. virulent) the ability of a pathogen to cause disease in a specific host

viruses ultramicroscopic, infectious particles, consisting of a nucleic acid core (RNA or DNA) and a protein coat; they have no metabolism of their own and are dependent on the machinery of living host cells for their reproduction

white rot type of wood decay, characterized by degradation of lignin

wilt (-disease) the collapse of those parts of a plant that are mechanically supported by cell turgidity; caused mainly by fungi or bacteria

wound an injury to tissues that is caused mechanically and happens suddenly; in certain cases the term is also used to describe wound-like disease injuries of parasitic origin

wound parasite a parasitic and pathogenic organism which gains entry into the host via a wound or damaged area

wound periderm a secondary layer of tissue which walls off a part of a plant that has previously become diseased or wounded

wound rot a wood decay in a tree stem, arising from an injury in that tissue, though which a causal agent has gained entry

zoospore a motile spore produced in a zoosporangium, i.e. one having flagella

Bibliography

1 Allen, M.C., and C.M. Haenseler. 1935. Antagonistic action of *Trichoderma* on *Rhizoctonia* and other soil fungi. *Phytopathology* 25: 244–252.

2 Anagnostakis, S. L. 1987. Chestnut blight. The classical problem of an introduced pathogen. *Mycologia* 78:23–37.

3 Anderson, N.A. 1982. The genetics and pathology of *Rhizoctonia solani*. *Ann. Rev. Phytopath*. 20:329–347.

4 Balder, H. 1990. Untersuchungen zur Revitalisierung streusalzgeschädigter Straßenbäume. *Gesunde Pflanzen*, Heft 10:356–361.

5 Banerjee, S. 1962. An oak (*Quercus robur* L.) canker caused by *Stereum rugosum* (Pers.) Fr. *Trans. Br. mycol. Soc.* 39:267–277.

6 Barklund, Pia. 1987. Occurrence and pathogenicity of *Lophodermium piceae* appearing as an endophyte in needles of *Picea abies*. *Trans. Br. mycol. Soc.*, 89:307–313.

7 Barklund, Pia, and J. Rowe. 1981. *Gremmeniella abietina (Scleroderris lagerbergii)*, a primary parasite in a Norway spruce die-back. *Eur. J. For. Path*. 11:97–108.

8 Barr, M.E. 1978. The Diaporthales in North America. *Mycol. Memoir* No. 7. New Yorker Bot. Garden. Cramer, Lehre; 232 pp.

9 Barrett, D.K., and B.J.W. Greig. 1985. The occurrence of *Phaeolus schweinitzii* in the soils of Sitka spruce plantations with broadleaved or non-woodland histories. *Eur. J. For. Path*. 15:412–417.

10 Bavendamm, W. 1944. *Valdensia heterodoxa*, ein neuer Buchenschädling. *Forstwiss. Centrabl. u. Tharandt. Forstl. Jahrb*.:54–60.

11 Bazzigher, G. 1973. Zunehmende Verbreitung einer Nadelschüttekrankheit an Föhren. *Bündner Wald* 26:194–196.

12 Bergmann, W. 1976. *Ernährungsstörungen bei Kulturpflanzen*, 183 pp. VEB Fischer, Jena;

13 Bernatzky, A. 1978. Tree ecology and preservation. Series: *Developments in Agriculture and managed-forest ecology*, Elsevier pp. 357, Amsterdam;

14 Beyer-Ericson, L., E. Damm, and T. Unestam. 1991. An overview of root dieback and its causes in Swedish forest nurseries. *Eur. J. For. Path*. 21:439–443.

15 Binns, W.O., G.J. Mayhead, and J.M. Mackenzie. 1980. Nutrient deficiencies of conifers in British Forests; an illustrated guide. *Leafl. For. Comm*. No. 76.

16 Björkman. E. 1963. Resistance to snow blight (*Phacidium infestans* Karst.) in different provenances of *Pinus sylvestris* L. *Stud. For. Suecica* 5:16 pp.

17 Blumer, S. 1963. *Rost- und Brandpilze auf Kulturpflanzen*, 379 pp. Fischer, Jena.

18 Blumer, S. 1967. *Echte Mehltaupilze (Erysiphaceae)*, 436 pp. Fischer, Jena.

19 Bondier, B. 1986. Essai de mise au point de méthode de lutte contre *Didymascella thujina*. *Rept. Ministère de l'Agric.*, Service de Prot. des Végétaux, France

20 Bonnet-Massimbert, M., and C. Muller. 1975. La conservation des faines est possible. *Rev. Forest.* France 27:129–138.

21 Booth, C. 1959. Studies of Pyrenomycetes: IV. *Nectria* (Part I). *Comm. Mycol. Inst., Mycol. Pap*. 73, 115 pp.

22 Booth, C. 1966. The genus *Cylindrocarpon*. *Comm. Mycol. Inst., Mycol. Pap*. 104, 56 pp.

23 Booth, C. 1977. *Fusarium*. Comm. Mycol. Inst., Kew; 58 pp.

24 Brasier, C.M. (1983). The future of Dutch elm disease in Europe. In: *Research on Dutch elm disease in Europe*, D.A. Burdekin (Ed.) *Forestry Comm. Bulletin* 60, 96–104. HMSO, UK.

25 Brasier, C.M. 1991. *Ophiostoma novo-ulmi* sp. nov., causative agent of current Dutch elm disease pandemics. *Mycopathologia* 115:151–161.

26 Brasier, C.M., J. Lea, and M.K. Rawlings. 1981. The aggressive and non-aggressive strains of *Ceratocystis ulmi* have different temperature optima for growth. *Trans. Br. mycol. Soc.* 76:213–218.

27 Brasier, C.M., and R.G. Strouts. 1976. New records of *Phytophthora* on trees in Britain. I. Phytophthora root rot and bleeding canker of Horse chestnut (*Aesculus hippocastanum* L.). *Eur. J. For. Path.* 6:129–136.

28 Breitenbach, J., and F. Kränzlin. 1984. *Fungi of Switzerland*. Vol. 1, *Ascomycetes*. Ed. Mycologia, Luzern; 313 pp.

29 Breitenbach, J., and F. Kränzlin. 1986. *Fungi of Switzerland*. Vol. 2. *Non-gilled fungi*. Ed. Mycologia, Luzern; 412 pp.

30 Bruck, R.I., Z. Solel, I.S. Ben-Ze'ev, and A. Zehavi. 1990. Disease of Italian cypress caused by *Botryodiplodia theobromae* Pat. *Eur. J. For. Path.* 20:392–396.

31 Burdekin, D.A. 1979. Common decay fungi in broadleaved trees. *Forestry Comm. Arboriculture Leafl.* No. 5, HMSO, London 41 pp.

32 Burdekin, D.A. (Ed.) 1983. Research on Dutch elm disease in Europe. *Forestry Comm. Bull.* 60, HMSO, UK.

33 Butin, H. 1957. Die blatt- und rindenbewohnenden Pilze der Pappel unter besonderer Berücksichtigung der Krankheitserreger. *Mitt. Biol. Bundesanst. Land-u. Forstw.* Berlin-Dahlem, H. 91:64 pp.

34 Butin, H. 1960. Die Krankheiten der Weide und deren Erreger. *Mitt. Biol. Bundesanst. Land-u. Forstw.* Berlin-Dahlem, H. 98:46 pp.

35 Butin, H. 1973. Morphologische und taxonomische Untersuchungen an *Naemacyclus niveus* (Pers. ex Fr.) Fuck. ex Sacc. und verwandter Arten. *Eur. J. For. Path.* 3:146–163.

36 Butin, H. 1977. Taxonomy and morphology of *Ascodichaena rugosa* gen. et sp. nov. *Trans. Br. mycol. Soc.* 69:249–254.

37 Butin, H. 1981. Über den Rindenbranderreger *Fusicoccum quercus* Oudem. und andere Rindenpilze der Eiche. *Eur. J. For. Path.* 11:33–44.

38 Butin, H. 1986. Endophytische Pilze in grünen Nadeln der Fichte (*Picea abies*). *Z. Mykologie* 52:335–346.

39 Butin, H. 1991. Mykologische Untersuchungen an vergrauten Holzoberflächen im Gebirge. *Holz als Roh-u. Werkstoff* 49:235–238.

40 Butin, H. 1992. Effect of endophytic fungi from oak (*Quercus robur* L.) on mortality of leaf inhabiting gall insects. *Eur. J. For. Path.* 22:237–246.

41 Butin, H., and H. Dohmen. 1981. Eine neue Rindenkrankheit der Roteiche. *Forst-u. Holzwirt* 36:97–99.

42 Butin, H., and I. Kappich. 1980. Untersuchungen über die Neubesiedlung von verbrannten Waldböden durch Pilze und Moose. *Forstwiss. Centralbl.* 99:283–296.

43 Butin, H., and T. Kowalski. 1983. Die natürliche Astreinigung und ihre biologischen Voraussetzungen I. Die Pilzflora der Stieleiche (*Quercus robur* L.). *Eur. J. For. Path.* 13:428–439.

44 Butin, H., and M. Paetzoldt. 1974. Schäden an *Juniperus virginiana* L. durch *Phomopsis juniperovora* Hahn. *Nachrichtenbl. Deut. Pflanzenschutzd. (Braunschweig)* 26:36–39.

45 Cannon, P.F., D.L. Hawksworth, and M.A. Sherwood-Pike. 1985. *The British Ascomycotina. An annotated checklist*. Comm. Mycol. Inst., Kew, 302 pp.

46 Caspari, C.-O., and H. Sachsse. 1990. Rißschäden an Fichte. *Forst und Holz* Nr. 23:685–688.

47 Cassagrande, F. 1969. Ricerche biologiche e systematiche sũ particũlari ascomiceti pseũdosferiali. *Phytopath. Z.* 66:97–135.

48 Cellerino, G.P. 1979. Marssoninae dei pioppi. *Cellulosa Carta* 30:3–23.

49 Christiansen, E., and H. Solheim. 1990. The bark beetle-associated blue-stain fungus Ophiostoma polonicum can kill various spruce and Douglas fir. *Eur. J. For. Path.* 20:436–446.

50 Cooper, J.I. 1979. *Virus diseases of trees and shrubs*. Oxford, Inst. Terrestrial Ecol.; 74 pp.

51 Darker, G.D. 1932. The Hypodermataceae of conifers. *Contrib. Arnold Arbor.* Harvard Univ. 1:1–131.

52 Day, W.R. 1938. Root-rot of sweet chestnut and beech caused by species of *Phytophthora*. *Forestry* 12:101–116.

53 De Kam, M. 1981. The identification of the two subspecies of *Xanthomonas populi in vitro*. *Eur. J. For. Path.* 11:25–29.

54 Défago, G. 1937. *Cryptodiaporthe castanea* (Tul.) Wehmeyer, parasite du chataignier. *Phytopath. Z.* 10:168–177.

55 Delatour, C. 1978. Recherche d'une méthode de lutte curative contre le *Ciboria batschiana* (Zopf) Buchwald chez les glands. *Eur. J. For. Path.* 8:193–200.

56 Delatour, C. 1986. Le problème de *Phytophthora cinnamomi* sur le chêne rouge (*Quercus rubra*). Bull. OEPP/EPPO 16:499–504.

57 Dennis, R.W.G. 1978. *British Ascomycetes*. 585 pp. J. Cramer, Vaduz.

58 Dhingra, O.D., and J.B. Sinclair. 1978. Biology and pathology of *Macrophomina phaseolina*, 166 pp. Imp. Univ. Federal Vicosa. Vicosa, Brasil.

59 Diamandis, S. 1979. 'Top-dying' of Norway spruce, *Picea abies* (L.) Karst., with special reference to *Rhizosphaera kalkhoffii* Bubák. VI. Evidence related to the primary cause of 'top-dying'. *Eur. J. For. Path.* 9:183–191.

60 Dickenson, S., and B.E.J. Wheeler. 1981. Effects of temperature, and water stress in sycamore, on growth of *Cryptostroma corticale*. *Trans. Br. mycol. Soc.* 76:181–185.

61 DiCosmo, F., H. Peredo, and D.W. Minter. 1983. *Cyclaneusma* gen. nov., *Naemacyclus* and *Lasiostictis*, a nomenclatural problem resolved. *Eur. J. For. Path.* 13:206–212.

62 Dimitri, L. 1975. Die Wundfäule – ein aktuelles Problem der Forstwirtschaft. *Holz-Zentralbl.* 101:803–805.

63 Dobson, M. C. 1991. Deicing salt damage to trees and shrubs. *Forestry Comm. Bull.* 101, 64 pp. HMSO. London.

64 Domsch, K.H., and F.J. Schwinn. 1965. Nachweis und Isolierung von pflanzenpathogenen Bodenpilzen mit selektiven Verfahren. *Zentralbl. Bakt. Parasitenkd.* Abt.I, Suppl.1:461–485.

65 Donaubauer, E. 1964. Untersuchungen über die anfälligkeit verschiedener Pappelklone für *Septotinia populiperda* Waterman et Cash. *Phytopath. Z.* 50:134–143.

66 Donaubauer, E. 1972. Distribution and hosts of *Scleroderris lagerbergii* in Europe and North America. *Eur. J. For. Path.* 2:6–11.

67 Eden-Green, S.J., and E. Billing. 1974. Fireblight. *Rev. Pl. Path.* 53:353–365.

68 Ehrlich, J. 1934. The Beech Bark Disease, a Nectria disease of *Fagus* following *Cryptococcus fagi* (Baer). *Can. J. Res.* 10:593–692.

69 Ellis, B., and J. Pamela Ellis. 1985. *Microfungi on land plants. An identification handbook*, 818 pp. Croom Helm, London.

70 Ellis, B., and Pamela Ellis. 1990. *Fungi without gills*. Chapman Hall, London; 329 pp.

71 Ellis, M.B. 1971. *Dematiaceous Hyphomycetes*. Comm. Mycol. Inst., Kew; 608 pp.

72 Ellis, M.B. 1976. *More Dematiaceous Hyphomycetes*. Comm. Myc. Inst., Kew; 507 pp.

73 European and Mediterranean plant protection Organization. 1987. Guideline for the biological evaluation of fungicides. *Phacidium infestans. EPPO Bull.* 17:395–400.

74 Evans, H.C. 1984. The genus *Mycosphaerella* and its anamorphs *Cercoseptoria, Dothistroma* and *Lecanosticta* on pines. *Comm. Mycol. Inst., Mycol. Pap.* 153, 102 pp.

75 Ferdinandsen, C., and C.A. Jørgensen. 1938/39. *Skogvtrae ernes Sygdomme*. Gyldendal, Kopenhagen; 570 pp.

76 Flack, N.J., and T.R. Swinburne. 1977. Host range of *Nectria galligena* Bres. and the pathogenicity of some Northern Ireland isolates. *Trans. Br. mycol. Soc.* 68:185–192.

77 Francis, Sheila, M. 1986. Needle Blights of Conifers. *Trans. Br. mycol. Soc.* 87:397–400.

78 Frank, A.B. 1885. Über die auf Wurzelsymbiose beruhenden Ernährung gewisser Bäume durch unterirdische Pilze. *Ber. Deutsch. Bot. Ges.* 3:128–145.

79 Freyer, K. 1976. Untersuchungen zur Biologie, Morphologie, Verbreitung und Bekämpfung von *Herpotrichia parasitica* (Hartig) E. Rostrup (vormals *Trichosphaeria parasitica* Hartig). *Eur. J. For. Path.* 6:152–166 u. 222–238.

80 Gäumann, E. 1959. Die Rostpilze Mitteleuropas. *Beitr. Kryptogamenfl. Schweiz* 12, 1407 pp.

81 Gibbs, J.N., and C.M. Brasier. 1973. Correlation between cultural characters and pathogenicity in *Ceratocystis ulmi* from Europe and North America. *Nature* 241:381–383.

82 Gourbiere, F., and M. Morelet. 1979. Le genre *Rhizosphaera* Mangin et Hariot. *Rev. Mycologie* 43:81–94.

83 Gregory, P.H., S. Waller. 1951. *Cryptostroma corticale* and sooty bark disease of sycamore (*Acer pseudoplatanus*). *Trans. Br. mycol. Soc.* 34:579–597.

84 Gregory, S.C. 1982. Bark necrosis of *Acer pseudoplatanus* L. in northern Britain. *Eur. J. For. Path.* 12:157–167.

85 Greig, B.J.W. 1981. Decay fungi in conifers. *Forestry Comm. Leaflet* No. 79, HMSO, London.

86 Greig, B.J.W., and C.C. Gulliver. 1976. Decays in oaks in the Forest of Dean. *Forestry* 49:157–159.

87 Greig, B.J.W., S.C. Gregory, and R.G. Strouts. 1991. Honey fungus. *Forestry Comm. Bull.* No. 100, HMSO, London.

88 Gremmen, J., and M. de Kam. 1970. *Erwinia salicis* as the cause of dieback in *Salix alba* in the Netherlands and its identity with *Pseudomonas saliciperda. Neth. J. Path.* 76:249–252.

89 Gremmen, J., and M. de Kam. 1974. Research on poplar canker (*Aplanobacter populi*) in the Netherlands. *Eur. J. For. Path.* 4:175–181.

90 Guillaumin, J. J., C. Bernard, C. Delatour, and M. Belgrand. 1983. Le dépérissement du Chêne à Tronçais: pathologie racinaire. *Rev. For. France* 35:415–424.

91 Habermehl, A., and H.W. Ridder. 1979. Neues Verfahren zum Nachweis der Rotfäule. *Holz-Zentralbl.* 105:383–384.

92 Hahn, Gl., G. 1943. Taxonomy, distribution and pathology of *Phomopsis occulta* and *P. juniperovora. Mycologia* 35:112–129.

93 Hanisch, B., and E. Kilz. 1991. *Monitoring of forest damage. Spruce and pine*. Helm Publ. Bromley, Kent; 334 pp.

94 Harley, J.L., and S.L. Smith. 1983. *Mycorrhizal symbiosis*, 483 pp. Academic Press, London – New York.

95 Harrington, T.C., and F.W. Cobb (Eds.) 1988. *Leptographium root diseases on conifers*, 149 pp. APS Press, St. Paul, MN.

96 Hartig, R. 1900. *Lehrbuch der Pflanzenkrankheiten*, 3. Aufl. 324 pp. Springer, Berlin.

97 Hartley, C. 1921. Damping-off in Forest Nurseries. *Bull. U.S. Dep. Agric.*, No. 934, 99 pp.

98 Hartmann, G., R. Blank, and S. Lewark. 1989. Oak decline in Northern Germany. *Forst u. Holz* 44:475–487.

99 Hartmann, G., F. Nienhaus, and H. Butin. 1988. Farbatlas Waldschäden. *Diagnose von Baumkrankheiten*, 256 pp. Ulmer, Stuttgart.

100 Hawksworth, D.L. 1972. *Hypoxylon mediterraneum* (de Not.) Ces. & de Not. Comm. Mycol. Inst. Descript. No. 359.

101 Hawksworth, D.L., and F. Rose. 1970. Qualitative scale for estimating Sulphur Dioxide air pollution in England and Wales using epiphytic lichens. *Nature* 227:145–148.

102 Hawksworth, D.L., B.C. Sutton, and G.C. Ainsworth. (Eds.) 1983. *Ainsworth & Bisby's dictionary of the fungi*, 412 pp. Comm. Mycol. Inst., Kew.

103 Heiniger, U., and M. Schmid. 1989. Association of *Tiarosporella parca* with needle reddening and needle cast in Norway spruce. *Eur. J. For. Path.* 19:144–150.

104 Hendry, S.J., L. Boddy, and D. Lonsdale. 1990. Strip cankering of beech (*Fagus sylvatica*) – a preliminary survey. *Fourth Intern. Mycol. Congr. Regensburg* 1990, IIF-350/2 (Abstract).

105 Herring, T.F. 1962. Host range of the violet root rot fungus *Helicobasidium purpureum* Pat. *Trans. Br. mycol. Soc.* 45:488–393.

106 Hilber, O. 1991. Some aspects on the revitalisation of a damaged spruce stand by use of the ectomycorrhizal fungi. In: *Science and cultivation of edible fungi*. Maher (Ed.) pp. 625–633. Balkema Rotterdam.

107 Houston, D.R., E.J. Parker, R. Perrin, and K.J. Lang. 1979. Beech Bark Disease: A comparison of the Disease in North-America, Great Britain, France and Germany. *Eur. J. For. Path.* 9:199–211.

108 Illingworth, K. 1973. Variation in the susceptibility of Lodgepole Pine provenances to Sirococcus shoot blight. *Can. J. For. Res.* 3:585–589.

109 Jahn, H. 1971. Stereoide Pilze in Europa (Stereaceae Pil. emend. Parm. u. a. *Hymenochaete*) mit besonderer Berücksichtigung ihres Vorkommens in der Bundesrepublik Deutschland. *Westfäl. Pilzbriefe* 8:69–176.

110 Jahn, H. 1990. *Pilze an Bäumen*. 2. Aufl. 272 pp. Patzer Verlag, Berlin u. Hannover.

111 Jalaluddin, M. 1967. Studies on *Rhizina undulata*. I. Mycelial growth and ascospore germination. *Trans. Br. mycol. Soc.* 50:449–459.

112 Janse, J. D. 1981a. The bacterial disease of ash (*Fraxinus excelsior*), caused by *Pseudomonas syringae* subsp. *savastanoi* pv. *fraxini*. I. History, occurrence and symptoms. *Eur. J. For. Path.* 11:306–315.

113 Janse, J. D. 1981b. — II. Etiology and taxonomic considerations. *Eur. J. For. Path.* 11:425–438.

114 Jansen, E. C. 1969. The watermark disease, a severe danger for the white willow (*Salix alba* L.) *Ned. Bosb. Tijdschr.* 41:118–126.

115 Jülich. W. 1972. Monographie der Athelieae (Corticiaceae, Basidiomycetes). *Willdenowia, Beih.* 7, 283 pp.

116 Käärik, A. 1965. The identification of the mycelia of wood-decay fungi by their oxidation reactions with phenolic compounds. *Stud. Forest. Suecc.* 31, 80 pp.

117 Kechel, H.G., and E. Böden. 1985. Resistensprüfungen an Pappeln aus Gewebekultur. *Eur. J. For. Path.* 15:45–51.

118 Kistler, B.R., and W. Merrill. 1977. Etiology, symptomatology, epidemiology and control of Naemacyclus needlecast of Scotch pine. *Phytopathology* 68:267–271.

119 Korhonen, K. 1978a. Inter-sterility of *Heterobasidion annosum*. *Comm. Inst. For. Fenn.* 94, 24 pp.

120 Korhonen, K. 1978b. Interfertility and clonal size in the *Armillaria mellea* complex. *Karstenia* 18:31–42.
121 Laurence, J.A. 1981. Effects of air pollution on plant-pathogen interactions. *Z. Pflanzenkrankh. Pflanzensch.* 87:156–172.
122 Leith, I.D., D. Fowler. 1988. Urban distribution of *Rhytisma acerinum* (Pers.) Fr. (tar-spot) on sycamore. *New Phytologist* 108:175–181.
123 Lindeijer, E.J. 1932. De bacterie-ziekte van den Wilg veroorzaakt door *Pseudomonas saliciperda* n. sp. *Proefschr. Amsterdam*; 82 pp.
124 Lobanow, N.W. 1953. Mycotrophism of woody plants. (Rus.) *State Publ. Sov. Sci.*, 232 pp. Moscow.
125 Lonsdale, D. 1980. *Nectria* infection of beech bark in relation to infestation by *Cryptococcus fagisuga* Lindinger. *Eur. J. For. Path.* 10:161–168.
126 Lonsdale, D., and D. Wainhouse. 1987. Beech bark disease. *Forestry Comm. Bull.* 69, 15 pp. HMSO, London.
127 Luque, J., and J. Girbal. 1989. Dieback of cork oak (*Quercus suber*) in Catalonia (NE Spain) caused by *Botryosphaeria stevensii. Eur. J. Forest Path.* 19:7–13.
128 McDermott, J.M., and R.A. Robinson. 1989. Provenance variation for disease resistance in *Pseudotsuga menziesii* to the Swiss needle-cast pathogen *Phaeocryptopus gaeumannii. Can. J. For. Res.* 19:244–246.
129 Malençon, G. and J. Marion. 1951. Un parasite des suberaies Nord. Africaines L'*Hypoxylon mediterraneum* (D.Ntrs) Ces. et D.Ntrs. *Revue Forest. Française* 11:681–686.
130 Manners, J. G. 1953. Studies on Larch canker. I. The taxonomy and biology of *Trichoscyphella willkommii* (Hartig) Nannf. and related species. *Trans. Br. mycol. Soc.* 36:362–374.
131 Martinsson, O. 1985. The influence of pine twist rust (*Melampsora pinitorqua*) on growth and development of Scots pine (*Pinus sylvestris*). *Eur. J. For. Path.* 15:103–110.
132 Miller, J.H. 1961. *A monograph of the world species of* Hypoxylon, 158pp. Univ. Georgia Press, Athens.
133 Minter, D.M. 1981. *Lophodermium* on pines. *Comm. Mycol. Inst., Mycol. Pap.* 147, 54 pp.
134 Mitchell, C.P., and C.S. Millar. 1976. Effect of the needle-cast fungus *Lophodermella sulcigena* on growth of Corsican pine. *Forestry* 49:153–158.
135 Mitchell, C.P., C.S. Millar, and B. Williamson. 1978. The biology of *Lophodermella conjuncta* Darker on Corsican pine needles. *Eur. J. For. Path.* 8:108–118.
136 Mittal, R.K., R.L. Anderson, and S.B. Mathur. 1990. Microorganisms associated with tree seeds: world checklist 1990. *Petawa Nat. For. Inst., Forestry Canada Inform. Rep.* PL-X-96; 57 pp.
137 Mix, A.J. 1969. A Monograph of the Genus *Taphrina*. *Bibl. Mycologica* 18, 167 pp.
138 Monod, M. 1983. Monographie taxonomique des Gnomoniaceae. *Sydowia*, IX. Beih.; 315 pp.
139 Morelet, M. 1980. La maladie à *Brunchorstia*. 1. Position systématique et nomenclature du pathogène. *Eur. J. For. Path.* 10:268–277.
140 Müller, K. 1912. Über das biologische Verhalten von *Rhytisma acerinum* auf verschiedenen Ahornarten. *Ber. Deut. Bot. Ges.* 30:385–390.
141 Münch, E. 1913. Hitzeschäden an Waldpflanzen. *Naturwiss. Z. Forst- u. Landwirtsch.* 11:557–562.
142 Neely, D., and E.B. Himelick. 1963. *Aesculus* species susceptible to leaf blotch. *Plant. Dis. Reptr.* 47:170.
143 Nienhaus, F. 1979. Lärchen-Degeneration durch Rickettsien-ähnliche Bakterien. *Allgem. Forstz.* 34:130–132.

144 Nienhaus, F. 1989. *Laubbaumvirosen*. Waldschutz-Merkblatt Nr.14, Parey, Hamburg; 6 pp.
145 Oberwinkler, F. 1977. Das neue System der Basidiomyceten: *Beitr. Biologie der niederen Pflanzen*. Fischer, Stuttgart.:59–105.
146 Osorio, M., and B.R. Stephan. 1991. Life cycle of *Lophodermium piceae* in Norway spruce needles. *Eur. J. For. Path.* 21:152–163.
147 Parker, A.K., and J. Reid. 1969. The genus *Rhabdocline* Syd. *Can. J. Bot.* 47:1533–1545.
148 Patrick, K.N. 1990. Watermark disease of willows. *Arboriculture Research Note* 87/90/PATH, Forestry Comm. UK.
149 Pawsey, R.G. 1962. Leaf blotch of horse-chestnut. *Pl. Path.* 11:137– 138.
150 Peace, T.R. 1962. *Pathology of trees and shrubs*, 735 pp. Oxford University Press.
151 Perrin, R. 1976. Clef de détermination des *Nectria* d'Europe. *Bull. Soc. Mycol. Franc.* 92:335–347.
152 Perrin, R. 1979. La pourriture des faines causée par *Rhizoctonia* Kühn: Incidence de cette maladie après les fainées de 1974 et 1976. Traitement curatif des faines en vue de la conservation. *Eur. J. For. Path.* 9:89–103.
153 Peterson, G.W. 1967. Dothistroma needle blight of Austrian and Ponderosa pines: epidemiology and control. *Phytopathology* 57:437–441.
154 Petrini, L.E., and E. Müller. 1966. Haupt- und Nebenfruchtformen europäischer *Hypoxylon*-Arten (Xylariaceae, Sphaeriales) und verwandter Pilze. *Mycologia Helvetica* 7:501–627.
155 Phillips, D.H., and C.W.T. Young. 1976. Group dying of conifers. *Forestry Comm. Bull. Leafl.* No. 5. HMSO, London. 7 pp.
156 Pinon, J. 1973. Les rouilles du peuplier en France. *Eur. J. For. Path.* 3:221–228.
157 Pinon, J. 1979. Origine et principaux caractères des souches françaises d'*Hypoxylon mammatum* (Wahl.) Miller. *Eur. J. For. Path.* 9:129–142.
158 Prillwitz, H.-G. 1986. Verticilliose, eine häufige Krankheit an Steinobst und Ziergehölzen. *TASPO* 1/2:36–37.
159 Raddi. P., and A. Panconesi. 1981. Cypress canker disease in Italy: biology, control possibilities and genetic improvement for resistance. *Eur. J. For. Path.* 11:340–347.
160 Redfern, D.B., S.C. Gregory, J.E. Pratt, and G.A. MacAskill. 1987. Foliage browning and shoot death in Sitka spruce and other conifers in northern Britain during winter 1983–84. *Eur. J. For. Path.* 17:166–180.
161 Redfern, D.B., J.T. Stoakley, H. Steele, and D.W. Minter. 1989. Dieback and death of larch caused by *Ceratocystis laricicola* sp. nov. following attack by *Ips cembrae*. *Pl. Path.*, 36:467–480.
162 Reichard, M., and A. Bolay. 1986. La maladie de l'encre du châtaignier dans le canton de Genève. *Rev. suisse de viticulture arboriculture horticulture* 18.
163 Ridé, M., and S. Ridé. 1978. *Xanthomonas populi* (Ridé) comb. nov. (syn. *Aplanobacter populi* Ridé) specificité, variabilité et absence de relations avec *Erwinia cancerogena* Ur. *Eur. J. For. Path.* 8:310–333.
164 Rishbeth, J. 1979. Modern aspects of biological control of *Fomes* and *Armillaria*. *Eur. J. For. Path.* 9:331–340.
165 Robak, H. 1952. *Phomopsis pseudotsugae* Wilson – *Discula pinicola* (Naumov) Petrak, as a saprophyte on coniferous woods. *Sydowia* (Annales Mycologici Ser. II.) 6:378–382.
166 Rohde, T. 1937. Über die 'Schweizer' Douglasienschütte und ihren vermuteten Erreger *Adelopus* sp. Mitt. *Forstwirtsch. u. Forstwiss.* 8, 28 pp.
167 Rohmeder, E. 1972. *Das Saatgut in der Forstwirtschaft*, 273 pp. Parey, Hamburg-Berlin.
168 Roll-Hansen, F. 1965. *Pucciniastrum areolatum* on *Picea engelmannii*. Identification by spermogonia. *Medd. Norske Skogfors.* 76:391–397.

169 Roll-Hansen, F. 1972. *Scleroderris lagerbergii*: Resistance and differences in attack between pine species and provenances. *Eur. J. For. Path.* 2:26–39.
170 Roll-Hansen, F. 1985. The *Armillaria* species in Europe – a literature review. *Eur. J. For. Path.* 15:22–31.
171 Roll-Hansen, F. 1989. *Phacidium infestans*. A literature review. *Eur. J. For. Path.* 19:237–250.
172 Roll-Hansen, F., and H. Roll-Hansen. 1969. *Neofabraea populi* on *Populus tremula* ×*Populus tremuloides* in Norway. *Medd. Norske Skogfors.* 27:217–226.
173 Rose, D.R. 1989a. Marssonina canker and leaf spot (anthracnose) of weeping willow. *Arboriculture Research Note* 79/89/PATH, Forestry Comm., UK.
174 Rose, D.R. 1989b. Scab and black canker of willow. *Arboriculture Research Note 78/89/PATH*, Forestry Comm., UK.
175 Ryvarden, L. 1976. The Polyporaceae of North Europe. Vol. 1: *Albatrellus – Incrustoporia. Fungiflora*, Oslo.:7–214.
176 Ryvarden, L. 1978. The Polyporaceae of North Europe. Vol. 2: *Inonotus - Tyromyces. Fungiflora*, Oslo.:219–507.
177 Sachsse, H. 1982. Schäden an lebenden 'Muhle-Larsen'-Pappeln. *Holz als Roh- und Werkstoff* 40:461–369.
178 Scheffer, R.J. 1990. Mechanisms involved in biological control of Dutch elm disease. *J. Phytopathology* 130:265–276.
179 Schmelzer, K. 1977. Zier-, Forst- und Wildgehölze. In: Klinkowski, M.; *Pflanzliche Virologie* 4:276–405.
180 Schmidt-Vogt, H. 1989. *Die Fichte. Krankheiten-Schäden-Fichtensterben*. Band II/2; Parey, Hamburg; 607 pp.
181 Schneider, R., and J.A. von Arx. 1966. Zwei neue, als Erreger von Zweigsterben nachgewiesene Pilze. *Kabatina thujae* n.g., n.sp. und *K. juniperi* n.sp. *Phytopath. Z.* 57:176–182.
182 Schönhar, S. 1958. Bekämpfung der durch *Meria laricis* verursachten Lärchenschütte. *Allgem. Forstz.* 13:100.
183 Schulze, E.-D., O.L. Lange, and R. Oren. 1989. Forest decline and air pollution. A study of spruce (*Picea abies*) on acid soils. *Ecol. Stud.* 77, 475 pp. Springer, Berlin-Heidelberg.
184 Schütt, P. 1977. Das Tannensterben. *Forstwiss. Centralbl.* 96:177–186.
185 Schütt, P. 1985. Control of root and butt rots; limits and prospects. *Eur. J. For. Path.* 15:357–363.
186 Seemüller, E., and W. Lederer. 1987. MLO-associated decline of *Alnus glutinosa*, *Populus tremula* and *Crataegus monogyna*. *J. Phytopathology* 121:33–39.
187 Shigo, A.L. 1979. Tree decay. An expanded concept. *U.S. Dep. Agric. Inf. Bull.* 419, 73 pp.
188 Shigo, A.L. 1980. Branches. *J. Arboriculture* 6:300–304.
189 Shigo, A.L. 1986. *A new tree biology*, 596 pp. Shigo and Trees Ass., Durham/New Hampshire.
190 Shigo, A.L., and H. Butin. 1981. Radial shakes and 'Frost cracks' in living oak trees. *USDA Forest Service, Res. Paper* NE-478; 21 pp.
191 Shigo, A.L., and H.G. Marx. 1977. Compartmentalization of decay in trees. *USDA Inf. Bull.* 405, 73 pp.
192 Shigo, A.L., W. Shortle, and P. Garrett. 1977. Genetic control suggested in compartmentalization of discolored wood associated with tree wounds. *Forest Science* 23:179–182.
193 Sieber, T. 1988. Endophytic fungi in the needles of healthy-looking and diseased Norway spruce (*Picea abies* [L.] Karsten). *Eur. J. For. Path.* 18:321–342.
194 Sieber, T., C. and Hugentobler. 1987. Endophytische Pilze in Blättern und Ästen

gesunder und geschädigter Buchen (*Fagus sylvatica* L.). *Eur. J. For. Path.* 17:411–425.

195 Siepmann, R. 1976. *Polyporus schweinitzii* Fr. und *Sparassis crispa* (Wulf. in Jacq.) ex Fr. als Fäuleerreger in einem Douglasienbestand (*Pseudotsuga menziesii* (Mirb.) Franco) mit hohem Stammfäulenanteil. *Eur. J. For. Path.* 6:203–210.

196 Siepmann, R. 1977. *Fomes annosus* (Fr.) Cke. und andere Stammfäuleerreger in einem Douglasienbestand, *Pseudotsuga menziesii* (Mir.) Franco. *Eur. J. For. Path.* 7:287–296.

197 Sinclair, W.A., H.H. Lyon, and W.T. Johnson. 1987. *Diseases of trees and shrubs*, 575 pp. Comstock Pub. Ass., Cornell Univ. Press, Ithaca and London.

198 Smith, I.M., J. Dunez, R.A. Lelliott, D.H. Phillips, and S.A. Archer. (Eds.) 1988. *European handbook of plant diseases*, 583 pp. Blackwell Scientific. Publications.

199 Smith, W.H. 1981. *Air pollution and forests*, 379 pp. Springer, New York-Heidelberg-Berlin.

200 Smith, W.H. 1989. *Air pollution and forests. Interactions between air contaminants and forest ecosystems*. 2nd ed., Springer, New York; 618 pp.

201 Stalpers, J.A. 1978. Identification of wood-inhabiting Aphyllophorales in pure culture. *Stud. Mycol.* No. 16, 248 pp.

202 Stephan, B.R., and H. Butin. 1980. Krebsartige Erkrankung an *Pinus contorta*-Herkünften. *Eur. J. For. Path.* 10:410–419.

203 Stipes, R.J., and R. Campana. 1981. *Compendium of Elm Diseases*, 96pp. Americ. Phytopath. Soc.

204 Strobel, G.A., and G.N. Lanier. 1981. Dutch Elm Disease. *Scientific American* 245:40–50.

205 Strouts, R.G. 1973. Canker of cypress caused by *Coryneum cardinale* Wag. in Britain. *Eur. J. For. Path.* 3:13–24.

206 Strouts, R.G. 1990. Coryneum canker of Monterey cypress and related trees. *Arboriculture Research Note* 39/90/PATH, Forestry Comm., UK.

207 Strouts, R.G. 1991. Anthracnose of London plane (*Platanus X hispanica*). *Arboriculture Research Note* 46/91/PATH, Forestry Comm., UK.

208 Sutton, B.C. 1980. *The Coelomycetes*. Comm. Mycol. Inst., Kew; 696 pp.

209 Tamblyn, N., and E.W.B. Da Costa. 1958. A simple technique for producing fruit bodies of wood-destroying Basidiomycetes. *Nature* 181:578–579.

210 Tippet, J.T., and A.L. Shigo. 1981. Barriers to decay in conifer roots. *Eur. J. For. Path.* 11:51–59 .

211 Upadhyay, H.P. 1981. A monograph of *Ceratocystis* and *Ceratocystiopsis*, 176 pp. Univ. Georgia Press. Athens.

212 Vajna, L. 1986. Branch canker and dieback of sessile oak (*Quercus petraea*) in Hungary caused by *Diplodia mutila*. *Eur. J. For. Path.* 16:223–229.

213 Vaucher, H. (Comp.) 1986. *Elseviers's dictionary of trees and shrubs*, 413 pp. Elsevier, Amsterdam, Oxford, New York.

214 Van Haut, H., and H. Stratmann. 1970. *Farbatlas über Schwefeldioxid-Wirkung an Pflanzen*, 206 pp. Giradet, Essen.

215 Veldeman, R. 1971. *Chalaropsis* sp., a new parasitic fungus on poplar, the cause of bark lesions. *Med. Rijksfac. Landb. Wetenschappen Gent* 36:1001–1005.

216 Viennot-Bourgin, G. 1949. *Les champignon parasites des plantes cultivées*, 1850 pp. Masson, Paris.

217 Vigouroux, A. 1979. Une méthode simple de recherche de *Ceratocystis fimbriata* f. *platani* sur arbre en place. *Eur. J. For. Path.* 9:316–320.

218 Von Arx, J.A. 1970. A revision of the fungi classified as *Gloeosporium*. *Bibliotheca Mycol.* 24, 203 pp.

219 Von Arx, J.A. 1981. *The genera of fungi sporulating in pure culture*, 424 pp. J. Cramer, Vaduz.

220 Von Tubeuf, C. 1895. *Pflanzenkrankheiten durch kryptogame Parasiten verursacht*, 599 pp. Springer, Berlin.

221 Von Tubeuf, C. 1923. *Monographie der Mistel*, 831 pp. Berlin.

222 Waldie, J.S.L. 1930. Oak seedling disease caused by *Rosellinia quercina* Hartig. *Forestry* 4:1–6.

223 Waterhouse, G. M. 1970. The genus *Phytophthora* de Bary. *Comm. Mycol. Inst., Mycol. Pap.* 122.

224 White, O.E. (1948). Fasciation. *Bot. Rev.* 14:319–351.

225 Wilkins, W.H. 1936. Studies in the genus *Ustulina* with special reference to parasitism. II. A disease of the common lime (*Tilia vulgaris*) caused by *Ustulina*. *Trans. Br. mycol. Soc.* 20:133–156.

226 Wilson, M., and D.M. Henderson. 1966. *British rust fungi*, 384 pp. Cambridge University Press, Cambridge.

227 Wulf, A., R. Kehr. 1991. Bark beetle hazards following storm damage. *Mitt. Biolog. Bundesanst. Land- u. Forstw.* H. 267, 227 pp.

228 Zentmeyer, G.A. 1980. *Phytophthora cinnamomi* and the disease it causes. *APS Monograph* No. 10. Americ. Phytopath. Soc., St. Paul.

Index

Page numbers in bold type refer to illustrations, (In entries for pathogens, host names are shown only when they are not apparent from the trivial names of the pathogens; e.g. *piceae* for pathogens of *Picea*.)